Lecture Notes in Energy

Volume 45

Lecture Notes in Energy (LNE) is a series that reports on new developments in the study of energy: from science and engineering to the analysis of energy policy. The series' scope includes but is not limited to, renewable and green energy, nuclear, fossil fuels and carbon capture, energy systems, energy storage and harvesting, batteries and fuel cells, power systems, energy efficiency, energy in buildings, energy policy, as well as energy-related topics in economics, management and transportation. Books published in LNE are original and timely and bridge between advanced textbooks and the forefront of research. Readers of LNE include postgraduate students and non-specialist researchers wishing to gain an accessible introduction to a field of research as well as professionals and researchers with a need for an up-to-date reference book on a well-defined topic. The series publishes single- and multi-authored volumes as well as advanced textbooks.

Indexed in Scopus and EI Compendex The Springer Energy board welcomes your book proposal. Please get in touch with the series via Anthony Doyle, Executive Editor, Springer (anthony.doyle@springer.com)

Saffa Riffat · Mardiana Idayu Ahmad · Aliff Shakir

Sustainable Energy Technologies and Low Carbon Buildings

Springer

Saffa Riffat
Department of Architecture and Built Environment
University of Nottingham
Nottingham, UK

Mardiana Idayu Ahmad
School of Industrial Technology
Universiti Sains Malaysia
Gelugor, Pulau Pinang, Malaysia

Aliff Shakir
School of Industrial Technology
Universiti Sains Malaysia
Gelugor, Pulau Pinang, Malaysia

ISSN 2195-1284 ISSN 2195-1292 (electronic)
Lecture Notes in Energy
ISBN 978-3-031-78852-9 ISBN 978-3-031-78853-6 (eBook)
https://doi.org/10.1007/978-3-031-78853-6

This Springer imprint is published by the registered company Springer Nature Switzerland AG
The registered company address is: Gewerbestrasse 11, 6330 Cham, Switzerland

Preface

Energy cannot be created or destroyed, it can only be transferred or transformed from one form to another.
—Albert Einstein

This book is intended to provide a comprehensive guide and foundational understanding of sustainable energy technologies and their critical role in reducing energy consumption in buildings across various climatic conditions. It has been crafted to inspire further research and innovation, particularly within the realm of green technology for building applications. The book can serve as a valuable resource for senior undergraduate and graduate students, engineers, professionals, practitioners, scientists, researchers, planners, technologists, and employees in the fields of engineering, technology, and both pure and applied sciences. Furthermore, it can be used as a university reference book, meeting the growing demand for advanced courses on sustainable materials and low-carbon technologies at technical and general universities alike.

This book represents the culmination of many years of research and case studies, spanning from the earliest developments to the most recent advances in sustainable energy systems, with a focus on energy-efficient building technologies. The content is organized into six chapters, each offering in-depth technical insights into various facets of sustainable energy technologies and their application in green buildings. By exploring these topics, readers will gain a deeper understanding of how innovations in energy systems can contribute to global sustainability goals and reduce the environmental footprint of the built environment.

Chapter 1 provides an overview of the global energy landscape, focusing on the rapid urbanization and rising living standards that have driven a marked increase in global energy demand. This chapter introduces the critical need for integrated solutions that link energy management with sustainable development, a concept reinforced by the United Nations 2030 Agenda and the Sustainable Development Goals (SDG). Particularly SDG 7, which emphasizes access to affordable, reliable, and sustainable energy. It highlights how reducing energy consumption through green innovations, in the form of sustainable energy technologies and low carbon buildings, is vital for meeting global sustainability objectives. The chapter sets the stage for the overall theme of the book by discussing the significance of energy-related technological innovations in the broader context of sustainable development.

Chapter 2 explores into the importance of building energy efficiency and the advancement of green technologies, specifically focusing on thermal energy-efficient systems. This chapter provides guidance and a foundational understanding of energy consumption in buildings and strategies to enhance energy efficiency. These systems address the common mismatch between energy supply and demand in buildings through the use of renewable energy technologies like solar collectors and high-efficiency heat pumps. The chapter presents a detailed discussion on the design, development, and validation of low-cost, modular energy-efficient systems, which have been proven effective in reducing both energy consumption and operational costs. Emphasis is placed on understanding the technical terms and principles behind energy recovery technology, with a focus on building applications. This chapter sets the foundation for understanding the mechanisms and theory behind energy recovery systems and how they contribute to lower energy costs, reduced greenhouse gas emissions, and improved environmental impact in the building sector.

Chapter 3 examines the integration of sustainable energy generation technologies in energy-efficient buildings, with a focus on improving energy performance through advanced electrical systems and the use of renewable energy sources. It explores the critical role of smart grids, energy management systems, and dynamic pricing models in optimizing energy usage by employing predictive algorithms and real-time monitoring tools. The chapter also highlights the incorporation of renewable energy technologies such as solar photovoltaics (PV), wind turbines, and geothermal heat pumps, stressing their importance in boosting energy self-sufficiency and ensuring grid stability. Additionally, it emphasizes the potential of building-integrated photovoltaics (BIPV) and microgrids to generate power on-site and contribute to sustainable urban infrastructure. Innovative applications like solar-powered smart windows, thin-film solar cells, and photovoltaic thermal collectors (PV/T) are introduced as groundbreaking solutions that could revolutionize energy efficiency in modern buildings. By integrating these technological advancements into sustainable building designs, this chapter presents strategies for significantly reducing energy consumption, cutting operational costs, and lessening the environmental impact of the built environment.

Chapter 4 provides an in-depth exploration of energy-efficient lighting technologies designed for building applications, with a focus on sustainable solutions that reduce energy consumption and enhance indoor environmental quality. The chapter covers natural lighting methods, such as solar daylighting, alongside advanced electric lighting technologies like compact fluorescent lamps (CFLs), light-emitting diodes (LEDs), and cold cathode fluorescent lamps (CCFLs). The benefits of each technology, including energy conservation, extended lifespan, and improved light quality, are discussed comprehensively. The chapter also highlights the integration of smart lighting systems and sensors for optimized lighting control, as well as hybrid solar lighting systems, which collectively contribute to substantial energy savings. By combining daylighting techniques with energy-efficient lamps and advanced lighting controls, the chapter outlines effective strategies for minimizing energy use while enhancing visual comfort and promoting environmental sustainability in residential, commercial, and institutional buildings.

Chapter 5 explores into the innovative architectural designs and technological advancements propelling the development of zero-energy and low-carbon buildings. It examines key strategies aimed at cutting energy consumption while advancing environmental sustainability, with an emphasis on energy-efficient insulation, passive solar heating, and the integration of renewable energy sources like solar and wind power. The chapter also looks at the role of smart grids and building energy management systems (BEMS) in optimizing energy use, enabling these buildings to interact with larger energy networks. With a global viewpoint, it reviews how regions such as North America, Europe, and Asia are adopting zero-energy practices, supported by robust energy policies and incentives. In addition, the chapter highlights the economic and health benefits associated with zero-energy and low-carbon buildings, including reduced operational costs, enhanced property values, and better indoor air quality. As more cities embrace sustainable development, these buildings are becoming central to urban planning, helping create resilient, energy-efficient infrastructures worldwide.

Chapter 6 discusses the concept of eco-cities, underscoring the growing need for sustainable urban development in the face of increasing global energy demands and rapid urbanization. The chapter outlines key strategies like the integration of renewable energy, zero-waste management, and water conservation to address the environmental challenges posed by urban expansion. It also explores the significance of innovative urban planning, which prioritizes resilience, circular economies, and carbon-neutral growth, helping to break the link between economic growth and resource depletion. The chapter further examines the difficulties of implementing eco-city initiatives in fast-growing urban areas, particularly in developing countries, where the pressures of urbanization are most intense. It concludes with an emphasis on the importance of collaboration among governments, industries, and communities in promoting sustainable urban living, with a focus on strategies to advance sustainable cities globally.

In the process of preparing and writing this book, the support from various individuals and institutions has been invaluable. We would like to extend our sincere gratitude to the editorial team at Springer Nature and Springer International Publishing for their contributions in various capacities, helping bring this book to completion. Our heartfelt thanks also go to the School of Industrial Technology, Universiti Sains Malaysia, and the Department of Architecture and Built Environment, University of Nottingham, UK, for their assistance in gathering the materials and information necessary for the publication. Additionally, we are deeply grateful to all the reviewers for dedicating their time to reviewing the content of this book. Finally, we express our appreciation to our families for their patience and unwavering support throughout the writing process.

The work presented in this book is part of the outcomes of several funded research projects, and we would like to take this opportunity to express our gratitude to the sponsors for their financial and technical support. Our appreciation goes to the International Centre for Genetic Engineering and Biotechnology, ICGEB Grant Number of 304/PTEKIND/6501382.I151 (R504-LR-GAL008-0006501382-I151). Any opinions, findings and conclusions or recommendations expressed in this material are those of the authors and do not necessarily reflect the views of the MOE and USM.

Nottingham, UK — Saffa Riffat
Penang, Malaysia — Mardiana Idayu Ahmad
Penang, Malaysia — Aliff Shakir
September 2024

Contents

About the Authors

Saffa Riffat is recognized globally as Leading Expert in sustainable technologies and eco-buildings. He is Chair of Sustainable Energy and heads the Architecture, Climate, and Environment Research Group at the University of Nottingham, UK. Additionally, he holds the position of Professor in the Department of Architecture and Built Environment at the University of Nottingham and serves as President of the World Society of Sustainable Energy Technologies (WSSET). His research expertise spans renewable energy, sustainable technologies, eco-cities, energy-efficient buildings, heat pumps, cooling systems, energy storage, and thermally powered energy cycles. With over 650 peer-reviewed publications and 30 international patents to his name, he was also awarded a Doctor of Science (D.Sc.) degree by the University of Oxford for his pioneering contributions to the field of heat pumps and ventilation technologies. e-mail: saffa.riffat@nottingham.ac.uk

Mardiana Idayu Ahmad received her Ph.D. in Engineering Science (Sustainable Energy Technologies) at the Department of Architecture and Built Environment, Faculty of Engineering, University of Nottingham, UK, in 2011. She is currently Associate Professor in Energy and Environmental Technology, Universiti Sains Malaysia. Her research interests cover a broad spectrum, focusing on sustainable energy technologies, clean technology and bio-based nano and micro materials for energy and environmental applications. She is committed to integrating these fields through her work, which has led to the publication of more than 100 national and international contributions, including journal articles, research books, book chapters, conference papers, and various other scholarly works. e-mail: mardianaidayu@usm.my

Aliff Shakir is currently Postdoctoral Fellow at Universiti Sains Malaysia. He earned his Ph.D. in Environmental Technology, with research focused on sustainable materials, biomass recovery, and bio-composites. His expertise spans natural fiber processing, biofuel production, clean technology, nanomaterials and renewable energy technologies. His work has contributed to several international research collaborations, including projects in marine biotechnology, bioenergy, and sustainable materials. His research bridges the gap between academia and industry, promoting sustainable energy solutions and green materials for low-carbon technologies. He continues to explore innovative approaches to bio-based product development, with a commitment to driving environmentally responsible practices in energy and materials science. e-mail: maliffshakir@usm.my

Chapter 1
Introduction: Sustainable Energy Technologies and Low Carbon Buildings Towards Sustainability

Contents

Abstract Due to the swift progression of urbanization and enhancements in living standards, global energy demand is anticipated to increase markedly in the forthcoming decade. The rise in energy consumption requires an integrated solution that connects energy management with sustainable development. During the United Nations (UN) Sustainable Development Summit in September 2015, global leaders ratified the 2030 Agenda, which delineated 17 Sustainable Development Goals (SDG). SDG 7 is particularly significant as it seeks to guarantee access to affordable, reliable, sustainable, and modern energy for everyone. Attaining this objective necessitates coordinated efforts to advance sustainable energy technologies and low carbon building for the conservation and management of natural resources. A primary strategy in this initiative is to diminish energy consumption while promoting environmentally sustainable technological innovations. This chapter offers a concise summary of energy-related green innovations and their significance in fulfilling the global demand for sustainable development.

S. Riffat et al., *Sustainable Energy Technologies and Low Carbon Buildings*, Lecture Notes in Energy 45, https://doi.org/10.1007/978-3-031-78853-6_1

1.1 Overview of World Energy Production and Consumption

Energy constitutes the cornerstone of contemporary human existence, propelling economic expansion and industrial advancement. Current global energy production and consumption patterns are predominantly unsustainable, heavily dependent on fossil fuels that are both expensive and detrimental to the environment. The global energy landscape has encountered a brief alleviation from the urgent pressures of the energy crisis; however, substantial challenges persist. Although the significant surges in fossil fuel prices observed in 2022 have abated, the global energy markets remain marked by tension and volatility (Liu 2023).

Mardiana and Riffat (2020) project that global energy demand will rise by 50–80% in the next decade, primarily driven by urbanization and improving living standards (Ahmad 2020). The increase in energy demand heightens concerns regarding energy security, which is characterized by prompt investments in energy infrastructure that correspond with economic growth and environmental sustainability. Furthermore, the past two decades have experienced a significant rise in primary energy consumption and CO_2 emissions, by 49% and 43%, respectively (Salehabadi et al. 2020).

The impacts of climate change are becoming more evident. The global average surface temperature has increased by roughly 1.2 °C above pre-industrial levels, resulting in extreme weather events including heatwaves, wildfires, and floods. This warming trend is inherently connected to the ongoing increase in greenhouse gas emissions, which have not yet peaked (Allan et al. 2021). The energy sector is the primary contributor to air pollution, with more than 90% of the global population subjected to hazardous levels of air contaminants. This exposure directly accounts for over 6 million premature deaths each year (Organization 2022). Initiatives to enhance access to vital energy services, including electricity and clean cooking fuels, have encountered obstacles in certain areas, worsening social disparities and hindering advancements toward global sustainability objectives (Liu 2023). Confronting the interrelated issues of energy consumption, technological advancement, and sustainable development necessitates a comprehensive strategy that incorporates various energy sources, financial investments, and policy initiatives to guarantee enduring environmental sustainability.

A glimmer of hope can be found in the expansion of the clean energy sector, which is occurring in the midst of this turbulent scenario. Solar photovoltaic (PV) technology and electric vehicles (EV) are experiencing rapid expansion, which is a sign that a new energy economy that is more environmentally friendly is emerging. The imperative to reduce greenhouse gas emissions, in addition to favourable economic conditions for mature renewable energy technologies, has been a driving force behind the forty percent increase in global investment in clean energy technologies that has occurred since the year 2020 that has occurred. It is becoming increasingly important for nations that are highly dependent on fuel imports to place a higher priority on energy security. The transition to clean energy is seen as a strategic move that can enhance industrial capabilities and generate green jobs. The transition to clean energy, on the other hand, is not without its difficulties. The supply chain of certain environmentally friendly technologies, such as wind power, is experiencing pressures that could potentially impede their growth. However, there are discernible indications that progress is being made. The percentage of electric vehicles that are sold worldwide has seen a significant increase, going from one in every 25 vehicles sold in the year 2020 to one in every five vehicles sold by the year 2023. In addition, it is anticipated that the capacity of renewable energy sources from around the world will increase by a staggering 500 gigawatts (GW) in the year 2023. Investment in solar power is particularly noteworthy, as more than one billion dollars is allocated each day to the deployment of solar energy. As an additional point of interest, the manufacturing capacity for essential components of clean energy, such as solar photovoltaic modules and electric vehicle batteries, is rapidly expanding (Liu 2023).

1.2 Building Energy Consumption

Building energy consumption denotes the aggregate energy utilized by residential, commercial, and industrial structures to sustain their operations. This encompasses energy for heating, cooling, illumination, appliances, and equipment, in addition to energy utilized for water heating and ventilation. The constructed environment is a primary consumer of global energy, representing approximately 40% of total energy consumption worldwide and substantially contributing to greenhouse gas (GHG) emissions (Florini 2011). The rising energy demand in buildings is propelled by population growth, urbanization, and enhanced living standards. This trend poses a considerable challenge for climate change mitigation, as the energy utilized in buildings primarily derives from fossil fuel sources, which are significant contributors to global carbon dioxide (CO_2) emissions (Ürge-Vorsatz et al. 2015). Energy consumption in buildings is intrinsically connected to carbon emissions, attributable to the prevalent utilization of carbon-intensive energy sources like coal, oil, and natural gas. Fossil fuels are frequently utilized in power generation, supplying the electricity and heat necessary for building operations. The energy utilized in buildings substantially contributes to global CO_2 emissions, intensifying the greenhouse effect and accelerating climate change.

A significant fraction of the energy usage in a building pertains to its operational phase, encompassing lighting, heating, ventilation, air conditioning (HVAC), and the utilization of electrical devices. In the United States, buildings contribute to 39% of CO_2 emissions, primarily due to their energy consumption habits (Thomas et al. 2019). Furthermore, the construction of edifices termed embodied energy, contributes to carbon emissions due to the necessity of energy-intensive materials such as cement and steel (Pomponi and Moncaster 2016).

1.2.1 Strategies to Mitigate Energy Consumption in Buildings

Mitigating building energy consumption is crucial to reducing carbon emissions and achieving global climate goals, such as those outlined in the Paris Agreement. Strategies to reduce energy consumption in buildings can be categorized into several key approaches: improving energy efficiency, incorporating renewable energy sources, and promoting behavioural changes.

(a) **Improving energy efficiency in buildings**

Energy efficiency refers to reducing the amount of energy required to provide the same level of building service, such as heating, cooling, or lighting. Several strategies can be implemented to enhance energy efficiency in buildings:

i. **High-performance building envelope**
   Enhancing the thermal efficiency of a building envelope (walls, windows, roofs) can substantially decrease energy requirements for heating and cooling. High-performance insulation materials and energy-efficient windows, including double or triple-glazed options, diminish heat loss during winter and curtail heat gain in summer.
ii. **Energy-efficient HVAC systems**
   Enhancing to energy-efficient HVAC systems, such as heat pumps and high-efficiency air conditioners, can diminish the energy usage linked to indoor climate regulation. Intelligent thermostats and energy management systems can enhance HVAC efficiency.
iii. **Lighting efficiency**
   Adopting energy-efficient lighting technologies, such as light-emitting diodes (LED), can significantly decrease electricity usage. LED utilize up to 75% less energy than conventional incandescent bulbs and possess a longer lifespan, rendering them a more sustainable option for residential and commercial structures.
iv **Building automation systems (BAS)**
   The implementation of Building Automation Systems (BAS) automates control mechanisms for HVAC, lighting, and security, thereby optimizing energy consumption by modulating energy use according to occupancy levels and external weather conditions.

(b) **Incorporating renewable energy sources**

An additional effective strategy to diminish building energy consumption and related carbon emissions is the integration of on-site renewable energy technologies. These technologies enable buildings to produce their own clean energy, diminishing dependence on grid-supplied, fossil fuel-based electricity.

i. **Solar photovoltaic (PV) systems**

Installing solar panels on rooftops enables buildings to capture solar energy for the generation of on-site electricity. Solar photovoltaic systems can substantially diminish carbon footprint of the building by substituting conventional electricity generation.

ii. **Geothermal heat pumps**

Geothermal heat pumps utilize the consistent temperature of the earth to deliver heating and cooling for structures, thereby minimizing the energy needed for these functions. This technology is especially efficient in climates characterized by extreme temperatures, where heating and cooling demands are substantial.

iii. **Wind energy integration**

Small-scale wind turbines can be deployed in specific locations to produce electricity for building consumption. This technology, while less prevalent than solar PV, can serve as a valuable alternative in regions with significant wind potential.

(c) Promoting behavioural change

It is essential to make behavioural adjustments in addition to technological interventions in order to reduce the amount of energy that is consumed in buildings. It is possible to achieve significant reductions in energy consumption by providing residents with information on energy conservation techniques and encouraging the adoption of habits that have a low impact on the environment.

i. **Energy conservation through smart use**

Encouraging people who live in a building to turn off lights, appliances, and electronic devices when they are not in use can help save energy. In order to effectively mitigate "phantom loads" caused by devices that are in standby mode, the utilization of smart plugs and power strips can be advantageous.

ii. **Increasing awareness of thermal comfort**

Adjusting the settings on the thermostat to reduce the amount of heating and cooling that is required can significantly cut down on energy consumption. It is possible to achieve significant savings on energy costs by adjusting thermostats to a higher setting during the summer and a lower setting during the winter.

iiii. **Water conservation**

Reducing the amount of hot water used by installing low-flow fixtures and water heaters that are energy-efficient can significantly cut down on the amount of energy required for water heating, which is a primary component of the energy consumption of buildings.

1.3 Global Warming and Greenhouse Gases

Since the onset of the Industrial Revolution, the global average annual temperature has increased by slightly over 1 °C, approximately 2 °F. Accurate records commenced in 1880, and from that year until 1980, the temperature rose by an average of 0.07 °C

Table 1.1 Average composition of the atmosphere up to an altitude of 25 km

Gas name	Chemical formula	Percent volume (%)
Nitrogen	N_2	78.08
Oxygen	O_2	20.95
Water[a]	H_2O	0 to 4
Argon	Ar	0.93
Carbon dioxide[a]	CO_2	0.0360
Neon	Ne	0.0018
Helium	He	0.0005
Methane[a]	CH_4	0.00017
Hydrogen	H_2	0.00005
Nitrous Oxide[a]	N_2O	0.00003
Ozone[a]	O_3	0.000004

[a]Variable gases

(0.13 °F) every decade. Since 1981, the rate of increase has more than doubled: over the past four decades, the average global temperature has risen by 0.18 °C, equivalent to 0.32 °F. Over time, the energy received by the surface of the Earth is approximately equal to the energy re-emitted into space as infrared radiation, thus maintaining a relatively constant surface temperature. The Earth temperature is significantly affected by the presence, density, and composition of its atmosphere, as indicated in Table 1.1

Carbon dioxide is the primary contributor to anthropogenic global warming, representing 66% of the greenhouse effect (Pryck 2021). The principal sources of CO_2 emissions are the combustion of fossil fuels, deforestation, and industrial operations. Methane, a significant greenhouse gas, is predominantly released through agricultural practices, such as livestock husbandry and rice production. Nitrous oxide, though less prevalent, is a potent greenhouse gas released during fossil fuel combustion and the application of nitrogen-based fertilizers.

Extensive human activity has resulted in global warming since the onset of industrialization, disrupting the natural equilibrium. The rise in average global temperatures caused by the greenhouse effect is referred to as global warming. Examples of greenhouse gases include water vapor, carbon dioxide (CO_2), methane (CH_4), nitrous oxide (N_2O), chlorofluorocarbons (CFC), hydrofluorocarbons (HFC), perfluorocarbons (PFC), and sulfur hexafluoride (SF6). Their presence is advantageous as it maintains the average global temperature at a comfortable 60°F (15.6 °C). However, when the concentration of these gases escalates, complications arise. Table 1.2 depicts the worldwide atmospheric concentrations of significant greenhouse gases, emphasizing the substantial increase in their levels since pre-industrial eras.

In addition to human-made sources, the majority of greenhouse gases have considerable natural sources, and there are tremendous natural mechanisms for removing them from the atmosphere. Yet, the continued increase in atmospheric concentrations demonstrates that for each of the major greenhouse gases, more gas is emitted

Table 1.2 Global Atmospheric Concentrations of Selected Greenhouse Gases

Gas	Pre-industrial atmospheric concentration (ppm)	1992 Atmospheric concentration	Percent change (Recent years) (%)
Carbon Dioxide (CO_2)	278	356	0.4
Methane (CH_4)	0.700	1.714	0.6
Nitrous Oxide (N_2O)	0.275	0.311	0.25
Sulphur Hexafluoride (SF6)	0	32 ppt	0.6

annually than is absorbed. This indicated that, natural absorption systems are falling behind. It is necessary to analyses each of greenhouse gases below:

1.3.1 Water Vapour

Water vapour is by far the most common, with an atmospheric concentration of nearly 1%, compared with less than 0.04% for carbon dioxide. Water vapour is the most common greenhouse gas present in the atmosphere. It is emitted into the atmosphere in enormous volumes through natural evaporation from oceans, lakes, and soils and is returned to Earth in the form of rain and snow. Water vapour is so plentiful in the atmosphere that additional emissions are unlikely to absorb any significant amount of infrared radiation. It is also likely that the amount of water vapour held in the atmosphere is generally in equilibrium, and that increasing emissions of water vapour would not increase atmospheric concentrations. According to currently available information, anthropogenic water vapour emissions at the surface of the Earth are unlikely to be an important element in either causing or ameliorating climate change (Li et al. 2024).

1.3.2 Carbon Dioxide CO_2

Carbon dioxide is the main contributor responsible for 80% of emissions from industrialized countries. The gas is released into the atmosphere primarily from burning fossil fuels: coal, oil, petrol and natural gas as well as forest fire. A growing population coupled with increasing demands on transport and energy, has led to emissions increasing at dramatic rates. CO_2 contributes about 66% to the greenhouse effect (Lamb et al. 2021). Most of the CO_2 could be absorbed by the ocean creatures and all kinds of plants on land. CO_2 is taken up by minute sea creatures or dragged to the ocean depths by the circulation of water. But desertification and deforestation have been becoming serious due to human activities, weakening the conditioning capability of the Earth itself.

1.3.3 Methane CH_4

Methane is released during coal-mining activities, oil exploration and when vegetation is burnt during land clearance. The main source of methane is agricultural activity. It is released from wetlands such as rice paddies and from animals, particularly cud-chewing species like cows. An increasing population coupled with increasing agricultural production must lead to the increase of methane emission. Since the 1960s the amount of methane in the air has increased by 1% per year—twice as fast as the build-up of CO_2.

1.3.4 Nitrous Oxide

Nitrous oxide (N_2O) is a colourless, non-flammable gas with a sweetish odour, commonly known as "laughing gas", and sometimes used as an anaesthetic. Nitrous oxide comes from both natural and man-made processes, as shown in Table 1.2. Nitrous oxide can be naturally produced by oceans and rainforests. Man influenced sources, which represent around 45% of output to the atmosphere, mainly include fossil fuel combustion in power stations, the use of fertilizers in agriculture, burning rain forests and human and animal waste (organic matter), nylon and nitric acid production, etc. Nitrous oxide is broken down in the atmosphere by chemical reactions that involve sunlight.

1.3.5 Halocarbons (CFC, HFC, and HCFC, PFC (Including CF_4, C_2F_6, and C_3F_8)) and Sulfur Hexafluoride (SF6)

These man-made engineered chemicals are from human industrial activities. Chlorofluorocarbons (CFC), hydrofluorocarbons (HFC) and hydrochlorofluorocarbons (HCFC) found in fridges and air conditioners etc. are extremely effective greenhouse gases. Although there are lower concentrations of CFC in the atmosphere than CO_2 they trap more heat. A CFC molecule is 10,000 times more effective in trapping heat than a CO_2 molecule. In addition, CFCs are ozone-depleting greenhouse gas. A portion of the CFCs find their way into the stratosphere, where they can be destroyed by sunlight. This reaction, however, releases free chlorine atoms into stratosphere, and the free chlorine atoms tend to combine with stratospheric ozone (O_3), which protects the surface of the Earth from certain wavelengths of potentially damaging solar ultraviolet radiation (e.g. causes human and animal skin cancers).

Beyond the halocarbons (CFC, HFC, and HCFC), there are a range of engineered chemicals, produced in relatively small quantities, such as halocarbons perfluorocarbons (PFC), which include CF_4, C_2F_6, and C_3F_8. PFCs are emitted as by-products of aluminium smelting. Other chemicals such as sulfur hexafluoride (SF6), used as an insulator in utility-scale electrical equipment. All of these chemicals are potent greenhouse gases with very long atmospheric lifetimes.

1.4 Energy and Sustainable Development Goals

During the United Nations (UN) Summit for Sustainable Development, which took place on September 25, 2015, the adoption of Transforming Our World: the 2030 Agenda for Sustainable Development was a momentous occasion in the history of global efforts to advance sustainability. In order to combat the growing threat of climate change and eradicate poverty by the year 2030, this ambitious agenda established a comprehensive framework consisting of seventeen Sustainable Development Goals (SDG). These goals are designed to promote social justice, reduce inequality, and eradicate poverty overall. According to the United Nations, the SDG are founded on the idea of promoting balanced progress across economic, social, and environmental dimensions. This is done in recognition of the fact that these three domains are interconnected and necessary for achieving long-term prosperity and well-being for all populations.

As a natural extension and enhancement of the earlier Millennium Development Goals (MDG), which included eight targets specifically focused on reducing poverty, hunger, and disease, as well as promoting education, gender equality, and environmental sustainability, the Sustainable Development Goals were conceived of as a natural extension and enhancement of the MDG. The Millennium Development Goals, which were adopted in the year 2000, made significant strides in reducing extreme poverty and addressing other critical global issues. However, the MDG were limited in scope and did not fully encompass the broader and more interconnected challenges of the twenty-first century (Sachs 2012). As a response, the Sustainable Development Goals intend to build on the progress that was made under the Millennium Development Goals, while also introducing a framework that is more comprehensive and inclusive. This framework places an emphasis on the necessity of addressing poverty, environmental degradation, and social injustices in a coordinated manner.

The overarching vision of this new global initiative is articulated in the Preamble of the 2030 Agenda, which describes the following:

> "This Agenda is a plan of action for people, planet and prosperity. It also seeks to strengthen universal peace in larger freedom. We recognize that eradicating poverty in all its forms and dimensions, including extreme poverty, is the greatest global challenge and an indispensable requirement for sustainable development" (United Nations, 2015).

The relationship between climate change and sustainable development has been a topic of discussion in the field of climate economics for a considerable amount of time. During their Special Working Session in 1987, the World Commission on Environment and Development (WCED) provided a definition of sustainable development. This definition states that sustainable development is defined as "development that meets the needs of the present without compromising the ability of future generations to meet their own needs" (Brundtland and Mansour 1987). Global warming can lead to intergenerational disparities in the distribution of climatic resources, which could potentially compromise the ability of future generations to meet their needs. This definition brings to light the primary challenge that is posed by global warming.

As a result of the emissions of greenhouse gases (GHG) that are caused by modern economic activities, it is anticipated that these emissions will have a negative impact on the climate conditions of the future, which will in turn threaten the concept of sustainable development across generations. The implementation of a carbon budget is something that climate economists advocate for as a response to the challenges that have been presented. According to estimates of global warming thresholds that the planet is able to withstand without experiencing severe consequences (Nordhaus 1994), this budget establishes limits on the total amount of carbon dioxide that can be emitted.

The implementation of carbon-based taxes is intended to be one of the primary mechanisms that will be utilized in order to enforce this carbon budget. The internalization of the external costs of greenhouse gas emissions would be accomplished through the implementation of such taxes, which would set the price of carbon emissions at a rate that reflects the time-discounted cost of the damage they impose on future generations (Woodward and Bishop 1997). In order to come up with the appropriate discount rate for these taxes, it is necessary to have an accurate measurement of the amount of time that passes between the emission of greenhouse gases and the material impacts that they have on global warming.

This measurement is made more difficult by the complex feedback loops and interactions that occur within climate subsystems. As a result, it is challenging to accurately model these effects using the conventional physical modelling techniques. As a consequence of this, there is a growing reliance on methods that are driven by data in order to identify and monitor warming trends. According to the Intergovernmental Panel on Climate Change (IPCC), these trends serve as essential benchmarks for adjusting the baseline climate conditions against which future warming is measured. The research that has been done on climate tipping points, which are critical thresholds at which even slight changes in temperature can lead to irreversible damage to climatic subsystems, has further highlighted the significance of accurately tracking warming trends. The identification of these tipping points is of the utmost importance because in the event that they are exceeded, the effects of global warming will become significantly more severe and could potentially have catastrophic consequences (Lenton 2013).

The term "social cost of carbon" (SCC) refers to the economic damages that are associated with an incremental increase in carbon emissions in a given year. Recent studies, have quantified the SCC. Through the process of monetizing the long-term damages that are caused by each additional ton of carbon dioxide that is emitted, the SCC is an essential instrument in the field of climate economics (Dietz et al. 2016). It is responsible for linking carbon budgets to carbon pricing strategies. The existence of this connection highlights the significance of conducting thorough warming trend analysis as a basis for developing efficient climate policy and economic strategies with the goals of reducing the effects of global warming and fostering progress toward sustainable development.

The year 2012 was designated by the General Assembly of the United Nations as the International Year of Sustainable Energy for All. This designation highlights the growing recognition on a global scale of the significant role that energy plays in the development of economies and also in the mitigation of climate change. Immediately after the official launch of the Sustainable Energy for All (SE4All) initiative in September 2011, which was designed to address the pressing need for access to sustainable energy, this designation was bestowed upon the organization. The initiative reflected the understanding that energy plays a fundamental role in driving economic growth, improving quality of life, and enabling the transition to low-carbon economies, all of which are essential to mitigating the effects of climate change (Mahama 2012). The initiative was a reflection of this understanding.

The SE4All initiative outlined three interconnected and lofty goals that were to be accomplished by the year 2030. These goals were to be pursued throughout the Decade of Sustainable Energy for All, which was declared to be in effect from 2014 to 2024. All of these goals are cantered on making sure that everyone has access to contemporary energy services. There has been a global increase in energy efficiency that is twice as fast as before. Increasing the proportion of renewable energy sources in the overall energy mix of the world (Mahama 2012).

The implementation of this initiative marked a significant milestone in the governance of global energy, as it reflected a collective recognition of the importance of a coordinated approach to the achievement of sustainable energy systems. For the purpose of fostering long-term sustainability, the framework that focuses on renewable energy, energy efficiency, and equitable access to energy serves as an essential component. The United Nations Sustainable Development Summit, which took place in September 2015, ultimately decided to adopt the 2030 Agenda for Sustainable Development, which included a total of seventeen Sustainable Development Goals (SDG). This decision was made in accordance with the objectives of the SE4All initiative. These objectives are intended to address a wide variety of global challenges, such as poverty, inequality, climate change, and environmental degradation, among others. Although the Sustainable Development Goals are based on earlier frameworks such as the Millennium Development Goals (MDG), they place an emphasis on the necessity of an integrated approach to sustainable development that strikes a balance between the economic, social, and environmental dimensions.

When it comes to these objectives, Sustainable Development Goal 7 (SDG 7) specifically addresses the requirement to "ensure access to affordable, reliable, sustainable, and modern energy for all" by the year 2030 (Nations 2030). It is believed that having access to energy that is both clean and reliable is a key factor in the accomplishment of other Sustainable Development Goals, particularly in areas such as the elimination of poverty, the improvement of health, the advancement of education, and the availability of clean water. Furthermore, according to the United Nations Development Programme (UNDP), energy access is a crucial component in the process of driving industrialization, fostering economic growth, and enhancing resilience to climate change. If we did not have access to energy that is both affordable and sustainable, we would be significantly hampered in our efforts to reduce global poverty and achieve climate goals.

According to the World Commission on Environment and Development (WCED), the Brundtland Report from 1987, which was formerly referred to as Our Common Future, established the basis for this agenda by defining sustainable development as "development that meets the needs of the present without compromising the ability of future generations to meet their own needs". The interdependence of environmental sustainability and human well-being is brought to light by this principle, which continues to serve as the foundation for global efforts to achieve sustainable energy and development. The concentration on energy within the Sustainable Development Goals (SDG), in particular SDG 7, serves to emphasize the crucial role that energy plays in the accomplishment of broader social and economic goals, as well as in addressing the climate crisis. The SDG7 have listed the following targets to be achieved by 2030:

(a) Ensure universal access to affordable, reliable and modern energy services.
(b) Increase substantially the share of renewable energy in the global energy mix.
(c) Double the global rate of improvement in energy efficiency.
(d) Enhance international cooperation to facilitate access to clean energy research and technology, including renewable energy, energy efficiency and advanced and cleaner fossil-fuel technology, and promote investment in energy infrastructure and clean energy technology.
(e) Expand infrastructure and upgrade technology for supplying modern and sustainable energy services for all in developing countries, in particular least developed countries, Small Island developing States, and land-locked developing countries, in accordance with their respective programmes of support.

In addition to the specific emphasis placed on energy access within Sustainable Development Goal 7, both Sustainable Development Goal 9 and Sustainable Development Goal 13 place an emphasis on critical areas related to decarbonisation and sustainability within the context of economic and industrial development. According to the United Nations, Sustainable Development Goal 9 (SDG 9) encourages the development of infrastructure that is resilient, the advancement of sustainable industrialization, and the promotion of innovation, particularly in the field of environmentally friendly technologies (Nations 2030). The achievement of this objective highlights the significance of making investments in infrastructure that is capable of withstanding changes in the environment while simultaneously fostering economic growth in a manner that is sustainable. In addition to this, it emphasizes the significance of technological innovation in the achievement of energy efficiency and the reduction of emissions.

Particularly in terms of fostering responsible consumption and production, as outlined in Sustainable Development Goal 12, energy development plays a crucial role in indirectly supporting the achievement of all Sustainable Development Goals (SDG). Because energy is one of the primary natural resources, it is essential that it be managed effectively and utilized prudently in order to reduce the negative effects on the environment. According to the International Renewable Energy Agency (IRENA), it is essential to make the transition to high-efficiency renewable energy technologies such as solar, wind, geothermal, and hydropower in order to guarantee maximum energy output, safety, and energy security at reduced economic and environmental costs.

For instance, solar energy harvesting in areas that are characterized by prolonged exposure to sunlight and high solar irradiance can maximize the amount of energy that is produced. On the other hand, the entire process of energy production, from the extraction of resources to the consumption of energy, results in the generation of a number of different types of waste, such as emissions of greenhouse gases, pollution of water, and waste heat. According to previous study, it is essential to make selective use of clean energy technologies that have lower waste profiles (Panwar et al. 2011). Some examples of these technologies include geothermal energy, hydropower, wind energy, and artificial kinetic energy. These renewable technologies generate very little waste and contribute to the development of a circular economy by lowering the amount of resources that are depleted and stifling the production of harmful by-products.

In contrast to the widespread belief, solar energy and nuclear energy, despite the fact that they provide a significant amount of energy output, can present difficulties in terms of waste management and thermal pollution. The production and disposal stages of solar energy systems, particularly photovoltaic cells, result in the generation of hazardous waste. On the other hand, nuclear power plants continue to face challenges that are associated with radioactive waste and thermal discharge (Fernández-Arias et al. 2023). In light of this, the implementation of these technologies needs to be subjected to a thorough analysis with regard to the environmental consequences they produce. In addition, the incorporation of renewable energy sources into the residential, commercial, and industrial sectors contributes to the reduction of fossil fuel consumption, as well as the assurance of affordability and dependability. According to previous report, the substitution of conventional cooking fuels with electric induction cookers and the promotion of hybrid or electric vehicles are two examples of actions that have the potential to significantly reduce carbon emissions and reduce reliance on fossil fuels (Dincer and Acar 2015).

A significant contribution to the improvement of water management systems is also made by the development of energy. This aligns with Sustainable Development Goal 6 (Clean Water and Sanitation) and promotes sustainable urbanization and industrialization as envisioned in Sustainable Development Goal 11 (Sustainable Cities and Communities) and Sustainable Development Goal 9 (Industry, Innovation, and Infrastructure). Advanced energy systems make it possible to efficiently treat and pump water, which guarantees a consistent supply of clean water. In the end, responsible energy development helps to foster economic growth, social well-being, and equity, all while protecting the biodiversity of both terrestrial and aquatic environments (Carlsen and Bruggemann 2022).

In a similar vein, Sustainable Development Goal 13 (SDG 13) places an emphasis on critical climate action, more specifically the requirement for robust strategies to mitigate and adapt to climate change. In particular, this objective places an emphasis on the significance of lowering emissions of greenhouse gases and increasing resilience against the effects of climate change, particularly for ecosystems and communities that are particularly vulnerable (Adopted and Climate change, 2014). In order to accomplish the overarching goals of the 2030 Agenda for Sustainable Development, Sustainable Development Goal 13 (SDG 13) is an essential component. This is because it encourages climate resilience and low-carbon development pathways.

The Sustainable Development Goals (SDG) need to be addressed in a comprehensive and integrated manner because their goals are interconnected. For instance, if we concentrate on sustainable infrastructure (SDG 9) without taking into account the wider climate impacts (SDG 13), we might be unable to make significant headway toward achieving long-term sustainability. In a similar vein, the climate actions that are outlined in Sustainable Development Goal 13 are heavily reliant on the sustainable transformation of key sectors such as industry, energy, and transportation, which are at the core of Sustainable Development Goal 9. As a result, pursuing these objectives in isolation would be detrimental to the interconnectedness of these objectives as well as the overall effectiveness of efforts to promote global sustainability (Biermann et al. 2017).

Over the course of the past few years, nearly all sectors within the environmental and energy community have increasingly embraced the concept of "a more sustainable future" (Raimi et al. 2023). This includes everything from transportation and agriculture to waste management, industrial production, and building construction. This transition toward sustainability is largely driven by the necessity of reducing reliance on fossil fuels, improving energy efficiency, and promoting the conservation of natural resources and sustainable management of those resources by promoting sustainable management. It is imperative that industries, governments, and other stakeholders work together to devise and put into action strategies that promote clean energy and low-carbon technologies in order to accomplish these lofty Sustainable Development Goal targets from the United Nations.

When it comes to shaping a future that is both decarbonized and sustainable, the commitment of the industrial sector, in particular, is absolutely necessary. For the purpose of driving the necessary transformations in energy systems, infrastructure, and production processes, it is essential for industry leaders, governments, and organizations representing civil society to work together. It will be necessary to make significant investments in renewable energy, green innovations, and policies that promote energy efficiency and reduce emissions in order to achieve a sustainable future that is decarbonized, affordable, and secure. Redoubled efforts are required to ensure equitable access to clean energy while simultaneously reducing the carbon footprint in countries with high energy consumption and significant energy access deficits (Raimi et al. 2023; Citaristi 2022). This is especially important for countries that have a high energy consumption and significant energy access deficits.

1.5 Sustainable Energy Technologies and Low Carbon Buildings as Green Innovation

A crucial international policy framework that was established in 1992 with the intention of addressing climate change was established by the United Nations Framework Convention on Climate Change (UNFCCC) in response to the growing environmental crisis that was occurring at the time. In order to "prevent dangerous anthropogenic interference with the climate system" (Dawson and Spannagle 2008), the United Nations Framework Convention on Climate Change (UNFCCC) has set as its primary objective the stabilization of greenhouse gas concentrations in the atmosphere at levels that would be effective. In order to accomplish this objective, 192 nations have dedicated themselves to the implementation of policies that are centred on the reduction of energy consumption and the development of energy practices that are more sustainable. The goal of these efforts is to reduce the most severe consequences of climate change and to slow the rate at which temperatures are rising around the world.

One of the most important strategies for accomplishing these goals is to encourage environmental-friendly innovation, which is also known as innovation that is focused on sustainability. The term "green innovation" refers to a variety of strategies that are applied with the intention of lowering carbon emissions and improving environmental overall efficiency. The development of products and processes that have a minimal impact on the environment is the focus of this concept, which is frequently used interchangeably with other terms such as eco-innovation, environmental innovation, or sustainable innovation (Kemp and Pearson 2008). The low-carbon and circular economy frameworks, both of which aim to decouple economic growth from carbon emissions and resource depletion, are both advanced by green innovation, which plays a crucial role in the advancement of these various frameworks.

Environmental technologies, eco-efficiency, low-carbon design, and energy-efficient technologies are some of the concepts that fall under the umbrella term known as "green innovation. Green innovation is an umbrella term that involves a wide range of concepts. It is the realization of the integration of technological advancements with socially responsible practices with the intention of accomplishing the objectives of sustainable development. The environmental footprint of industries, particularly in sectors such as energy production, manufacturing, and construction, can be significantly reduced through the implementation of these innovations, which are highly important. For example, the development of energy-efficient technologies in the building sector has been identified as a key area where significant reductions in CO_2 emissions can be achieved (Canton 2021). This is one of the key areas where significant advancements can be made.

The term "green innovation" can be defined in a straightforward manner as the development of new products or processes that have the dual purpose of minimizing negative effects on the environment and fostering sustainable economic growth. Therefore, it is essential for policymakers, industries, and researchers to concentrate on the development and deployment of technological solutions that are contemporary, innovative, and environmentally friendly, particularly in crucial areas such as energy systems and building technologies (Desai 2020). There is a possibility that these innovations will significantly lessen the carbon footprint of these industries, while simultaneously enhancing their resistance to climate change and making a contribution to the long-term sustainability of the environment.

1.5.1 Sustainable Energy Technologies

In the context of energy, the term "sustainable energy technologies" refers to innovations and systems that provide energy in ways that are sustainable from an economic, social, and environmental perspective. The ability of sustainable energy technologies to fulfil the energy requirements of the present without jeopardizing the capacity of future generations to fulfil their own energy requirements is the essence of environmentally friendly energy technologies. The primary objective of these technologies is to lessen the negative effects on the environment that are caused by conventional energy sources such as fossil fuels. In addition, these technologies frequently make use of renewable resources such as solar, wind, and geothermal energy (Panwar et al. 2011).

Energy systems that are designed and implemented to minimize environmental degradation, improve energy efficiency, and foster a transition towards renewable energy sources are examples of sustainable energy technologies. These technologies can be defined as following the aforementioned definition. Priority is given to renewable resources with an emphasis on the reduction of carbon emissions, pollution, and resource depletion (Owusu and Asumadu-Sarkodie 2016). Such an approach is in line with the more comprehensive objectives of sustainable development. It is necessary for a sustainable energy technology to fulfil the following requirements:

(a) **Capacity for regeneration**

It should obtain its power from sources that are naturally replenished on a timescale that is comparable to that of humans, such as the sun, the wind, or the changes in the hydrological cycle.

(b) **The effects on the community**

In order to mitigate or eradicate harmful environmental effects, such as emissions of greenhouse gases, the destruction of habitats, and pollution of air or water, it is necessary to take action.

(c) **Effectiveness**

The output of energy should be maximized in relation to the input, while the amount of waste produced should be minimized.

(d) **The possibility of social and economic impact**

It ought to be easily accessible and reasonably priced, thereby assisting the energy requirements of a wide variety of communities and contributing to the growth of local economies (Vallecha et al. 2021).

In accordance with the kind of energy source that they are able to harness, sustainable energy technologies can be classified into a number of different categories. Among the most common and extensively researched categories are the following:

i. **Technology utilizing solar energy**

Through the use of photovoltaic (PV) cells, concentrated solar power (CSP), or solar thermal systems, solar energy technology is able to convert sunlight into either electricity or heat. The ability of solar photovoltaic (PV) technology to generate electricity from direct sunlight makes it a leading solution for energy sustainability (Ren21 2021). This technology is one of the most rapidly advancing fields of sustainable energy. When it comes to reducing reliance on the grid for electricity, solar photovoltaic panels, for instance, are frequently utilized in both residential and commercial settings. For instance, the Bhadla Solar Park in India, which is the largest solar farm in the world, generates more than 2245 megawatts of electricity, which demonstrates the scalability of solar technology for the production of mass energy (Koutoudjian et al. 2021).

ii. **Technology used for wind energy**

Through the use of turbines, wind energy technology is able to harness the kinetic energy of wind in order to generate electricity. The relatively low operational costs and zero emissions that wind farms produce during the process of energy generation have contributed to their rise to prominence as a significant component of the global energy mix. Wind farms can be found both onshore and offshore. The designs of wind turbines are constantly being improved in order to improve energy capture and reliability (Heier 2014). Wind energy is one of the most efficient forms of renewable energy. For instance, the Gansu Wind Farm in China, which is also referred to as the Jiuquan Wind Power Base, is the largest onshore wind farm in the world. It has a capacity of more than 20 gigawatts (GW), which demonstrates the potential of wind energy for the production of large amounts of energy (Council and Global offshore wind report 2020).

iii. **Technology utilizing geothermal energy**

It is possible to generate electricity or provide direct heating through the utilization of geothermal energy technology, which draws heat from the interior of the Earth. Considering that the Earth is constantly producing heat through the process of radioactive decay, this source of energy is not only dependable but also sustainable. According to previous study, geothermal plants are extremely efficient, meaning that they provide base load energy while having a minimal impact on the environment (Lund et al. 2011). Geothermal energy is used to heat approximately 90% of homes in Iceland, and geothermal power plants are responsible for producing approximately 25% of the electricity in the country (Fridleifsson 2001). Examples of countries that are at the forefront of geothermal energy include Iceland.

iv. **Technology relating to hydropower**

Hydropower is one of the most established and well-established forms of renewable energy technology ever developed. Utilizing the energy that is contained within falling or flowing water, it is able to generate electricity. It is possible to obtain this energy from dams, rivers, or the movements of the tides. Hydropower plants can operate on either a large or small scale, and they produce only a small amount of greenhouse gas emissions. However, the environmental impact that these plants have on aquatic ecosystems needs to be carefully managed. As an illustration, the Three Gorges Dam in China is the largest hydropower plant in the world. It generates 22.5 gigawatts of electricity, which is equivalent to the output of twenty coal-fired power stations (Magee 2015).

Specifically, Sustainable Development Goal 7 (SDG 7) calls for affordable, reliable, sustainable, and modern energy for all people (Nations 2030). Sustainable energy technologies are an essential component in the process of achieving the Sustainable Development Goals (SDG) of the United Nations International. The transition to a low-carbon economy, the mitigation of climate change, and the reduction of global dependence on fossil fuels are all accomplished through the utilization of these technologies. Improvements in energy efficiency that are driven by sustainable energy technologies help reduce the overall demand for energy, which in turn leads to a reduction in the amount of carbon dioxide emissions that are produced on a global scale. Additionally, the implementation of renewable energy sources has the potential to stimulate local economies by generating employment opportunities in the areas of technology development, installation, and maintenance, as well as by helping underserved communities gain access to energy (Maradin et al. 2017).

Not only does the successful implementation of sustainable energy technologies provide direct benefits to the environment, but it also promotes social equity by significantly increasing access to energy in regions that are still in the process of developing. In addition to this, it helps to strengthen economic resilience by diversifying energy portfolios and decreasing reliance on volatile markets for fossil fuels (Vallecha et al. 2021). In order to effectively combat climate change and environmental degradation on a global scale, the implementation of sustainable energy technologies is an essential component. These technologies provide energy solutions that are cleaner, more efficient, and more equitable. They do this by harnessing renewable energy sources such as solar, wind, geothermal, and hydropower. As technological advancements continue to advance, the role of sustainable energy technologies will only grow in importance in the process of shaping a future that is more sustainable and resilient for everyone.

1.5.2 Sustainable and Low Carbon Buildings

Buildings that are both environmentally friendly and low in carbon emissions have become an important issue for both developing countries and developed countries in the twenty-first century. According to the United Nations Department of Economic and Social Affairs (UN DESA), since it is anticipated that the global population will increase from 7.7 billion in 2019 to approximately 9.7 billion by 2050, and then further increase to 0.9 billion by the end of the century, the demand for essential resources such as water, energy, and raw materials will significantly increase. We anticipate that this surge in resource consumption will put a strain on biological systems and contribute to the deterioration of natural ecosystems through its effects. There is a significant contributor to this trend that is not sustainable, and that contributor is the built environment, specifically through its increased energy consumption. Previous research has demonstrated, time and time again, that buildings are among the most significant consumers of energy, accounting for close to thirty percent to forty percent of the total energy consumption across the globe (Adopted and Climate change, 2014; Canton 2021; Desai 2020).

In addition, the building industry is a significant contributor to greenhouse gas (GHG) emissions. It is responsible for more than one-third of the total global emissions, which makes it an essential sector in the fight against climate change and global warming (Skillington et al. 2022). Buildings are the primary source of these emissions because they are the primary source of energy consumption, which in turn leads to the production of waste and harmful atmospheric pollutants that contribute to the worsening of environmental degradation. It has become an important strategy for mitigating the negative effects that urbanization has on the environment, and one of the ways that this can be accomplished is by encouraging sustainable building practices. According to the United Nations Environment Programme (UNEP), sustainable buildings incorporate multidimensional approaches, concentrating on environmental, economic, social, ecological, technical, and technological pillars in order to achieve success over the long term.

Buildings that are environmentally friendly and sustainable provide solutions that can help reduce the adverse effects on the environment, the economy, and society that are characteristic of conventional building practices. These structures are designed to reduce greenhouse gas emissions through the implementation of a variety of strategies, such as the utilization of renewable energy sources, materials that are efficient in their use of resources, and advanced energy management systems. According to the United Nations Environment Programme (UNEP), achieving sustainability in buildings requires lowering the amount of embodied energy and emissions, increasing the utilization of natural resources, recycling materials, and meeting standards for indoor environmental quality (IEQ) (Desai 2020). The term "sustainable buildings" has been defined by a number of specialists as those that are able to reduce greenhouse gas emissions in a more efficient manner than conventional buildings and that contribute to the achievement of the Net-Zero Carbon Buildings Commitment (Sánchez Cordero et al. 2019).

Efficiency in energy use is an essential component in both the construction and operation of environmentally responsible buildings. As stated by the International Energy Agency (IEA) in 2020, it is considered the fundamental concept for the development of energy-efficient, low-energy, ultra-low-energy, and zero-energy buildings. One of the quickest and most cost-effective solutions to mitigate the environmental, economic, and social impacts of the building sector is the implementation of energy efficiency measures, which are widely recognized as being among the strategies. A wide variety of approaches, such as energy performance benchmarking, energy-saving technologies, building-integrated photovoltaic (BIPV) systems, advanced building envelopes, and innovative resource-efficient materials, can be utilized to accomplish the goal of achieving high energy efficiency in buildings (Adopted and Climate change 2014; Canton 2021; Desai 2020).

Using life cycle assessment (LCA) and cost–benefit analysis techniques, sustainable buildings have the potential to generate significant cost savings from an economic point of view. This is accomplished by lowering operational expenses associated with the building, such as those related to energy and maintenance (Technical Committee and TC 207 2006). Long-term financial sustainability can be achieved through the implementation of economic optimization models that incorporate solar power generation, energy storage systems, and life cycle cost minimization, as stated by the International Organization for Standardization (ISO) in 2018. The implementation of local rating systems that take into account cultural, social, and environmental dimensions is a socially beneficial measure that helps to ensure that sustainable buildings are able to cater to the requirements of a wide range of communities. In order to improve the well-being of occupants and their productivity, it is necessary to give priority to factors related to the quality of the indoor environment, such as the lighting, the air quality, and the thermal comfort (Desai 2020).

References

Adopted I (2014) Climate change 2014 synthesis report. IPCC: Geneva, Switzerland, pp 1059–1072

Ahmad MI et al (2020) Introduction: energy, green innovation and sustainable future. Energy Recovery Technology for Building Applications: Green Innovation towards a Sustainable Future pp 1–4

Allan RP et al (2021) Intergovernmental Panel on Climate Change (IPCC). Summary for Policymakers, in Climate change 2021: the physical science basis. Contribution of working group I to the sixth assessment report of the intergovernmental panel on climate change. Cambridge University Press, pp 3–32

Biermann F, Kanie N, Kim RE (2017) Global governance by goal-setting: the novel approach of the UN sustainable development goals. Current Opin Environ Sustainabil 26:26–31

Brundtland GH, Mansour K (2010) World commission on environment and development (WCED). 1987. Our common future, 2010: p 2010

Canton H (2021) International energy agency—IEA. The Europa Directory Int Org 2021. Routledge, pp 684–686

Carlsen L, Bruggemann R (2022) The 17 United Nations' sustainable development goals: A status by 2020. Int J Sust Dev World 29(3):219–229

Citaristi I (2022) United Nations Development Programme—UNDP. The Europa directory of international organizations 2022. Routledge, pp 183–188

Council GWE (2020) Global offshore wind report 2020. GWEC: Brussels, Belgium, 19:10–12

Dawson B, Spannagle M (2008) United Nations framework convention on climate change (Unfccc). The complete guide to climate change. Routledge, pp 392–403

De Pryck K (2021) Intergovernmental expert consensus in the making: the case of the summary for policy makers of the IPCC 2014 Synthesis Report. Global Environmental Politics 21(1):108–129

Desai BH (2020) 14. United Nations Environment Programme (UNEP). Yearbook of Int Environ Law 31(1):319–325

Dietz S et al (2016) 'Climate value at risk'of global financial assets. Nat Clim Chang 6(7):676–679

Dincer I, Acar C (2015) A review on clean energy solutions for better sustainability. Int J Energy Res 39(5):585–606

Fernández-Arias P, Vergara D, Antón-Sancho Á (2023) Global review of international nuclear waste management. Energies 16(17):6215

Florini A (2011) The international energy agency in global energy governance. Global Pol 2:40–50

Fridleifsson IB (2001) Geothermal energy for the benefit of the people. Renew Sustain Energy Rev 5(3):299–312

Heier S (2014) Grid integration of wind energy: onshore and offshore conversion systems. John Wiley & Sons

Kemp R, Pearson P (2008) MEI project about measuring eco-innovation. Final report

Koutoudjian G et al (2021) About IRENA The International Renewable Energy Agency (IRENA) is an intergovernmental organisation that supports countries in their transition to a sustainable energy future, and serves as the principal platform for international co-operation, a centre of excellence, and a repository of policy, technology, resource and financial knowledge on renewable energy. IRENA promotes the. IRENA promotes the, 2021: p. 24.

Lamb WF et al (2021) A review of trends and drivers of greenhouse gas emissions by sector from 1990 to 2018. Environ Res Lett 16(7):073005

Lenton TM (2013) Environmental tipping points. Annu Rev Environ Resour 38(1):1–29

Li X, Peachey B, Maeda N (2024) Global warming and anthropogenic emissions of water vapor. Langmuir 40(14):7701–7709

Liu D (2023) International energy agency (IEA). The Palgrave Encyclopedia of Global Security Studies. Springer, pp 830–836

Lund JW, Freeston DH, Boyd TL (2011) Direct utilization of geothermal energy 2010 worldwide review. Geothermics 40(3):159–180

Magee D (2015) Hydropower and end-use electrical efficiency in china: State support and potential contribution to low-carbon development. Copenhagen J Asian Stud 33(1):64–89

Mahama A (2012) 2012 international year for sustainable energy for all: African Frontrunnership in rural electrification. Energy Policy 48:76–82

Maradin D, Cerović L, Mjeda T (2017) Economic effects of renewable energy technologies. Naše Gospodarstvo/our Economy 63(2):49–59

Nordhaus WD (1994) Managing the global commons: the economics of climate change, vol 31. MIT press Cambridge, MA

Organization W.H., *WHO ambient air quality database, 2022 update: status report*. 2023: World Health Organization.

Owusu PA, Asumadu-Sarkodie S (2016) A review of renewable energy sources, sustainability issues and climate change mitigation. Cogent Eng 3(1):1167990

Panwar NL, Kaushik SC, Kothari S (2011) Role of renewable energy sources in environmental protection: a review. Renew Sustain Energy Rev 15(3):1513–1524

Pomponi F, Moncaster A (2016) Embodied carbon mitigation and reduction in the built environment–What does the evidence say? J Environ Manage 181:687–700

Raimi D et al (2023) Global energy outlook 2023: sowing the seeds of an energy transition. Resources for the future

Ren21 R (2021) Global Status Report, REN21 Secretariat), Paris, 2021

Sachs JD (2012) From millennium development goals to sustainable development goals. The Lancet 379(9832):2206–2211

Salehabadi A et al (2020) Overview of energy, society, and environment towards sustainable and development. Energy, Society and the Environment, Solid-State Hydrogen Storage Materials, pp 1–8

Sánchez Cordero, A., S. Gómez Melgar, and J.M. Andújar Márquez, *Green building rating systems and the new framework level (s): A critical review of sustainability certification within Europe.* Energies, 2019. **13**(1): p. 66.

Skillington K et al (2022) A review of existing policy for reducing embodied energy and greenhouse gas emissions of buildings. Energy Policy 168:112920

Technical Committee ISO/TC 207, E.M., *Environmental management-life cycle assessment-principles and framework.* 2006: International Organization for Standardization.

Thomas RS et al (2019) The next generation blueprint of computational toxicology at the US Environmental Protection Agency. Toxicol Sci 169(2):317–332

Nations U (2015) Transforming our world: the 2030 agenda for sustainable development. United Nations, Department of Economic and Social Affairs, New York, 1, p 41

Ürge-Vorsatz D et al (2015) Heating and cooling energy trends and drivers in buildings. Renew Sustain Energy Rev 41:85–98

Vallecha H et al (2021) Evaluation of barriers and enablers through integrative multicriteria decision mapping: Developing sustainable community energy in Indian context. Renew Sustain Energy Rev 138:110565

Woodward RT, Bishop RC (1997) How to decide when experts disagree: uncertainty-based choice rules in environmental policy. Land Economics, pp 492–507

Chapter 2
Thermal Energy-Efficient Systems for Building Applications

Contents

S. Riffat et al., *Sustainable Energy Technologies and Low Carbon Buildings*, Lecture Notes in Energy 45, https://doi.org/10.1007/978-3-031-78853-6_2

Abstract The increasing need for building energy efficiency, the advancement of green technology, and the development of thermal energy-efficient systems are crucial in world today. These systems optimize the use of renewable energy and provide an effective solution for the common mismatch between building energy supply and demand. This chapter focuses on the design, development, validation, and demonstration of low-cost, modular thermal energy-efficient systems for various building applications, including heating, cooling, and domestic hot water generation, utilizing technologies such as solar collectors and highly efficient heat pumps. The primary goal for researchers and engineers is to develop energy-efficient thermal systems and processes with optimized designs to address rising energy costs and stringent greenhouse gas emission targets. Improving energy efficiency is essential for businesses to stay competitive, leading to substantial efforts in enhancing both overall systems and individual components. These improvements ultimately reduce energy consumption, costs, and environmental impact. This chapter provides a comprehensive discussion on thermal energy-efficient systems, detailing their description, working mechanisms, and applications.

2.1 Solar Water Heater

Energy is essential for the life of humans, and the impressive accomplishments of civilization have mostly been enabled by the progressively efficient and widespread use of different energy sources. An eminent obstacle in the twenty-first century is the development of techniques for producing and utilizing energy that can satisfy the escalating needs while safeguarding the worldwide ecology. Renewable energy sources, such as solar, wind, biomass, and geothermal energy are plentiful, limitless, and eco-friendly (Shakir et al. 2023). Out of all the available options, solar energy is particularly noteworthy as a very promising type of renewable energy that has the potential to fulfill the energy requirements of the entire world. A solar water heater is a thermal device that utilizes water to capture solar thermal energy, making it one of the most effective implementations of solar thermal technology (Ahmed et al. 2021). This technology is efficient, dependable, and economical for utilizing solar energy to meet 50–80% of the demand for hot water. The use of solar water heating technology is widespread in several countries, aiming to decrease utility expenses in both commercial and residential domains. Solar water heaters have become highly efficient in all geographical and climatic regions due to technological improvements.

In addition to being used domestically, solar water heaters have a wide range of commercial and industrial applications. These include usage in swimming pools, laundries, hotels, and restaurants, as well as in sectors such as food and beverage, processing, and textiles. On a global scale, the residential sector consumes 15–25% of its total energy on water heating. Water heating accounts for 18% of domestic energy consumption in the United States and 23% in the United Kingdom (Yildiz et al. 2021). Within the industrial sector, the process of heating water can contribute to a significant amount of energy consumption. Water heating in the textile industry can contribute to as much as 65% of the overall energy consumption during processes such as dyeing, finishing, and drying (Kumar et al. 2021a). In addition, solar water heaters have the capability to supply energy to district heating networks and assist in powering huge industrial loads.

2.1.1 System Description and Working Mechanism

A solar water heater generally consists of a solar thermal collector, a water tank, and other accessory facilities such as supporting stands, connecting pipes, and instrumentation. In any solar water heating system, two essential elements are there: one is the storage tank, and the next is a solar collector (Lazzarin 2020). The collector panel is designed to optimize the capture of solar radiation by utilizing highly efficient materials and coatings that enhance heat absorption. Modern collectors employ selective coatings that significantly reduce heat loss, allowing for greater energy capture even in cooler climates. These technological improvements have increased the thermal efficiency of solar collectors by up to 20% compared to earlier models, making them more viable for a wider range of geographical locations (Evangelisti et al. 2019). The collectors have the absorber on a dark absorbing surface. Hence, solar radiations are collected and the collected heat energy is given up through a heated fluid, generally water. In some advanced systems, other fluids such as glycol mixtures are used as heat transfer fluids to enhance performance in colder climates where water may freeze. This adaptability of the heat transfer medium allows solar water heaters to be operational year-round in diverse environmental conditions. The tank with heated water is insulated to not let heat go out. High-quality insulation materials such as polyurethane foam or vacuum insulation are used in modern tanks to minimize heat loss. These tanks are often designed to maintain water at elevated temperatures for extended periods, reducing the need for additional heating energy and ensuring a continuous supply of hot water.

Solar water heating has been put into two categories, basically: active and passive systems. Active systems require external energy sources, typically electricity to operate pumps or controllers that circulate the heat transfer fluid. This approach will make them more efficient in terms of energy output but also more expensive to install and maintain. These systems are ideal for regions with colder climates where passive systems might struggle to maintain optimal performance. In comparison, passive solar water heating systems are far less expensive than active systems. However, their simplicity offers greater reliability and longer service life, as passive systems contain fewer mechanical components that can break down (Patel 2023). This makes them especially popular in regions where maintenance services are limited or where cost considerations are paramount. Compared to active solar water heating systems, these passives are less effective but tend to be more reliable and durable. Despite the lower efficiency, recent advancements in passive system design, such as improved thermosiphon circulation and better insulation materials, have increased their effectiveness, making them a viable option for a greater range of applications.

(a) **Active Solar Water Heating Systems**

Active solar water heating systems are more technologically advanced than passive systems and rely on mechanical devices such as pumps and controllers to circulate the heat transfer fluid. This makes them more efficient in terms of heat capture and distribution, allowing them to function well even in climates where solar radiation is less intense. These systems also offer greater flexibility in terms of installation as the placement of collectors and storage tanks is not constrained by natural circulation methods. Additionally, active systems can integrate various sensors and smart controls that optimize performance based on real-time weather conditions and hot water demand, further enhancing energy efficiency.

i. **Direct Circulation Techniques**

Such systems operate by circulating water through the collectors into a dwelling using pumps. In a direct circulation system, water from the household is pumped through the solar collectors where it is heated by solar energy and then returned to a storage tank or directly used. This design is straightforward and effective, particularly in regions with abundant sunlight and mild climates (Sarbu and Sarbu 2021a). However, since the fluid being heated is the same water that is used for consumption, these systems are susceptible to freezing in colder climates, which can damage pipes and other components. Therefore, they are best suited for climates where temperatures do not drop below freezing. Such systems are also ideal for a climate with no frost. In regions where frost might occur, additional mechanisms such as freeze protection valves or temperature sensors can be installed to drain the system or circulate warm water during cold periods to prevent freezing.

ii. **Indirect Distribution Networks**

Active systems which use a pumping method that circulates a non-freezing fluid to transport the collected heat energy from the heated collectors to a heat exchanger in which water entering the residence is heated with the transferred heat. The use of non-freezing fluids such as glycol, in these systems makes them highly effective in extremely cold climates. The fluid absorbs heat from the solar collectors and transfers it to the household water through a heat exchanger, ensuring that the system continues to operate efficiently even in sub-zero temperatures. This method also minimizes the risk of freezing within the pipes system and collectors. Indirect systems can be combined with other heat sources such as conventional boilers or electric heaters to ensure a continuous supply of hot water even during periods of limited sunlight (Chauhan et al. 2021). Additionally, modern indirect systems often come with automatic control units that regulate the fluid flow based on weather conditions, optimizing the efficiency of heat transfer and minimizing energy consumption. These systems are pervasive in areas of extremely low temperatures. In such environments, indirect systems are often preferred because they require less maintenance and are more resilient to the harsh conditions that could otherwise compromise the longevity of a solar heating system.

(b) **Passive Solar Water Heating Systems**

Passive solar water heating systems do not require pumps or mechanical devices to circulate the heat transfer fluid. Instead, they rely on natural convection or gravity to move the water between the collector and the storage tank. This simplicity makes them more cost-effective and easier to maintain compared to active systems, although they may not be as efficient in regions with colder climates or frequent cloud cover. Passive systems are most suitable for regions with moderate climates, where the temperature does not frequently drop below freezing.

i. **Passive Integrated Collector-Storage Systems**

These are systems with a see-through water storage tank, whereby the water is directly heated by the sun. In integrated collector-storage (ICS) systems, the water tank itself acts as both the solar collector and the storage vessel. The design typically consists of a dark-colored storage tank, which is placed inside an insulated, transparent enclosure. Solar radiation passes through the transparent surface and heats the water directly inside the tank. This simple design allows the system to store heat during the day and provide hot water during the night, without the need for pumps or controllers (Bilardo et al. 2020). However, ICS systems are generally limited by their storage capacity, and they may not perform as well in regions with long, cold winters due to heat loss during the night. The heated water within the tank then joins the plumbing system. They work very well for households with high use of hot water during the day and at night. These systems are particularly beneficial in areas with mild climates, where cold weather is infrequent and households with consistent daily demand for hot water. They offer a cost-effective solution for residential hot water needs, as they can provide between 30–70% of the hot water supply in such conditions (Rao and Somwanshi 2022). Therefore, they work exceptionally well in areas where rare cold snaps can be found. In regions with occasional cold snaps, insulated storage tanks or freeze protection features may be added to enhance the durability and efficiency of the system.

ii. **Thermosyphon Techniques**

It is such a system with a collector mounted on the roof to heat water. Thermosyphon systems rely on the principle of natural convection to circulate water between the collector and the storage tank. In these systems, the storage tank is positioned higher than the collector usually on the roof, so that as the water in the collector is heated it naturally rises into the tank. Cooler water from the bottom of the tank flows down into the collector to be heated creating a continuous cycle of circulation (Koholé and Tchuen 2020). This design eliminates the need for pumps, making it highly reliable and requiring minimal maintenance. By merely opening a tap, water passes through the plumbing system. Thermosyphon systems are widely used in regions where electricity is unreliable or where simplicity and low maintenance are critical concerns. However, they are sensitive to the installation angle, as the tilt of the collector must be optimized for maximum solar exposure throughout the year. Most of these systems are 40 gallons. Larger systems can be designed for commercial or multi-family buildings, with capacities exceeding 100 gallons, although the performance may vary depending on the design and geographical location (Lin et al. 2022). Thermosyphon systems are an affordable and efficient solution for providing hot water in warm to moderate climates, with minimal operational costs and high reliability over the long term.

The water circulation in solar water heating systems can be driven by way of both passive (thermosyphon) and active (using a pump) modes (Xu et al. 2022). In passive systems, the water or heat transfer fluid circulates naturally due to the difference in density between hot and cold fluids. Whereas in active systems, mechanical pumps are used to circulate the fluid offering greater control and efficiency. The choice between these two modes depends on factors such as climate, the required efficiency, and the available budget for installation and maintenance. While active systems more expensive and complex, it allow for greater flexibility in terms of system design and installation. This is because s they are not dependent on the placement of the storage tank relative to the collector (Zhou et al. 2020). It can deliver heat for preparing domestic hot water, a swimming pool, or for providing heating in a heat-supplying circuit. The versatility of solar water heating systems extends beyond domestic use, with larger systems being implemented for industrial processes, district heating networks, and commercial applications. They can significantly reduce energy consumption in heating processes, particularly for large-scale hot water demand in hospitals, hotels, and industrial operations.

Type of solar thermal collectors There are several types of solar thermal collectors: flat-plate, evacuated tube, concentrating, parabolic, CPC, and solar tower. Each type of collector is designed to optimize heat capture and transfer for specific applications. Concentrating and parabolic collectors are commonly used in industrial settings where higher temperatures are required. While flat-plate and evacuated tube collectors are more suitable for residential or commercial hot water production. Solar towers are employed in large-scale solar thermal power plants, where they focus sunlight onto a central receiver to generate electricity. Mostly used for the application of domestic hot water preparation are flat-plate and evacuated tube solar collectors. These collector types are the most common in residential solar water heating systems due to their balance of cost, efficiency, and ease of installation. Flat-plate collectors are widely used in warmer climates while evacuated tube collectors with their superior insulation are preferred in colder regions where freezing is a concern.

(a) **Flat-Plate Collector**

Flat-plate solar thermal collectors are constructed from an insulated box that holds a black metal sheet containing pipes and is provided with a cover. The black metal sheet, known as the absorber plate is crucial as it absorbs the maximum amount of solar radiation and converts it into heat. The absorbed heat is then transferred to the fluid circulating through the pipes attached to the absorber. There are two general types of flat-plate collectors: glazed and unglazed (Tian et al. 2023). Glazed flat-plate collectors have one or more glass or polymer covers that create a greenhouse effect, trapping heat and improving the efficiency of the collector. The glazing also provides weatherproofing and durability, making glazed collectors more suitable for long-term installations in a variety of climates. Glazed flat-plate collectors mean a version that is insulated and weatherproof. They have a dark absorber plate enclosed with one or more often two glass or polymer covers (Ameri et al. 2021). Unglazed flat-plate collectors lack a cover and are typically used in applications where lower temperatures are sufficient, such as pool heating. Without the insulating cover, these collectors are less efficient in cooler weather but are cheaper to manufacture and install. Unglazed means that the type does not involve a cover or enclosure over the dark absorber plate, the latter being made of either metal or polymer and it serves mainly for heating swimming pools (Shemelin and Matuska 2022). In unglazed collectors, the absence of glazing reduces costs but also means that heat loss is higher, especially in cooler conditions. However, they are ideal for heating outdoor swimming pools where water temperatures do not need to be as high and where heat loss to the environment is less of a concern.

(b) **Integral Collector-Storage Systems**

More commonly called ICS or batch systems, these units comprise one or more black tanks or tubes inside an insulated, glazed box. The black tanks or tubes act as both the collector and the storage unit, allowing the system to absorb solar radiation directly into the stored water. The glazing over the tank enhances the greenhouse effect, trapping heat and raising the water temperature inside. The simplicity of this design makes ICS systems one of the most cost-effective options for solar water heating, as they require fewer components compared to more complex systems (Bilardo et al. 2020). However, the lack of active circulation limits their efficiency in certain climates. Cold water is pre-heated in these as it recirculates through the solar collector. In this process, the cold water enters the black tubes or tanks, where it absorbs heat from the sun. Once the water reaches a sufficiently high temperature, it is ready to be used or transferred to a conventional backup water heater. This mechanism will ensure a consistent supply of hot water. This integration with a backup system is particularly useful for cloudy days or during periods of increased hot water demand, providing a reliable solution for households.

It then passes to a more conventional backup water heater in a building so that hot water is always available. This configuration offers the advantage of ensuring hot water availability even during periods when solar heating alone is insufficient, such as during extended cloudy weather or winter months. However, because ICS systems do not actively circulate water, they can lose efficiency in colder environments as the heat collected during the day may be lost to the surroundings overnight. This type is used in warmer climates because freezing can be a problem. ICS systems perform best in regions where temperatures remain consistently above freezing, as the exposed water tanks can lose heat rapidly in cold weather. In freezing conditions, there is also a risk that the water inside the system could freeze, potentially causing damage to the tanks or tubes. To mitigate these risks, ICS systems are typically installed in climates with mild winters or in tropical regions where freezing is not a concern. Additionally, improvements such as insulating the storage tanks further or integrating freeze protection measures can expand their use to a wider range of climates, though these solutions may increase the cost.

(c) **Evacuated Tube Solar Collectors**

In general, evacuated tube collectors are parallel rows of optically transparent glass tubes. These tubes are arranged in such a way that they maximize exposure to sunlight, capturing solar radiation even when the angle of the sun is lower such as during winter months. This design makes evacuated tube collectors particularly efficient compared to other types of solar collectors, especially in regions with cooler climates or where sunlight is less intense. The vacuum inside the tubes minimizes heat loss, allowing the system to retain a higher proportion of the absorbed energy (Kumar et al. 2021b). An evacuated tube consists of two layers: the outer layer is usually made of glass and an absorber tube is made of metal to which a fin is attached. The outer glass tube provides durability and protection from the elements, while the vacuum between the outer layer and the absorber tube acts as an excellent insulator, reducing heat loss to the surrounding environment. The absorber tube typically made from copper or aluminum is coated with a selective material that enhances its ability to absorb solar radiation. The attached fin is designed to increase the surface area allowing for maximum heat absorption while minimizing radiant heat loss. This combination of high thermal absorption and low heat loss makes evacuated tube collectors far more efficient than flat-plate collectors in colder climates or during overcast days.

The design of the fin is such that it may absorb the solar radiation energy with the least possible loss of radiant heat energy. By optimizing the design of the fin, evacuated tube collectors ensure that most of the solar energy is transferred to the heat transfer fluid with minimal energy loss. This is particularly important in colder regions where the external temperature can significantly reduce the efficiency of other solar collector types. The vacuum layer effectively insulates the absorber tube from ambient temperatures allowing the system to maintain high efficiency even in sub-zero conditions (Kumar et al. 2021b). Today, countries are increasingly using evacuated tube collectors for commercial uses. Due to their high efficiency and adaptability, evacuated tube collectors are now widely implemented in large-scale commercial applications such as heating for hospitals, hotels, and industrial processes. In many countries, including China, Germany, and the United States, they have become a preferred choice for solar water heating and even for solar space heating in colder regions. Their modular design allows for easy scaling, enabling the use of evacuated tube systems in both small residential setups and large commercial installations. Additionally, ongoing research is improving the durability of evacuated tube collectors, making them even more viable for long-term use in various climates.

In extremely cold regions, all-glass evacuated solar collectors are susceptible to freezing and damage. Despite their high efficiency, traditional evacuated tube collectors that contain water are prone to freezing in very low temperatures. This poses a significant risk of damage to the glass tubes, leading to decreased performance and increased maintenance costs. To address this challenge, advancements in evacuated-tube technology have led to the development of heat pipe collectors which are better suited to harsh winter conditions. To mitigate this risk, solar evacuated-tube heat pipe collectors have been developed which avoid water in the evacuated tubes. Instead of water, a special working fluid is used within the heat pipe which ensures that the system remains operational even in sub-zero temperatures. This design eliminates the risk of water freezing within the tubes, thus preventing damage to the collector and improving its durability in extremely cold environments.

i. **Inner Black Tube**

The inner black tube design features an inner tube with a selectively coated outer surface, typically black. This coating is highly efficient at absorbing solar radiation, converting it into heat with minimal reflective losses. The black color is specifically chosen for its superior ability to absorb a broad spectrum of sunlight. The selective coating reduces radiated heat losses, ensuring that most of the captured solar energy is transferred to the heat pipe (Zhang et al. 2024). This coating efficiently absorbs solar energy and transfers the heat to a heat pipe inside the tube. The heat pipe, which contains a working fluid commonly a refrigerant or other heat-conductive liquid, undergoes phase changes to transport heat more effectively. When the working fluid inside the heat pipe is heated, it evaporates, turning into vapor and rising to the top of the pipe. At the top, the vapor releases its heat into a heat exchanger or manifold, where it condenses back into a liquid and flows back down to the bottom of the pipe. Then it is ready to repeat the process.

The heat pipe containing a working fluid evaporates and transports the heat to a manifold. This design is highly efficient because the phase change which is evaporation and condensation, allows for rapid heat transfer. This mechanism will make the system highly responsive to changes in solar radiation. Moreover, because the heat pipe is separated from the water in the storage tank, there is no risk of freezing making this system highly suitable for colder climates. This design is highly effective in absorbing and transferring heat, making it suitable for residential and commercial solar water heating systems in extremely cold regions by preventing freezing and damage. The inner black tube design has gained widespread popularity in regions with harsh winters, as it offers high thermal efficiency, minimal heat loss, and a robust solution to freezing problems that commonly plague other solar water heating systems.

ii. **Clear Tube**

The clear tube design utilizes a selectively absorptive coating applied to the clip fin sheet attached to the heat pipe. This selectively absorptive coating is engineered to maximize solar energy absorption while minimizing reflective losses. It works by capturing solar radiation across a broad spectrum and converting it into heat, which is then efficiently transferred to the heat pipe. The clear outer tube allows sunlight to pass through unobstructed, enhancing the overall energy capture. The use of a transparent tube also simplifies inspection and maintenance, as the internal components of collector are easily visible. This design is particularly advantageous in larger solar installations, where regular maintenance is required to ensure optimal performance.

Sunlight passes through the clear tube and is absorbed by the coating on the fin sheet. The clip fin sheet attached to the heat pipe serves as a critical component for transferring the absorbed heat to the heat pipe. The fin sheet increases the surface area available for heat absorption, improving the overall efficiency of the system. This ensures that a greater proportion of the sunlight that passes through the clear tube is captured and converted into usable heat energy (Tabarhoseini et al. 2022). The absorbed heat is then transferred to the heat pipe, which carries it to the manifold. As with the Inner Black Tube design, the heat pipe contains a working fluid that evaporates when heated and condenses after releasing its heat into the manifold. The heat is then transferred to the storage tank or used directly for heating purposes, while the cooled fluid recirculates within the heat pipe to continue the cycle. This efficient heat transfer process allows the Clear Tube design to maintain high performance, even in environments where sunlight may be less intense or temperatures are lower.

This design allows for better visibility and maintenance while ensuring efficient heat transfer. The transparent nature of the clear tube makes it easier to detect any issues such as blockages, dirt accumulation or damage to the internal components allowing for timely maintenance and cleaning. This feature is particularly beneficial in large-scale solar thermal installations, where system downtime can be costly. By facilitating easier maintenance, the Clear Tube design helps to ensure that the system operates at peak efficiency throughout its lifetime. It is ideal for larger solar thermal systems and provides a reliable solution for preventing freezing in extremely cold climates. Due to its robust design, the Clear Tube collector is well-suited for larger-scale commercial or industrial applications such as district heating or hot water supply for large buildings. The clear outer tube, in combination with the advanced heat transfer mechanism, ensures that the system continues to perform efficiently in extremely cold climates where freezing could otherwise be a concern. The separation of the heat pipe from the working fluid ensures that freezing water is never in direct contact with the collector, making it a reliable solution for harsh winter conditions.

The high cost of evacuated tubes for solar evacuated tube heat pipe collectors has led to the development and research of the integral collector-storage (ICS) system as an affordable alternative (Bilardo et al. 2020). An ICS system consists of one or more black-painted tanks or tubes housed in a glazed, insulated box as shown in Fig. 2.1. This simple construction makes ICS systems relatively inexpensive and

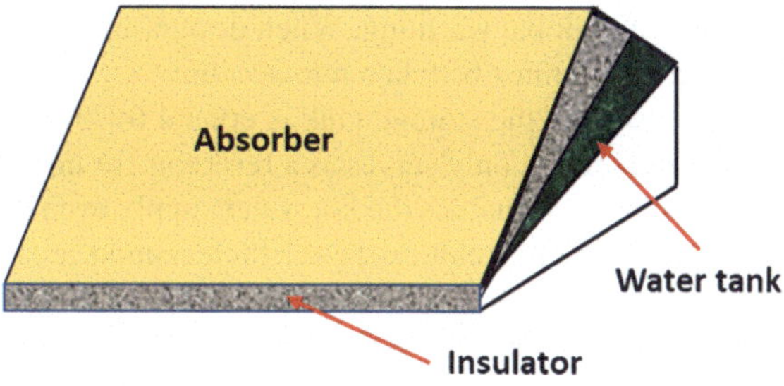

Fig. 2.1 Schematic diagram of integral collector-storage solar water heater with an inverted triangle water tank (Anupam Rao and Somwanshi 2022)

reliable. However, they lose a significant amount of heat from the top and can easily freeze in cold weather, requiring complete drainage during severe winters and reuse in the following season. To address these issues, a previous study designed a triangular water tank and enhanced the ICS system using transparent insulation materials (TIM) between the cover and the tank (Paneri et al. 2019). Experimental results with a transparent honeycomb structure between the cover and tank showed that water temperatures could rise by 18–20 °C, achieving a daily average efficiency of 55% under an average ambient temperature of 23 °C. This transparent honeycomb structure demonstrated superior thermal performance compared to a transparent corrugated (Vee-form) sheet.

However, conventional backup heating proves to be very much a necessity in view of the intermittent nature of solar energy. There is a need for the demand for backup heating for solar water heating systems to happen during rainy weather and also at periods of peak demand. This makes it not unusual for conventional storage water heaters to be offered as a package deal with the solar system. Besides that, the solar collectors themselves may have an integrated backup system, such as rooftop tanks with a system of thermosyphons (Amirgaliyev et al. 2021). ICS systems can be integrated into hybrid systems, which can also include a tankless or demand-type water heater that may be used for backup. Domestic water heating is among the big five end-uses for concern domestic hot water, especially with its daily use all around the year. The solar water heating system is expected to be one of the most customarily used systems in the whole world since the entire world is focusing majorly on a low-carbon economy and the energy crisis. China is one of the leading manufacturers and users of solar water heating systems (Pomianowski et al. 2020). By 2002, the installed base of solar domestic hot water systems had reached nearly 40 million m^2 with an annual production and sales volume of around 8 million m^2. Over 1000 manufacturers of solar thermal systems generated a turnover exceeding €1 billion. In China, the market for evacuated tube collectors is well established both nationally and internationally. In the United States, approximately 1 million m^2 of collector surface had been installed by 2001, primarily using unglazed absorbers for swimming pool heating (Kalogirou 2004). Europe saw 1.1 million m^2 of collector area installed by 2002 predominantly consisting of flat plate collectors with Germany contributing about 50% of this total. The European Union has set a target of 100 million m^2 of collector installations by 2010, emphasizing the crucial role of solar thermal systems in combating global warming. When designing a solar water heater (SWH), several key parameters must be taken into account.

The design of the storage tank is crucial for maintaining water at high temperatures. The tank not only serves as a reservoir for hot water but also acts as a thermal storage system. It enables the hot water supply even during periods without sunlight such as at night or on cloudy days. Efficient tanks are typically made from corrosion-resistant materials like stainless steel, copper, or polymer composites (Liu et al. 2021). Insulation usually in the form of polyurethane foam or vacuum panels minimizes heat loss, ensuring that water remains hot for extended periods. Thermal stratification within the tank occurs when the hotter water rises to the top and cooler water settles at the bottom, affecting energy storage capacity. Maintaining stratification is essential for efficient system performance, as it ensures a continuous supply of hot water at the desired temperature. Advanced tanks often incorporate internal partitions or baffles to preserve this layering, enhancing heat retention and reducing the need for auxiliary heating during periods of high demand or low sunlight. Maintaining a consistent overall temperature in the storage tank is vital for preventing a drop in performance. Fluctuating temperatures increase energy consumption as more energy is required to reheat cooler water. Proper insulation and technologies such as reflective coatings can help stabilize the temperature of the tank. In some systems, backup heating elements are integrated to ensure the water remains hot during periods of insufficient solar energy.

In the SWH system, the collector captures solar radiation, which is changed into heat. It is to achieve maximum heat absorption by the absorber and then transfer the absorbed heat to the working fluid with the lowest thermal loss possible. As the intensity of the solar irradiation is increased, so is the efficiency of the working fluid in holding the heat developed in the surface. This shall taper down to some extent when the temperature of the existing conditions in ambient is too low to carry it away. The tilt angle of the collector installation often referred to as the collector tilt angle plays a significant role in its efficiency. Generally, the tilt reported in the literature should be so designed that it delivers the maximum solar exposure during all days of the year. Equally important, one of the significant parameters in defining the capacity of the tank or reservoir is tilting for the tilt-collecting of the thermosyphon-type evacuated tube solar water heater (Govindasamy et al. 2024). The tilt angle should be set equal to the latitude of the location so that maximum collection of solar radiation through the year is achieved. The greater than 45° collector tilt will cause stagnation zones within the evacuated tube by the flow system, as the natural convection flows would form an incomplete flow loop or loop dissipate. Stagnation occurs every time the heat-induced convection is not present, and therefore the efficiency is reduced. Turbulence may occur within the evacuated tube when the night commences as the incoming more excellent fluid gets agitated with the surrounding high-temperature water, thereby eroding the thermally defined zone that restitutes itself later on.

The choice of working fluid is crucial as it directly impacts the overall performance and efficiency of the solar collector. In most solar water heating systems, the common working fluids include water, oil, or air each chosen based on the system design and the local climate. Water is the most widely used fluid due to its high heat capacity and availability. But it presents challenges in colder climates, where freezing can occur. In these cases, antifreeze solutions such as glycol mixtures are often added. Oil and air are used in systems designed for higher temperatures, although they are less common in residential applications. However, these fluids generally have low thermal conductivities, limiting their ability to transfer heat efficiently. In recent developments, nanofluids, base fluids enhanced with nanoparticles have emerged as a promising alternative for improving the thermophysical properties of working fluids (Alqaed et al. 2022). Nanoparticles made of metals, oxides, or carbon-based materials are suspended in traditional fluids like water or oil, significantly increasing their thermal diffusivity, conductance, and convective heat transfer coefficients. These enhanced properties allow nanofluids to absorb and transfer heat more effectively, thereby boosting the performance of the solar collector.

Studies have shown that even small concentrations of nanoparticles can lead to substantial improvements in heat transfer efficiency, making nanofluids a viable solution for increasing the output of solar thermal systems (Sheikholeslami and Farshad 2021). Nanomaterials also have the potential to improve the optical properties of solar collectors by enhancing the absorption of solar radiation. By increasing the fluid ability to capture and retain solar energy, nanofluids can reduce heat loss and optimize the system efficiency. While the use of nanofluids in solar thermal applications is still an area of active research, their potential for increasing system performance, particularly in extreme conditions. As more advanced nanomaterials are tested and integrated, the efficiency gains from using such fluids are expected to become more significant, making solar collectors even more effective.

2.1.2 Applications in Buildings

The solar water heating process is one of the lowest-cost renewable energy sources, with thousands of these systems installed globally every year. Solar water heaters are highly adaptable and can be used in a wide range of applications, from small domestic installations to large commercial and industrial systems. In residential settings, these systems provide an eco-friendly alternative to traditional electric or gas water heaters, helping homeowners reduce their energy bills and carbon footprints. Solar water heaters have also become an essential component in green building initiatives, offering a sustainable solution for hot water generation in homes, apartments, and even high-rise buildings. The simplicity of their design and the relatively low cost of operation make them an attractive option for households aiming to reduce energy costs and dependence on non-renewable energy sources. In larger installations, solar water heaters are increasingly used in community centers, hospitals, nursing homes, hotels, and restaurants (Ahmadi et al. 2021). These buildings typically have high hot water demand, and solar water heaters can be sized to meet these needs efficiently. Industrial applications include facilities such as dairy plants and swimming pools, where large amounts of hot water are required for processes like pasteurization or maintaining pool temperatures. In commercial settings, water can be heated to temperatures of 60–80 °C with high efficiency, providing a cost-effective alternative to electric boilers or gas-fired systems (Chinnasamy and Arunachalam 2023). The versatility of solar water heaters makes them suitable for a variety of applications, each tailored to the specific hot water demands of the facility.

Household solar water heater capacities range from as low as 100 gallons for small families to as high as 300 gallons for extended commercial and institutional applications. For example, a 300-gallon capacity system can meet the hot water needs of hotels, restaurants, and hospitals, where continuous availability is critical. The scalability of solar water heaters allows them to be customized according to the specific requirements of each application. This will ensure that both residential and commercial users can benefit from energy savings. Furthermore, in cases where the hot water demand exceeds the system capacity, solar water heaters can be integrated with conventional backup systems to ensure a steady supply of hot water, even during periods of low sunlight. A 100-L solar water heater can save approximately 1500 kWh of electricity annually, effectively replacing a conventional electric geyser. With widespread adoption, solar water heaters can significantly reduce peak demand on the electricity grid. For example, installing just 1000 solar water heaters could reduce peak demand by about 1 Megawatt, helping to alleviate pressure on energy suppliers, especially during high-demand periods. In addition to saving energy, these systems play a critical role in reducing greenhouse gas emissions, with each unit potentially preventing the release of 1.5 tons of carbon dioxide annually. This makes solar water heaters not only a cost-effective solution but also an environmentally sustainable one, contributing to national and global carbon reduction targets.

In industrial applications where steam is generated for processes like food production or sterilization, solar water heaters can reduce energy or fuel costs by 70–80%. In these scenarios, solar water heating systems are designed to integrate with existing steam boilers or steam generators allowing businesses to capitalize on solar energy during daylight hours and rely on conventional systems as a backup during periods of low sunlight. The energy savings from such systems are substantial, and the return on investment (ROI) is often realized within two to four years. For industries with high energy costs, this makes solar water heating a financially attractive option while also aligning with sustainability goals. System losses include draw-off pipe losses, storage tank losses, primary circulation pipe losses, and losses incurred during hot water generator heating. Accurately accounting for these losses during the design and installation phase is essential for optimizing system performance. Mathematical models such as the Hottel-Whillier-Bliss Equation are frequently used to simulate and assess the performance of solar water heating systems (Bouhdjar et al. 2021). These models help evaluate beneficial energy gain, transmission absorption product, absorber fin efficiency factor, removal factor, and overall collector performance. Solar water heaters generally require just a one-time investment. The actual payback period varies compared to the performance of the system, local weather, and hence, the solar insolation number available compared to the price of fossil fuels. Therefore, solar water heaters demand a substantially more significant initial investment than gas- or electric-fired water heaters. With quite a bit of help from the price of fuel, solar water heaters can often be cost-competitive over the life of heater compared to heating water with electricity, fuel oil, propane, or natural gas. Scientific reports note that the payback period is between 4 and 8 years for a professionally installed solar water heater (Muneer et al. 2006). All other costs, except the payback, are the savings one enjoys. The life span would range from 15 to 40 years, depending upon the maintenance.

2.2 Solar Air System

Solar air systems or solar air heating systems utilize solar energy to produce heat, which is subsequently conveyed by air for direct utilization or storage. These systems are efficient for warming or preheating air in buildings and can also be applied in industrial processes such as crop drying. Their function involves the use of collectors to capture solar radiation, which is then converted into thermal energy and transferred to the surrounding air (Almutairi et al. 2022). Solar air systems integrated into new residential structures are a cost-effective solution, as they eliminate the need for additional wall coverings and ventilation fans. When retrofitted into existing buildings, they enhance indoor air quality, improve ventilation, and reduce heating expenses, making them economically beneficial even in suboptimal conditions. Recent advancements in solar air system technology have focused on the integration of thermal storage units to improve performance and reliability. The efficiency of these systems is greatly enhanced by the use of both sensible and latent heat storage systems, particularly those that employ phase change materials (PCM) (Wang et al. 2022a). This allows for the storage of excess heat during peak sunlight, which can be utilized during periods of low solar radiation.

Innovations such as modern insulation materials and the application of nanotechnology have further improved thermal efficiency and overall system performance. These design improvements help solar air systems adapt to diverse environmental conditions, increasing their potential for widespread adoption in both residential and industrial applications. Solar air systems are a sustainable alternative for heating applications and can be easily integrated with existing infrastructure. Their adaptability to a range of environmental circumstances, combined with their cost-effectiveness, positions them as a key technology in the transition towards renewable energy sources. As the demand for renewable energy solutions continues to grow, solar air systems present a viable option for reducing dependency on fossil fuels while maintaining high efficiency. Further research and development, particularly in the areas of thermal storage and system design, will continue to drive their evolution, making them even more effective and widely applicable in the future.

2.2.1 System Description and Working Mechanism

The heating load is a major factor in the energy consumption of buildings, often accounting for a substantial portion of the overall energy use, particularly in colder climates. In both residential and commercial buildings, heating systems are crucial for maintaining comfortable indoor temperatures during colder months. However, traditional heating methods such as gas or electric heaters contribute significantly to energy consumption and greenhouse gas emissions. In this context, reducing the heating load is vital for improving energy efficiency and minimizing environmental impacts. To meet the objectives of green building principles, it is essential to adopt environmentally friendly and efficient solutions that fulfill the heating requirements of structures without relying on non-renewable energy sources. Solar Air Heaters (SAH) offer feasible alternatives to conventional heating systems that rely on fossil fuels. These systems harness solar energy, a renewable and abundant resource, converting it into thermal energy that can be used for space heating. Unlike traditional systems, SAH does not emit harmful pollutants or deplete finite energy resources, making it a sustainable choice for modern building designs. SAH provide several advantages, including increased energy efficiency and a reduction in dependence on fossil fuels.

By utilizing solar energy, buildings equipped with SAH systems can significantly reduce their operational costs related to heating, making these systems economically viable in the long term. Moreover, they contribute to lowering carbon footprints, which is critical in achieving global climate goals. Nevertheless, there is a scarcity of comprehensive evaluations of these systems, limiting the understanding of their full potential. A study conducted in 2012 analyzed solar air heating systems, both with and without storage units (Tyagi et al. 2012). However, this evaluation did not cover research carried out between 2012 and 2022, highlighting the need for an updated and current analysis for scholars and practitioners engaged in this field. Todays, there have been significant advancements in solar air heating technologies particularly in terms of materials, storage efficiency, and system integration. These innovations could potentially address some of the limitations noted in earlier systems such as inconsistent heat output during cloudy days or limited heat storage capabilities. In 2015, another analysis conducted specifically examined the thermodynamic features of SAH (Saxena and El-Sebaii 2015). This study focused on the energy conversion and efficiency of SAH systems. But like the previous review, it did not include the latest breakthroughs such as the use of phase change materials (PCM) or advanced control mechanisms that optimize performance. Therefore, it is imperative to conduct a modern and comprehensive evaluation that addresses these deficiencies and offers a clearer understanding of the current state and future potential of SAH technology. The main types of solar air collectors include:

(a) **Facade or Roof-Integrated Panels**

Facade or roof-integrated panels are a versatile and aesthetically appealing solution for solar air heating systems. These panels not only collect solar energy but also serve as the weatherproof exterior of the building, effectively integrating with the structure design and architecture (Nihadh and Alkaff 2021). By positioning the solar air collectors directly within the building envelope such as the roof or facade, these panels eliminate the need for additional space. This making them ideal for urban environments where rooftop real estate may be limited. Glazed panels offer higher efficiency as they are equipped with a transparent cover that creates a greenhouse effect, trapping more heat. This allows the system to capture solar radiation more effectively and protect the absorber surface from the elements, thereby extending its lifespan (Kalaiarasi et al. 2020). Unglazed panels are simpler and less costly, making them suitable for regions with mild climates where extreme weather protection is less of a concern.

In addition to their energy efficiency, facade or roof-integrated panels contribute to the overall aesthetic of the building. Modern architectural designs have begun incorporating solar technologies not just for functionality but also for visual appeal, and these integrated panels can be customized to blend seamlessly with various building styles. This design approach makes the panels nearly invisible, ensuring that the building retains its architectural integrity while still benefiting from renewable energy. Furthermore, integrating solar air systems into the exterior of the building offers enhanced weather protection by acting as an additional insulating layer. This not only reduces heating costs but also improves indoor comfort by preventing heat loss during colder months. The dual-purpose nature of these panels as both energy collectors and part of the building structure provides a practical and efficient solution for energy-conscious construction.

(b) **Double Windows with Internal Adjustable Blinds**

Double windows with internal adjustable blinds offer a multi-functional approach to solar air heating systems by combining energy efficiency with enhanced indoor comfort and privacy. These systems feature an absorber that is integrated within the window structure, allowing the blinds to serve dual purposes. Not only do they collect solar energy to heat the air, but they also act as a mechanism for glare control, adjusting to block or allow sunlight as needed throughout the day (Su et al. 2022). The blinds can be manually or automatically adjusted to optimize the amount of solar radiation entering the building, depending on the season or time of day. It provides an additional level of control over indoor climate conditions. This versatility makes them especially useful for buildings with large windows, where both privacy and energy savings are important considerations.

In addition to their energy-harvesting role, these systems significantly improve indoor comfort by reducing glare and regulating the amount of natural light. By allowing building occupants to adjust the blinds based on their preferences, these windows create a more comfortable living or working environment, without sacrificing energy efficiency. The absorber in the blinds captures solar energy converting it into heat, which can then be used to warm the interior spaces or be stored for later use. This design reduces the need for artificial heating while also mitigating the impact of excessive sunlight, which can cause overheating in warmer months. The combination of energy efficiency, privacy, and enhanced indoor comfort makes these double windows a highly effective component of modern green building designs, offering both functionality and aesthetic appeal.

(c) **Transparent Secondary Facades**

Transparent secondary facades are an innovative solution for retrofitting existing buildings, providing a dual function of adding insulation and harnessing solar energy. These collectors are installed over the building primary facade, forming a transparent or semi-transparent outer layer that traps heat while still allowing natural light to pass through. The gap between the original facade and the secondary transparent layer acts as an insulating barrier, reducing heat loss and improving the overall energy efficiency of the building. This setup not only enhances thermal performance during colder months but also reduces the need for additional heating, thus lowering energy consumption (Sathkumara et al. 2023). The solar energy collected by the facade can be used to preheat ventilation air or provide supplementary heat to the building, making it an effective retrofit option for older buildings that may lack energy-efficient features.

In addition to improving energy efficiency, transparent secondary facades contribute to the aesthetic appeal of a building without altering its architectural character. This design is particularly beneficial for historic or architecturally significant buildings, where preserving the original exterior is essential. The secondary facade allows for solar energy collection without compromising the building appearance. It will offer a sleek and modern upgrade to existing structures. The additional layer also acts as a buffer against weather elements, protecting the primary facade from environmental wear and tear. By incorporating solar air heating systems into the facade, these collectors provide a cost-effective and sustainable solution for enhancing building performance, particularly in urban areas where space for traditional solar installations may be limited.

(d) **Glazed Spaces**

Glazed spaces such as atriums, sunspaces, or attics are versatile areas that can function as natural solar air collectors while also serving as functional architectural elements. These spaces are typically designed with large glass surfaces that capture and trap solar radiation, creating a greenhouse effect that heats the air within. This warm air can be circulated throughout the building to pre-heat ventilation air or provide supplemental heating during cooler months, reducing the need for conventional heating systems (Krarti 2022). The integration of glazed spaces into building design not only improves energy efficiency but also creates inviting, light-filled areas that enhance the overall living or working environment. In colder climates, glazed spaces offer a passive solar heating solution, maximizing the use of available sunlight to maintain comfortable indoor temperatures.

Beyond their heating function, glazed spaces contribute significantly to the building aesthetic and utility. These spaces can serve as gathering areas, recreational zones, or even indoor gardens, adding a multifunctional aspect to the structure. The transparent surfaces allow for natural lighting, which reduces the need for artificial lighting and enhances the interior ambiance. Architecturally, glazed spaces are often used to blur the lines between indoor and outdoor environments, creating a seamless connection with nature. This dual-purpose approach makes them a highly efficient and attractive option in both residential and commercial buildings. By combining energy efficiency with spatial functionality, glazed spaces provide a sustainable and aesthetically pleasing solution that aligns with modern green building principles.

Thermal storage and air circulation mechanisms in solar air systems Thermal storage in solar air systems can be incorporated either inside the building structure itself or through the use of additional mediums. Some examples of energy-efficient building techniques include the use of hypocausts which are concrete floor slabs with channels for circulating sun-warmed air (Zotos 2021). Murocausts, which are hollow-core brick walls, rock beds positioned in the basement or building core, and phase-change materials (PCM) (Fort 2013). In an alternative approach, the building can receive sun-warmed air directly to fulfill its heating, fresh air, or dehumidification needs. Air circulation can happen through the natural convection of warm air ascending (passive mode), but it is more frequently achieved through the use of fans (active mode). It is worth mentioning that air has a significantly lower heat capacity compared to water 0.0003 kWh/m^3 K against 1.16 kWh/m^3 K. This means that a big volume of air needs to be moved in order to convey a small amount of heat. The system can be powered by photovoltaic (PV) panels, as solar power output is directly proportionate to solar heat generation, resulting in a self-regulating system.

Solar air systems bring distinct benefits for space heating and tempering ventilation air. Unlike water, air does not need freeze protection, and any leaks will not damage the building or its contents. Active air systems provide superior heat distribution compared to passive systems, improving comfort levels and optimizing the use of solar energy more effectively. In low-energy office buildings, over 60% of winter heat loss is due to ventilation, making solar air systems an excellent choice for mechanically ventilated offices. Solar air systems come in various designs, each with its own advantages and disadvantages, as detailed in Table 2.1. The choice of system depends significantly on the specific application and the requirements of customer. In the six systems described here, the air collector is independently separated, allowing for tailored solutions to meet diverse heating and ventilation needs.

The solar air collector is an essential component of solar air heating systems and can be designed in various configurations to suit different applications. These collectors are generally divided into two primary categories: unglazed air collectors which are used to heat ambient air in commercial, industrial, agricultural, and process settings, and glazed solar collectors, which are primarily utilized for space heating. Additionally, solar air collectors can be categorized by their air distribution patterns or materials, with types including through-pass, front-pass, back-pass, combination front and back-pass, as well as glazed and unglazed collectors (Li et al. 2017). Theoretical analyses of various types of solar air collectors have been extensively documented, with comparative studies. The studies indicated that corrugated collectors exhibit superior thermal performance compared to conventional parallel

Table 2.1 Six different solar air heating system

System name	Schematic diagram	Principle and Advantages	Disadvantages
Ambient collector/room	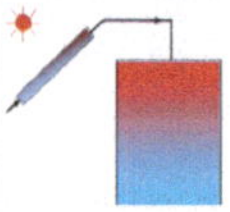	• High efficiency is achieved by circulating cool ambient air directly through a collector into the room, which makes the system both simple and cost-effective • During the summer months, the collector can be vented to the outside, enhancing its versatility for various applications, including vacation cottages and large industrial spaces	• Susceptible to overheating • Overhead heat distribution can cause discomfort
Collector/room collector (*Bara Costantini* system)	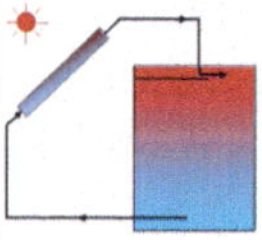	• Room air is circulated into the collector, where it is heated. It then rises and returns via a thermal storage ceiling back into the room through natural convection • The storage ceiling provides radiant heat after sunset • During the summer, the collector can be vented at the top to the outside, allowing room air to be replaced with cooler air from an earth storage device or open north-facing windows • Suitable for apartment buildings	• Prone to overheating • Overhead heat distribution can lead to discomfort

(continued)

Table 2.1 (continued)

System name	Schematic diagram	Principle and Advantages	Disadvantages
Collector/shell collector	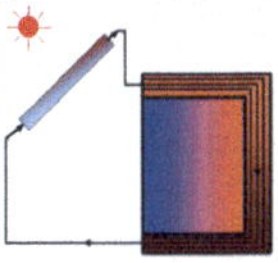	• Warmed air is circulated through a cavity between an outer insulated wall and an inner wall of the building envelope, creating a sun-buffered climate that reduces room heat losses through the building envelope • During the summer, the collector can be vented at the top to the outside, allowing room air to be replaced with cooler air from an earth storage device or open north-facing windows • This system is particularly suitable for retrofitting poorly insulated existing buildings	• Increased costs
Collector/storage collector	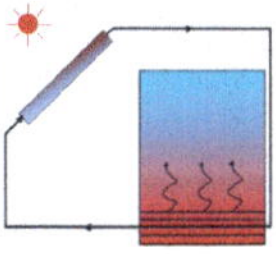	• Warmed air is circulated through channels in a massive floor or wall, which then radiates heat into the room after a delay of four to six hours • This system benefits from large radiating surfaces, thereby enhancing comfort. It is equipped with forced circulation, which ensures optimal system efficiency and output • This configuration has the advantage of keeping feet warm and heads cool	• Room temperature regulation remains challenging • High costs

(continued)

Table 2.1 (continued)

System name	Schematic diagram	Principle and Advantages	Disadvantages
Collector/storage collector plus room	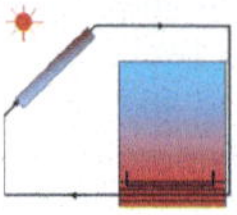	• This system actively discharges heat by circulating room air through separate channels in the storage device • Heat can be stored for longer periods and released when needed • The storage device can be located remotely from the rooms being heated	• High costs arise from the necessity of maintaining two independent systems of air channels and fans
Collector/heat exchanger/collector plus heat exchanger	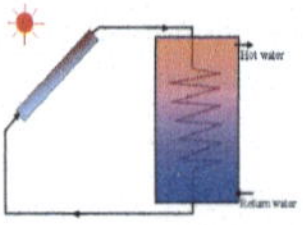	• This system combines the benefits of a solar air collector with the compactness of using water for heat distribution to the point of need • Conventional radiators, radiant floors, or walls can thus be utilized effectively • Additionally, the system can provide domestic hot water	• The system incurs high costs unless there is an existing air collector; otherwise, the use of a solar water collector is recommended

plate collectors, although they are outperformed by slate collectors (Tuncer et al. 2023). The corrugated-duct solar air collector is highly efficient due to its ability to absorb solar radiation through multiple reflections while minimizing convective losses, making it both simple and effective. However, transparent honeycomb insulation has been validated to provide superior thermal performance due to its enhanced radiation blockage and high solar transmission, making it particularly suitable for cold regions.

A solar chimney is a type of air heating system primarily designed to draw air upwards and facilitate ventilation. A solar chimney has been implemented to explore its capability in pre-heating air and removing volatile organic compounds (VOC) in the School of Built Environment at University of Nottingham. This innovative approach not only enhances ventilation efficiency but also contributes to improving indoor air quality by reducing harmful VOC levels. The research could be expanded to explore the possibility of extending the solar chimney to the full roof height of the house. A square duct made from a transparent honeycomb panel could be added to the existing solar chimney structure, as illustrated in Fig. 2.2. This modification would allow further investigation into the passive cooling and heating effects using an underground duct or heat exchanger during summer and winter.

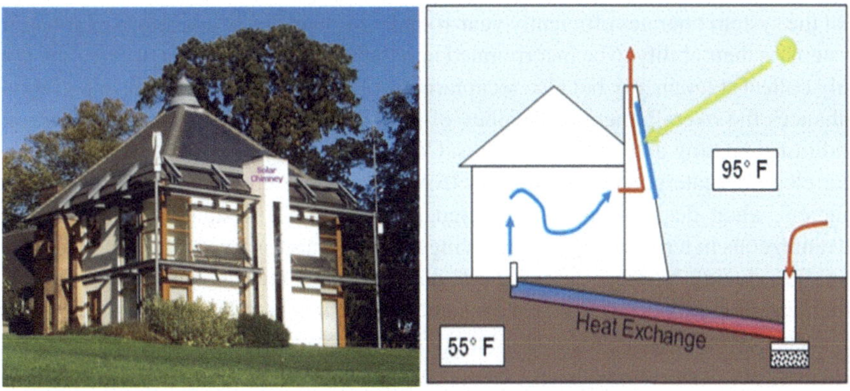

Fig. 2.2 Solar chimney in the eco-house at the University of Nottingham

2.2.2 Applications in Buildings

Solar air systems can be utilized in a wide range of contexts, including both building air heating and process air heating. These systems are applicable in new constructions as well as in retrofitting existing buildings, with the specific technology used depending on whether the building is industrial, commercial, or residential (Fudholi and Sopian 2019). In new buildings, solar air systems can be seamlessly integrated into the architectural design, while retrofitting older structures can provide significant energy savings by reducing reliance on fossil-fuel-based heating systems. The flexibility of solar air systems allows for tailored solutions that address the specific heating and ventilation requirements of each type of building. This adaptability is particularly valuable in industrial settings, where large spaces require efficient heating and in residential buildings, where comfort and cost-effectiveness are paramount. Ventilation heating technologies are specifically designed based on the type of building in which they are installed. In most residential and commercial buildings, ventilation air is essential for maintaining indoor air quality and comfort. Solar air systems can preheat this ventilation air before it is introduced into the building, significantly reducing the heating load (Hu et al. 2020). By using standard ductwork, an air-handling device draws ventilation air through a solar collector, heats it, and then distributes it throughout the building. This reduces the amount of energy required to heat the incoming fresh air, making solar ventilation air preheating devices an ideal solution for improving energy efficiency. The integration of solar air systems into ventilation not only lowers heating costs but also enhances indoor air quality by providing a steady supply of fresh air.

On colder days, solar collectors can heat the incoming air with the assistance of an air-handling unit, which provides any additional warmth as needed. Solar air systems can often meet the entire heating demand, making them a reliable source of warmth during the winter months. In contrast, during the summer, a bypass damper prevents the solar system from adding unnecessary heat to the air conditioning load, ensuring that the system operates efficiently year-round. One significant advantage of solar air systems is their ability to be incorporated into the facade of a building, where they not only collect solar energy but also recapture heat loss from the walls. This integration enhances the overall energy efficiency of the building and reduces the reliance on traditional heating and cooling systems. Compared to conventional heating systems like electric heaters, solar air heaters (SAH) are more cost-effective and energy-efficient when designed and configured properly. SAH systems are particularly advantageous in terms of long-term savings, as they harness free and renewable solar energy, reducing both operational costs and environmental impact. These systems, when installed properly, provide consistent and reliable heating with minimal maintenance, making them a practical solution for modern energy needs. Additionally, by offsetting the need for electricity or fuel-based heating, solar air heaters contribute to lowering carbon emissions, aligning with sustainability goals. The versatility of solar air systems, combined with their economic and environmental benefits, underscores their importance as a key technology in the shift towards greener and more energy-efficient buildings.

(a) **Space Heating**

Solar air heating panels are a highly effective solution for both residential and commercial space heating applications. These systems draw air from either the external environment or the building envelope, then pass it through a solar collector, where the air is warmed through conduction from the absorber. The heated air is then circulated into the living or working space either passively, by natural convection, or actively, with the aid of fans. This process significantly reduces the reliance on conventional heating systems, leading to energy savings and a lower carbon footprint. In colder climates, solar air heating can serve as a supplemental heat source. While in milder conditions, it may provide sufficient warmth on its own, even during the winter months. One of the key advantages of solar air heating systems is their ability to make use of solar energy without the need for complex mechanisms. Historically, buildings naturally retained heat from the sun during the day, especially before the widespread use of air conditioning. This passive heating effect kept indoor temperatures comfortable by capturing and storing solar radiation within the building structure.

In a similar way, solar air heating systems mimic this process by using collectors to harness the heat from the sun more efficiently. The technology is simple yet effective, and the integration of fans in modern systems allows for better heat distribution, ensuring that even large or multi-room spaces are warmed evenly. Solar air heating systems can be likened to the effect seen in vehicles, where the interior temperature can rise dramatically when exposed to sunlight, even with minimal ventilation. The air inside a car can exceed 50 °C on a sunny day, illustrating the potential of solar energy to heat enclosed spaces quickly and effectively. This same principle is applied in solar air heating systems, with collectors designed to optimize solar radiation capture. By taking advantage of natural processes, these systems not only reduce energy consumption but also offer a reliable, low-maintenance solution for maintaining indoor comfort. The simplicity and adaptability of solar air heating panels make them an attractive option for both residential and commercial buildings, particularly in areas where sunshine is abundant.

(b) **Process Heat Applications**

Solar air heating systems are not limited to space heating. They are also highly effective in various process heating applications, particularly in industries such as agriculture and manufacturing. In agricultural settings, solar air heating is commonly used for drying crops, where heated air is directed over harvested produce to reduce moisture content. This method offers a sustainable alternative to conventional drying techniques that often rely on fuel-based heating or electricity. By utilizing solar energy, farmers can significantly lower their operational costs while preserving the quality of their crops (Abdukarimov et al. 2021). Solar drying is especially beneficial for moisture-sensitive products like grains, fruits, and herbs, where precise control over drying conditions is necessary to maintain quality and prevent spoilage. The effectiveness of solar air heating in process heat applications extends beyond agriculture. For example, solar-heated air can be used to dry clothes in residential or commercial laundry settings, providing an eco-friendly and cost-efficient solution. In these systems, air heated by a solar collector is passed through the drying chamber, effectively removing moisture from clothing without the need for electricity or gas-powered dryers. This not only reduces energy consumption but also shortens drying times, making it a practical option for both individual households and larger-scale operations such as laundromats.

In industrial applications, solar air heating can be used for processes like drying wood, curing materials, or preheating air for combustion systems. It further demonstrating the versatility of solar technology in addressing various heating needs. Solar air systems offer a significant advantage in regions with high solar irradiance, where process heat demands are substantial. By integrating solar air heating into industrial processes, businesses can reduce their reliance on fossil fuels, decrease greenhouse gas emissions, and enhance energy efficiency. The scalability of solar air systems also makes them ideal for a range of process heat applications, from small-scale operations to large industrial plants. Additionally, the low maintenance requirements and durability of solar air systems make them a long-term, sustainable investment for businesses looking to improve their environmental footprint while maintaining productivity. The use of solar air heating for process applications underscores the growing importance of renewable energy in industrial and commercial sectors, where efficiency and cost savings are critical.

(c) **Ventilation Applications**

Solar air heating systems are also highly effective when applied to ventilation, providing a sustainable solution for maintaining indoor air quality while reducing heating loads. In many buildings, particularly commercial and industrial ones, fresh air intake is essential for proper ventilation. By using solar energy to preheat this incoming air, solar air systems can significantly reduce the amount of energy required for heating, especially during sunny periods (Li et al. 2020). As the fresh air is drawn through a well-designed solar collector, it is warmed before being introduced into the building, thereby reducing the need for additional heating. This not only improves energy efficiency but also contributes to a healthier indoor environment by ensuring a steady supply of fresh, warm air. One of the key technologies in ventilation applications is the use of transpired solar air collectors, which are designed to preheat incoming ventilation air. These collectors are mounted on the exterior of building, where they capture solar energy and transfer it to the air passing through small perforations in the collector surface. The preheated air is then drawn into the ventilation system of the building, reducing the workload on the heating system. Transpired collectors are particularly beneficial in colder climates, where the difference between outside and inside temperatures is significant. By preheating the air before it enters the building, these systems can lower energy consumption while maintaining indoor comfort, making them a practical solution for energy-conscious buildings.

As previously mentioned, another innovative approach to solar-assisted ventilation is the use of solar chimneys, which utilize the natural rise of heated air to create a ventilation effect. As solar energy heats the air within the chimney, the hot air is expelled through the top, creating a suction effect that draws in cooler air from the building or external environment. This process promotes natural air circulation, reducing the need for mechanical ventilation systems and cutting energy costs. Solar chimneys are particularly effective in buildings that require consistent air exchange, such as office spaces or industrial facilities, where maintaining air quality and temperature is critical. The combination of transpired collectors and solar chimneys demonstrates the versatility of solar air systems in providing both heating and ventilation solutions, further enhancing the overall energy efficiency of modern buildings.

(d) **Thermal Comfort and Drying**

Solar air heaters offer a sustainable and cost-effective solution for maintaining thermal comfort in buildings, as well as for drying applications in industrial and agricultural settings. These systems utilize solar energy to regulate indoor temperatures by providing supplemental heat during colder months or reducing the cooling load in the summer. By integrating solar air heaters into the building design, it is possible to achieve a more comfortable indoor environment while simultaneously lowering energy consumption and reducing reliance on fossil-fuel-based heating systems (Paraschiv et al. 2020). The performance of solar air heaters is influenced by various factors, including weather conditions and the operational design of the system. In regions with high solar irradiance, the heaters operate at peak efficiency, whereas in cloudy or colder climates supplemental heating may be required. The choice of insulating materials for the absorber is critical to minimizing heat loss and ensuring that the collected solar energy is effectively transferred to the air being heated. To further optimize performance, additional components such as turbulators can be incorporated to enhance heat transfer by creating turbulence in the airflow, ensuring more efficient energy capture and distribution. The integration of advanced materials like phase change materials (PCM) for thermal storage allows the system to store excess heat during the day and release it when sunlight is unavailable, providing continuous comfort and efficiency.

Solar air heaters are compatible with various other systems and technologies, which can further enhance their efficiency and performance. When combined with Trombe walls thermal mass walls designed to absorb and radiate heat or heat exchangers, solar air systems can provide a more consistent and reliable source of heating (Singh et al. 2021). Additionally, the use of tracking systems that follow the movement of the sun throughout the day can maximize solar energy capture, further boosting system efficiency. By incorporating these advanced technologies, solar air heaters can provide long-term energy savings while ensuring high levels of thermal comfort and improved drying capabilities in both residential and industrial applications. As a result, they offer a versatile and eco-friendly solution that aligns with modern energy efficiency standards and environmental goals.

2.3 Solar Heat Pump

A solar-assisted heat pump system combines an air-to-water heat pump with a solar collector panel, creating a unified refrigeration circuit where both components are directly interconnected. This synergy between the solar collector and heat pump allows the system to capture and utilize free solar energy from the environment, making it an efficient and environmentally friendly solution for both heating and cooling needs. The solar collector captures heat from the external environment including ambient air and sunlight, and converts it into a refrigerant gas (Zhang et al. 2023a). This process can occur even during periods of low sunlight or in colder climates as the system is designed to absorb heat from the air, ground, or other natural sources. The refrigerant gas is then compressed by the heat pump compressor significantly raising its temperature through the mechanical work of compression. This heated refrigerant can reach temperatures high enough to meet the heating demands of the home, whether it be for space heating or domestic hot water production. This elevated heat is subsequently delivered into the residence, often distributed through underfloor heating systems, radiators, or stored in a hot water cylinder for later use. By integrating a hot water storage unit, the system can also ensure a steady supply of hot water, which adds to its versatility and functionality throughout the year. The primary goal of the system is to optimize energy efficiency by achieving a high Coefficient of Performance (COP) while simultaneously reducing operational costs. The COP of a heat pump represents the ratio of useful heating or cooling provided to the energy consumed, and solar-assisted heat pumps tend to achieve significantly higher COP values compared to traditional systems (Humia et al. 2021).

By leveraging renewable energy sources like solar power, the system not only reduces the consumption of electricity but also significantly lowers greenhouse gas emissions. This eventually making the system as an attractive option for eco-conscious homeowners and businesses. By integrating renewable energy sources, the system enhances sustainability and reduces reliance on traditional fossil fuels. This approach aligns with global efforts to reduce carbon footprints and move toward more sustainable, energy-efficient solutions for heating and cooling. The heat pump can be paired with a variety of solar thermal panels, including flat plate, evacuated tube, roll-bond, heat pipe, and thermal plates, as well as hybrid panels such as monocrystalline, polycrystalline, and thin-film types (Ma et al. 2020). Each type of solar thermal panel offers distinct advantages depending on the climate, installation space, and specific energy requirements of the building. For example, evacuated tube collectors are highly efficient in colder climates, while flat-plate collectors are more suited for moderate conditions. The use of hybrid panels offers a distinct advantage by partially offsetting the electricity requirements of the heat pump. Thereby it lowering both energy consumption and the variable costs associated with the system. Hybrid panels which combine solar thermal and photovoltaic technology can generate electricity and heat simultaneously. This approach will further boosting the overall efficiency of the system. Furthermore, hybrid systems can improve overall system resilience by providing a supplementary power source, especially during peak demand periods or in regions with variable solar availability. This added flexibility makes solar-assisted heat pumps a reliable and adaptable solution for a wide range of climates and building types, from residential homes to large commercial buildings.

2.3.1 System Description and Working Mechanism

A heat pump is a device that upgrades heat energy from a low-temperature source, making it available as useful heat energy at a higher temperature through the application of external work or heat. Both heat pumps and refrigeration systems operate based on similar thermodynamic cycles and processes, though their primary objectives differ. The purpose of a heat pump is to supply additional heat to the high-temperature source, thereby raising its temperature. In contrast, the objective of a refrigeration system is to extract heat from the low-temperature source to lower its temperature, subsequently releasing the extracted heat to the high-temperature source as waste (Mota-Babiloni et al. 2020). Heat pumps can be classified according to the thermodynamic cycles they utilize, such as the vapor compression cycle, absorption cycle, adsorption cycle, and thermoelectric heating. Additionally, they can be classified based on the type of heat supplier at the low-temperature source. The most common heat suppliers include ambient air, the ground or groundwater, and solar energy. Some systems also recover waste heat from industrial processes or combine multiple sources known as Hybrid Heat Pumps to enhance efficiency and performance. Each heat supplier offers unique advantages depending on the environmental conditions and specific applications.

(a) **Vapor Compression Cycle Heat Pump**

The vapor compression cycle is the most commonly used mechanism in modern heat pump systems, particularly in residential and commercial applications. This cycle works by utilizing a refrigerant that absorbs heat at low temperatures and releases it at higher temperatures. The process starts with the refrigerant in a low-pressure, gaseous state. It is then compressed by a compressor which increases its pressure and temperature. The high-temperature, high-pressure gas is then passed through a condenser where it releases heat to the surrounding environment typically into a home heating system (Wu et al. 2021). The refrigerant cools and condenses into a liquid state before being throttled to a lower pressure and temperature, ready to absorb heat from the external environment again. This cycle repeats continuously, making it an efficient process for space heating, water heating, and cooling. Vapor compression systems are popular due to their high Coefficient of Performance (COP), meaning they provide more heating or cooling energy than the electrical energy they consume. One of the main advantages of this cycle is its adaptability to different environmental conditions as it can extract heat from air, water, or ground sources. However, these systems experience performance declines in extreme temperatures, particularly in very cold climates, where the efficiency of the compressor and the refrigerant decreases. This can lead to the need for supplemental heating systems or defrosting mechanisms to maintain efficiency. Still, the vapor compression cycle remains the most widely adopted technology in heat pumps due to its efficiency, reliability, and cost-effectiveness in a wide range of climates.

(b) **Absorption Cycle Heat Pump**

The absorption cycle differs from the vapor compression cycle in that it uses a liquid absorbent to facilitate the refrigeration process. In these systems, a refrigerant often ammonia or water is absorbed by a liquid absorbent such as lithium bromide or water. The process begins with the refrigerant absorbing heat from a low-temperature source, which causes it to evaporate. This vaporized refrigerant is then absorbed into the absorbent liquid, forming a solution. The solution is then heated typically using an external heat source such as solar energy, natural gas, or waste heat, to drive the refrigerant out of the absorbent, allowing it to be condensed back into liquid form (Zhang et al. 2020). This liquid refrigerant then re-enters the cycle, where it can absorb heat again from the low-temperature source. The external heat replaces the mechanical work that would typically be done by a compressor in vapor compression systems. Absorption heat pumps are less common in residential applications but are widely used in industrial settings where waste heat or renewable energy sources like solar thermal energy are available. They are particularly effective in situations where electricity is scarce or expensive, as they can operate using thermal energy rather than relying heavily on electrical power. One of the key advantages of absorption heat pumps is their ability to work in low-temperature environments, as they do not rely on mechanical compression. However, they tend to have lower COP than vapor compression systems, meaning they are generally less efficient. Additionally, absorption systems are often larger and more expensive, making them less practical for small-scale residential use but ideal for larger industrial applications where sustainability and energy savings are prioritized.

(c) **Adsorption Cycle Heat Pump**

The adsorption cycle heat pump operates similarly to the absorption cycle but relies on solid adsorbents rather than liquid absorbents. In this system, the refrigerant vapor is adsorbed onto the surface of a solid material such as silica gel or zeolites rather than being dissolved into a liquid. The process starts with the refrigerant evaporating at low temperatures and being adsorbed onto the solid surface. Heat is then applied to the solid material, which releases the refrigerant vapor. This vapor is subsequently condensed into liquid form, releasing heat to the environment (Caprì et al. 2020). The liquid refrigerant is then expanded and reintroduced to absorb heat from the low-temperature source. The cycle repeats, providing a continuous process of heat absorption and release. Adsorption heat pumps are known for their ability to operate efficiently with low-grade heat sources such as solar thermal energy, geothermal energy, or industrial waste heat. They are typically used in applications where conventional vapor compression systems may not be suitable such as in environments with fluctuating energy demands or where waste heat recovery is important. One of the major advantages of adsorption systems is their low maintenance and long operational life, as they do not require moving parts like compressors. Additionally, adsorption heat pumps are quiet and environmentally friendly since they can utilize natural refrigerants like water. However, like absorption systems, they tend to have lower efficiency compared to vapor compression systems, and their larger size and cost make them more suited to commercial or industrial applications.

(d) **Thermoelectric Heating**

Thermoelectric heating utilizes the Peltier effect, which occurs when an electric current is passed through the junction of two different conductors, causing heat to be absorbed on one side and released on the other. This effect allows thermoelectric heat pumps to directly convert electrical energy into a temperature gradient, with one side of the device becoming cold and the other side becoming hot. These systems do not rely on refrigerants or compressors, making them compact and environmentally friendly (Toan et al. 2020). Thermoelectric heat pumps are ideal for small-scale applications such as cooling electronic components, climate control in small devices, or heating and cooling small spaces. Their simplicity and solid-state nature mean there are no moving parts, which results in very low maintenance requirements. Despite their advantages, thermoelectric heat pumps are not widely used for large-scale heating or cooling due to their relatively low efficiency compared to other heat pump systems. Their COP is typically lower than that of vapor compression systems, meaning they require more electrical energy to produce the same amount of heating or cooling. As a result, thermoelectric heat pumps are primarily used in niche applications where their compact size, precision, and solid-state reliability are more important than overall energy efficiency. However, ongoing research into advanced thermoelectric materials may improve their efficiency in the future, potentially expanding their use in larger heating and cooling applications.

(e) **Heat Source for Heat Pump**

Heat pumps can utilize energy from air, water, or a combination of both with dual-source systems designed to switch between sources as needed. In many natural environments and specific climatic conditions, ambient air, natural water bodies, and

Table 2.2 Commonly used heat sources for heat pumps (Carroll et al. 2020; Sezen et al. 2021)

Source	Temperature range (in heating season) (°C)	Temperature range (in cooling season) (°C)
Ambient air	− 10 to 15	26–45
Exhaust air	17–23	26–34
Ground water	4–15	6–18
Lake water	0–15	10–20
River water	0–15	8–18
Sea water	4–15	10–25
Rock	0–15	10–20
Ground	0–15	10–20

geothermal heat from the earth can effectively supply the energy demands of a heat pump. Table 2.2 provides an overview of the most commonly utilized heat sources for these systems. Among these, solar thermal energy stands out as a particularly effective option for boosting the performance and efficiency of heat pumps.

2.3.2 *Applications in Buildings*

Heat pumps are highly versatile systems that can be employed in a wide range of settings from residential heating and cooling to large-scale industrial processes. Their ability to efficiently transfer heat from low-temperature sources and upgrade it for practical use makes them ideal for both small-scale applications, such as home heating, and more complex systems used in commercial or industrial settings. By drawing energy from renewable sources like air, water, and the ground, heat pumps reduce dependency on fossil fuels, lower greenhouse gas emissions, and offer significant energy savings compared to traditional heating and cooling systems.

(a) **Residential**

Heat pumps are increasingly used in residential settings due to their efficiency and environmental benefits. Air-source heat pumps (ASHP) are favored for home heating and cooling because they are easy to install and relatively affordable. These systems can extract heat from the outdoor air even in cooler climates, though they may require supplemental heating during extreme cold. Ground-source heat pumps (GSHP) provide more consistent performance throughout the year by utilizing the stable temperatures found beneath the Earth surface. GSHP systems are highly efficient, with lower operating costs, but their installation requires more space and higher upfront costs due to the need for ground loops or boreholes. Solar-assisted heat pumps (SAHP) are also gaining popularity in residential markets, especially in regions with abundant sunlight. These systems leverage solar energy to preheat the refrigerant, reducing electricity consumption and providing a more sustainable heating solution for homes. In residential settings, heat pumps are often used for space heating, cooling, and domestic hot water production. Underfloor heating systems are particularly well-suited for heat pumps as they allow for the efficient distribution of low-temperature heat throughout a home. Heat pumps can also be connected to traditional radiator systems, though adjustments may be required to optimize performance. The ability of heat pumps to provide both heating and cooling makes them a year-round solution for homeowners looking to reduce their energy consumption and environmental impact.

(b) **Commercial and Industrial**

Heat pumps are also widely used in commercial and industrial settings where the need for heating, cooling, and hot water is constant and often on a larger scale. In commercial buildings, heat pumps are integrated into HVAC systems to regulate indoor temperatures, improve air quality, and reduce energy costs. Air-to-water heat pumps are particularly useful for providing both space heating and cooling in office buildings, hotels, and shopping centers, as they can be adapted to work with large-scale air handling and distribution systems (Xiao et al. 2020). Hybrid systems, which combine different heat sources such as air and geothermal energy are becoming more common in commercial applications. This is because they offer improved efficiency and resilience, especially in climates with wide temperature fluctuations.

In industrial processes, heat recovery heat pumps are utilized to capture waste heat from manufacturing operations which can then be repurposed for other uses such as heating water or air. This helps to reduce overall energy consumption and operational costs in industries such as food processing, textiles, and chemicals. For large-scale industrial processes, ground-source or water-source heat pumps are often the most efficient, as they provide a stable and abundant heat source for year-round operations (Wang et al. 2020). Heat pumps can also be used in district heating systems, where they supply energy to multiple buildings. This approach will reduce the need for individual heating systems and providing centralized energy management. These applications not only contribute to reducing energy costs but also play a significant role in lowering carbon emissions in the industrial sector.

(c) **Agricultural and Process Heating**

In the agricultural sector, heat pumps are employed for a variety of tasks that require precise temperature control. Solar-assisted heat pumps (SAHP) are particularly effective for drying crops, herbs, and other agricultural products. By using solar energy to preheat the air, these systems can reduce moisture content while preserving the quality of the products being dried. Heat pumps offer a more energy-efficient alternative to conventional dryers that rely on electricity or fossil fuels, making them an attractive option for farmers looking to reduce costs and improve sustainability.

In greenhouses, heat pumps are used to maintain optimal growing temperatures year-round, which can significantly extend the growing season and improve crop yields. Ground-source heat pumps are often favored for these applications as they provide a stable heat source even in colder weather. In addition to heating, heat pumps can be used for cooling in summer months, ensuring that temperature-sensitive crops are not exposed to extreme heat. The versatility of heat pumps in agricultural settings extends to dairy farms and livestock facilities, where they are used to provide hot water for cleaning and sanitation purposes, reducing the reliance on traditional water heaters.

To further enhance the efficiency and adoption of heat pump systems, it is recommended to integrate renewable energy sources such as solar and geothermal energy which can significantly reduce operational costs and improve sustainability. Incorporating solar thermal panels or hybrid systems can optimize the performance of heat pumps, particularly in regions with extreme temperature fluctuations. Additionally, upgrading insulation, improving heat exchanger designs, and using advanced refrigerants can reduce energy losses and increase the overall efficiency of the systems. Government incentives, financial subsidies, and public awareness campaigns should be expanded to promote the widespread use of heat pumps, emphasizing the long-term savings and environmental benefits they offer. Educating building owners and operators on the proper sizing, installation, and maintenance of heat pumps is crucial to maximizing system efficiency and ensuring optimal performance throughout their lifespan.

2.4 Ground Source Heat Pump

Geothermal energy refers to the heat energy that is produced and retained within the earth. It is located within the rocks and fluids that occupy the fractures and pores in the crust of the earth. Calculations indicate that in the absence of an extra energy source other from solar input, the earth would have undergone cooling and solidification several millennia ago starting from a state of total moltenness. Geothermal energy is mostly derived from the radioactive decay that takes place deep inside the Earth (Mufti 1971). Geothermal heat often emerges at the surface in a dispersed fashion throughout most areas. Nevertheless, a variety of geological processes have led to the presence of very shallow geothermal resources in specific regions, encompassing substantial amounts of numerous western states. The resources can be categorized based on their temperature into three distinct groups: low temperature (below 90 °C or 194 °F), moderate temperature (between 90 °C and 150 °C or 194 and 302 °F), and high temperature (beyond 150 °C or 302 °F). The utilization of geothermal resources is mostly dictated by their temperature. The primary utilization of high-temperature resources is for the generation of electric power, with the current total geothermal electric power generation in the United States being around 2800 MW (Williamson et al. 2001). Low and moderate temperature resources are mostly used in two categories such as direct utilization and ground-source heat pumps.

Geothermal energy can be directly harnessed by using the heat from geothermal water for many purposes, including heating buildings, powering industrial processes, operating greenhouses, facilitating aquaculture, and supporting recreational resorts (Joao et al. 2022). This does not require the use of heat pumps or power plants.

Usually, these direct utilization projects function at resource temperatures ranging from 38 °C (100 °F) to 149 °C (300 °F). The direct use systems in the United States have an installed capacity of around 470 MW, which is enough to supply heating for approximately 40,000 residences of average size. Ground-source heat pumps utilize the earth or groundwater as a heat source in winter and as a heat sink in summer. These systems operate within a temperature range of 4 °C (40 °F) to 38 °C (100 °F). A heat pump, a device that facilitates the transmission of heat between two locations is employed to extract heat from the ground and transmit it into buildings during winter and vice versa during summer. Although there is a dearth of specific statistics regarding the number of ground-source heat pump systems that have been built, it is believed that between 10,000 and 40,000 units are installed each year.

2.4.1 System Description and Working Mechanism

The ground source heat pump (GSHP), commonly referred to as the geothermal heat pump (GHP) is a cutting-edge renewable energy technology that is becoming more and more popular for both residential and commercial structures GSHP are employed for the purpose of both room heating and cooling, as well as water heating. One significant benefit of this technology is its ability to harness and amplify pre-existing heat rather of relying on the burning of fossil fuels to generate heat. This system functions based on the notion of geothermal stability, which means that the temperature under surface of the earth remains generally consistent throughout the year. During winter, the temperature is higher than the air above, while during summer, it is lower, resembling the circumstances found within a cave. GSHP utilize the stable underground temperature to extract heat from the ground or groundwater and transport it into buildings during winter. Conversely, during summer, they remove heat from buildings and transfer it back into the ground. In essence, the ground functions as a heat source during winter and a heat sink during summer.

(a) **Ground Temperature**

The effectiveness of GSHP technology is primarily contingent upon the ground temperature. The geothermal energy derived from the interior of the earth is stored in the rocks of the earth crust. GSHP used in residential and commercial buildings generally make use of ground temperatures that fall between the range of 4–38 °C. It was indicated that by descend down into the earth interior, the temperature increases. The interior emits heat that propagates outward, while the crust acts as a protective barrier, shielding us from this internal heat. The mantle exhibits partial melting, the outer core is in a liquid state, and the inner core is in a solid state. A GSHP system is composed of three main components that work together to provide efficient heating and cooling. The first component is the underground geothermal heat exchanger subsystem, which harnesses thermal energy from the earth. The second component is the heat pump subsystem, which includes key elements such as the evaporator (a heat exchanger), compressor, and expansion valve. These components work in tandem to transfer and elevate heat. The third component is the heat distribution subsystem, typically involving a condenser section (another heat exchanger), which disperses the heated or cooled air or hot water through conventional ductwork or piping systems to maintain the desired temperature throughout the building.

The subterranean geothermal heat exchanger is a crucial and costly component of a ground source heat pump, especially when compared to an air source heat pump. GSHP systems are classified into three categories based on the heat sources or sinks they utilize: ground-coupled systems, groundwater systems, and surface water systems (Sarbu and Sarbu 2021b). These systems employ two primary loop configurations: closed-loop and open-loop, which can be installed horizontally, vertically, or in a body of water such as a pond or lake. The choice of installation method depends on the available land area and the specific soil and rock characteristics at the site, which determine the most cost-effective option for establishing the ground loop. The most common configuration is the closed-loop system, where high-density polyethylene (HDPE) pipes are buried either horizontally at depths of 4–6 feet or vertically at depths of 100–400 feet. These pipes house a heat exchanger filled with an environmentally friendly antifreeze and water solution. In winter, the fluid absorbs thermal energy from the earth and transfers it to the building. During summer, the process reverses, extracting heat from the building and transferring it to the cooler ground. The air supply ducts then distribute the heated or cooled air throughout the house, similar to conventional systems. Additionally, this system provides complimentary hot water during the summer and substantial hot water savings in winter. The choice of a horizontal, vertical, or pond/lake configuration depends on factors such as climate, soil conditions, land availability, and local installation costs, making these systems suitable for both residential and commercial applications.

i. **Horizontal Closed-Loop System**

The horizontal closed-loop system is one of the most cost-effective options for installing ground source heat pumps, particularly in residential settings where there is ample land available. This method is especially suited for new construction projects, as it requires sufficient space for trenching and pipe installation. Typically, trenches are dug to a depth of at least four feet, although deeper installations may be required depending on the design and ground conditions. The most common configuration involves laying two pipes at different depths at one at six feet and the other at four feet, within the same trench (Esmaeilpour et al. 2021). Alternatively, two pipes can be placed side by side at a depth of five feet in a two-foot-wide trench. This design allows for optimal heat exchange between the ground and the circulating fluid, ensuring efficient transfer of thermal energy into the system.

For areas where conventional horizontal configurations may not be feasible due to space constraints or land characteristics. As an alternative approach is to use coiled pipes, often referred to as slinky coils, in shorter trenches. This coiled configuration reduces the overall length of trenching required, making it possible to install horizontal closed-loop systems in smaller or irregularly shaped plots of land. By reducing the amount of digging and piping, this method lowers installation costs and makes horizontal closed-loop systems more accessible to a wider range of residential properties. Moreover, the flexibility of this installation method allows it to be adapted to different soil conditions and climates, ensuring that the system performs effectively throughout the year. Despite being less expensive than vertical systems, horizontal loops still deliver reliable heating and cooling by utilizing the consistent underground temperatures to transfer heat efficiently.

(b) **Vertical Closed-Loop System**

The vertical closed-loop system is commonly used in large commercial buildings, schools, and urban settings where land availability is limited. Unlike the horizontal system which requires extensive land for trenching. The vertical system uses deep boreholes to accommodate the piping making it ideal for areas with limited space. This method is also preferred in regions where the soil is too shallow for horizontal trenching or where minimal disruption to existing landscaping is important (Fallah et al. 2021). The vertical installation involves drilling small-diameter holes, typically around four inches in diameter, spaced about 20 feet apart. These boreholes can range in depth from 100 to 400 feet, depending on the specific site conditions, as well as the heating and cooling requirements for the building.

Once the holes are drilled, two pipes are inserted into each borehole and connected at the bottom with a U-bend to form a loop. These vertical loops are then linked together using horizontal pipes called manifolds, which are buried in shallow trenches and connected to the heat pump system within the building. The vertical design allows the heat pump to tap into the consistent temperatures found deep within the Earth, providing a stable and reliable heat source in the winter and a heat sink in the summer. Although the installation costs for vertical systems are higher than for horizontal loops, due to the specialized drilling equipment and labor required. This system has ability to operate efficiently in tight spaces makes it an attractive option for urban developments and large commercial projects.

(c) **Pond/Lake Closed-Loop System**

The pond/lake closed-loop system leverages naturally occurring water bodies as a medium for heat exchange, offering one of the most cost-effective options for ground source heat pumps in suitable locations. If a property has access to a lake or large pond that meets specific criteria for depth, volume, and water quality, this system can be a highly efficient alternative to traditional ground loops. An underground supply line is installed from the building to the water body, where the piping is coiled into circular loops and submerged at least eight feet below the surface of water to prevent freezing (Assad et al. 2022). The depth and placement of the coils ensure that the water remains at a relatively stable temperature, allowing the heat pump to efficiently extract or release heat throughout the year. This system works by transferring thermal energy between the water and the heat pump. During winter, the water in the pond or lake acts as a heat source. While in summer, it functions as a heat sink, absorbing heat from the building. To further enhance the efficiency of system and reduce heat loss, an insulating sheet or board can be placed on the surface of the water. In some cases, transparent insulation materials like honeycomb panels are used to increase heat retention. The pond/lake system is particularly appealing for properties with access to natural water bodies, as it requires less excavation and trenching compared to ground-based loops. However, the water source must meet specific environmental conditions to ensure long-term functionality and efficiency.

(d) **Open-Loop System**

The open-loop system is a unique ground source heat pump configuration that uses groundwater or surface water as the heat exchange fluid. Unlike closed-loop systems, which circulate a solution of water and antifreeze through buried pipes, open-loop systems directly extract water from a well, lake, or other water body. This water is pumped through the heat pump system where it absorbs or releases heat depending on the season (Violante et al. 2021). After circulating through the system, the water is returned to the environment through a discharge well, a recharge well, or directly into a surface water body. This makes the open-loop system ideal for areas with a plentiful and clean water supply, as well as for applications where local regulations permit the discharge of groundwater.

Open-loop systems offer many of the same benefits as closed-loop systems including energy efficiency and environmental friendliness. But they are typically less expensive to install due to the reduced need for excavation and piping. However, they are only viable in regions with sufficient water resources and where groundwater regulations allow for the safe discharge of water. Additionally, the water quality must be carefully monitored to prevent clogging, scaling, or corrosion of the heat pump system. While open-loop systems can be highly efficient, they require regular maintenance to ensure the water source remains clean and uncontaminated. This system is commonly used in regions where access to high-quality water is reliable and where a suitable method for water disposal is readily available.

2.4.2 Applications in Buildings

Numerous studies have explored the significant advantages of ground source heat pumps (GSHP), making them one of the most efficient and sustainable heating and cooling solutions available today. Their ability to harness the consistent temperature of the earth provides year-round comfort for residential, commercial, and industrial buildings. In the residential sector, GSHP systems can be installed in homes of any size, from small single-family houses to large multi-family units. These systems offer a versatile solution for providing both heating in the winter and cooling in the summer. In addition to space heating and cooling, GSHP systems can also supply domestic hot water, making them an all-in-one solution for energy-efficient home management. GSHP systems are adaptable to a variety of lot sizes which makes them suitable for both urban and rural homes. In urban areas where land space is limited, vertical closed-loop systems can be installed, allowing the system to utilize deep boreholes to access geothermal energy without requiring large areas for trenching. In suburban and rural settings, where more land is available, horizontal closed-loop systems are commonly used. These systems are installed in trenches under lawns, gardens, or even beneath driveways making efficient use of space without disrupting the landscape. This flexibility in installation options ensures that GSHP can be effectively used in a range of residential environments, regardless of the available land area or soil conditions.

Another key advantage of GSHP systems is their suitability for retrofitting existing homes. For older homes that rely on traditional heating and cooling systems, GSHP can be installed with minimal disruption to the home existing structure. The system can be integrated with the current ductwork or radiant floor heating systems, making it an ideal solution for homeowners looking to upgrade their energy systems without completely overhauling their infrastructure. By evaluating the existing ductwork, a qualified GSHP installer can recommend any necessary modifications to optimize the system efficiency, ensuring that the home benefits fully from the new geothermal technology. GSHP systems are also compatible with modern home energy management systems, making them a popular choice for homeowners interested in smart home technology. They can be integrated with programmable thermostats and smart controls, allowing for precise temperature management and further energy savings. Many homeowners appreciate the long-term financial benefits of GSHP systems as they significantly reduce energy bills by using renewable geothermal energy. Although the initial installation cost may be higher than traditional heating and cooling systems, the reduced operational costs and energy savings over time make GSHP systems a cost-effective investment.

Lastly, the environmental benefits of GSHP systems are considerable. By using the renewable energy stored in the earth, GSHP help reduce the reliance on fossil fuels, contributing to lower greenhouse gas emissions and a smaller carbon footprint for residential buildings. This makes GSHP systems not only a practical solution for homeowners but also an environmentally responsible choice for those looking to reduce their impact on the planet. In many cases, governments and energy providers offer incentives, tax credits, or rebates to encourage the adoption of GSHP systems, further enhancing their appeal to environmentally conscious homeowners.

2.5 Absorption Cooling System

The substantial increase in energy consumption across different applications has exerted immense strain on traditional energy sources, resulting in the imminent exhaustion of fossil fuels unless prompt measures are implemented. Moreover, the increasing need for energy has resulted in higher expenses, highlighting the pressing requirement to enhance energy production. Either by exploring new energy sources or by conserving existing ones through lower energy consumption. Recently, there has been a growing interest in using renewable energies, such as waste heat from industrial activities, wind, and solar energy, to power refrigeration systems. These energy sources are abundant and sustainable. Absorption refrigeration systems are a

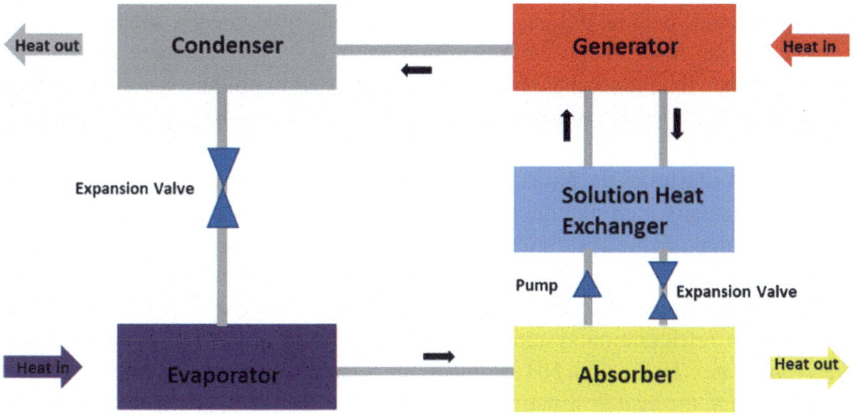

Fig. 2.3 General schematic of absorption system (Baroutaji et al. 2021)

viable alternative to conventional vapour compression refrigeration systems (VCRS), garnering significant scientific interest. These devices are driven by thermal energy and offer an efficient refrigeration mechanism.

Absorption refrigeration systems are comprised of four main components: the generator, the condenser, the evaporator, and the absorber as shown in Fig. 2.3. The application of heat to the generator results in the vaporization of the refrigerant. The absorption system, in contrast to conventional systems that utilize a compressor, utilizes an absorber to attract the refrigerant vapor into a diluted solution (Lima et al. 2020). Subsequently, the solution is sent to the generator, where thermal energy induces the refrigerant to undergo separation from the solution. The gaseous refrigerant transitions to the condenser, where it undergoes cooling and condensation, transforming into a liquid state. The liquid refrigerant flows via an expansion valve to decrease its pressure prior to entering the evaporator. In the evaporator, it absorbs heat and transforms into a gas, so delivering cooling. The highly concentrated solution is returned to the absorber by a heat exchanger, where it is prepared to absorb additional refrigerant vapour from the evaporator. This cycle perpetually recurs, facilitating effective cooling.

2.5.1 System Description and Working Mechanism

The vapor-compression cycle is categorized as a work-operated cycle since the increase in pressure of the vapor refrigerant is accomplished by a compressor that necessitates mechanical work. On the other hand, the absorption cycle is classified as a heat-operated cycle since its primary operating expense is related to supplying heat to remove the refrigerant vapor from the high-pressure liquid in the generator (Muhammad et al. 2021). The heat input can be obtained through the combustion of many sources such as oil, gas, natural gas, geothermal energy, solar energy, or industrial waste heat. While the absorption cycle does involve some mechanical labor to operate the pump, the quantity of work needed to produce a certain level of refrigeration is far lower compared to the work necessary in the vapor-compression cycle.

Figure 2.4 illustrates the basic absorption cycle. The components enclosed within the dotted frame, which include the absorber and generator, function as a thermal compressor, making the absorption cycle analogous to the vapor-compression cycle. In air-conditioning systems, the commonly used working pair is a solution of lithium bromide (LiBr) and water, where water serves as the refrigerant and LiBr as the absorbent. In refrigeration systems, the typical working pair consists of ammonia (NH_3) and water (H_2O), with NH_3 acting as the refrigerant and H2O as the absorbent.

In summary, the heat flow pattern to and from the four components in the absorption cycle can be described as follows: The generator receives heat at high temperatures, while the evaporator absorbs heat at low temperatures from the cooled air. Heat is expelled from the cycle at the absorber and condenser through circulating water or air, which is then discharged into the atmosphere. The Coefficient of Performance

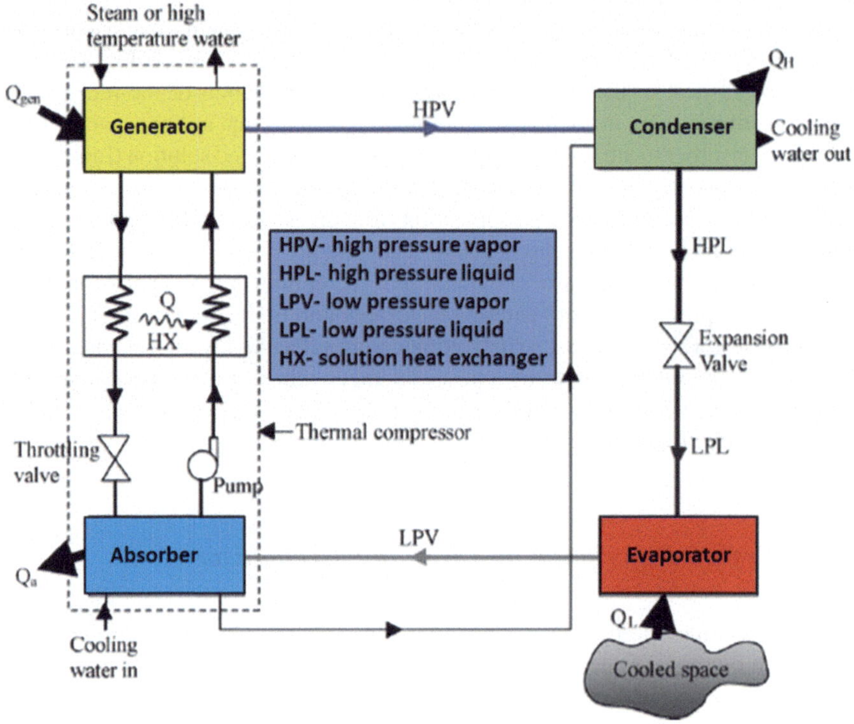

Fig. 2.4 An absorption system with heat exchanger (Bakthavatchalam et al. 2021)

(COP) of an absorption cycle is significantly lower than that of a vapor compression cycle. However, the absorption cycle primarily relies on less expensive thermal energy, unlike the vapor compression cycle, which depends on more costly electrical energy. This heat energy can be sourced from solar energy, industrial waste heat, geothermal energy, biomass combustion, and similar methods.

Although the absorption refrigeration system has its benefits, it also has certain limitations such as its large size, intricate design, and expensive initial investment. Refrigeration units of this kind often have capacities ranging from tens to hundreds of kilowatts. As a result, they are frequently employed in industrial settings and, more recently, in residential central air-conditioning systems. In the past, the first types of steam absorption chillers that used a single-stage and indirect-fired system had a COP ranging from 0.6 to 0.7. This low COP resulted in many absorption chillers being substituted with centrifugal chillers throughout the late 1970s and 1980s. Nevertheless, advancements and enhancements in the technology have persisted as a result of the utilization of low-temperature thermal energy and energy crises.

The absorption cycle is primarily utilized in cooling mode, but it can also be employed in heating mode as an absorption heat pump. In this configuration, the system extracts heat energy from the evaporator via the absorber, combines it with the heat input in the generator, and releases both to the hot water in the condenser for heating purposes. In Japan and Sweden, absorption heat pumps have been installed in industrial and district heating plants, utilizing industrial waste heat to provide hot water, typically at 74 °C, for winter heating or other applications, with a COP ranging between 1.2 and 1.7 (Johansson 2021; Jesper et al. 2021). Generally, the heat supplied to the generator can originate from solar thermal energy, geothermal energy, or waste heat. Solar absorption cooling systems have been extensively investigated due to the advantages of solar energy over fossil fuel combustion and its compatibility with summer climates. While geothermal energy presents a viable option, its application is limited to areas with local geothermal resources. Ideally, waste heat should be utilized in the generator of an absorption cooling system.

2.5.2 Applications in Buildings

Absorption refrigeration systems are widely applied in Europe, where they account for approximately 60% of all thermally powered refrigeration systems. This significant adoption is largely due to the numerous benefits absorption systems offer in various types of buildings, from residential to commercial and industrial structures. Their ability to operate using low-grade heat sources combined with their energy efficiency makes them an attractive option for sustainable cooling and refrigeration applications. As energy efficiency and environmental concerns become more pressing globally, absorption refrigeration systems present a viable alternative to conventional vapor compression systems, especially in buildings aiming to reduce their carbon footprint. One of the key advantages of absorption refrigeration systems is their reliance on low-grade thermal energy. These systems can be powered by heat from sources such as engine exhaust, industrial waste heat, and renewable energy sources like solar energy (Yadav et al. 2020). This flexibility in heat sourcing reduces the reliance on electricity and fossil fuels, contributing to lower CO_2 emissions and energy savings. Solar-powered absorption systems are particularly suitable for regions with abundant sunlight, making them a promising solution for eco-friendly buildings. Furthermore, the ability to tap into waste heat makes absorption systems highly efficient in industrial settings, where energy savings are a top priority.

Another notable benefit of absorption refrigeration systems is their use of eco-friendly refrigerants. In contrast to traditional vapor compression systems, which often use refrigerants that contribute to ozone depletion and global warming, absorption systems typically use natural refrigerants such as water or ammonia. These refrigerants have minimal environmental impact, making absorption systems a more sustainable choice for building applications that prioritize environmental responsibility (Karthick et al. 2022). The use of natural refrigerants also aligns with global efforts to phase out harmful substances like hydrofluorocarbons (HFC) further enhancing the system eco-friendliness. Quiet operation is another advantage that makes absorption refrigeration systems particularly well-suited for building applications. The absence of high-speed moving parts such as compressors, results in a quieter operation compared to traditional systems. This make them ideal for settings where noise pollution must be minimized such as hospitals, hotels, and residential complexes. Additionally, the lower mechanical complexity of these systems leads to reduced wear and tear, resulting in lower maintenance costs and a longer operational lifespan. Many absorption systems can last for 20–30 years with minimal maintenance, offering a durable and cost-effective cooling solution.

Another critical feature of absorption refrigeration systems is their ability to facilitate heat recovery, which improves overall energy efficiency. These systems can capture heat from virtually any source, including industrial processes or exhaust gases, and use it to power the cooling cycle. This capability makes absorption systems an attractive option for buildings that generate significant amounts of waste heat such as manufacturing facilities and data centers. The ability to recover and repurpose heat enhances energy savings and reduces the overall operational costs of the building cooling system. Lastly, absorption refrigeration systems offer the advantage of eliminating cycling loss, a common issue in vapor compression refrigeration systems (VCRS). In conventional systems, the frequent cycling on and off can generate excess heat and reduce efficiency. However, absorption systems maintain consistent performance without the need for cycling, resulting in higher energy efficiency and reduced waste heat generation. This makes absorption systems particularly well-suited for applications that require continuous cooling, such as refrigeration in large buildings, data centers, and industrial facilities.

2.6 Adsorption Cooling

Adsorption cooling and heat pump cycles operate on the principle of adsorbing a refrigerant gas into an adsorbent at low pressure followed by its release through the application of heat (Soni et al. 2021). The adsorbent, typically a porous material such as silica gel or activated carbon, functions as a "chemical compressor" driven by thermal energy, effectively acting as the system "pump." This method differs from traditional refrigeration cycles, as it does not require a mechanical compressor. Instead, it relies on heat sources, making it a more sustainable option. Solar thermal collectors, particularly in solar adsorption cooling, waste heat from industrial processes, and other readily available thermal sources are commonly used to power adsorption systems. By utilizing these heat sources, the system can operate efficiently especially in areas where waste heat or solar energy is abundant.

The advantages of adsorption cooling systems extend beyond energy efficiency. These systems also have fewer moving parts compared to conventional refrigeration systems, leading to lower maintenance costs and a longer operational lifespan (Chauhan et al. 2022). Furthermore, adsorption systems can be designed to operate with environmentally friendly refrigerants, reducing their impact on global warming and ozone depletion. This makes adsorption cooling an attractive option for regions with high ambient temperatures, where traditional air conditioning systems might struggle or become less efficient. Additionally, the ability to harness low-grade heat sources such as solar energy or industrial waste heat, provides a significant opportunity to reduce overall energy consumption and decrease reliance on conventional electricity-driven systems. As research and development in adsorption technology continue, we can expect further enhancements in efficiency and broader adoption across various industries and applications.

2.6.1 System Description and Working Mechanism

A solar adsorption chiller of intermittent nature exemplifies the practical application of adsorption cooling technology, as shown in Fig. 2.5. The system comprises three primary components: a solar collector, a condenser or heat exchanger, and an evaporator. The solar collector harnesses solar energy to elevate the temperature of the adsorbent, causing the release of the refrigerant which then condenses in the condenser. The condensed refrigerant subsequently evaporates within the evaporator, producing the desired cooling effect. The interior of the collector is lined with an adsorption bed filled with an adsorbent such as activated carbon, zeolite, silica gel, or their derivatives (Sinha 2020). These materials are capable of adsorbing refrigerants like methanol, ammonia, ethanol, and water. The insulated refrigeration box containing water allows the adsorbent to effectively absorb a significant amount of refrigerant vapor at room temperature and release it at elevated temperatures. During the day, solar irradiation heats the collector, causing the refrigerant to vaporize and

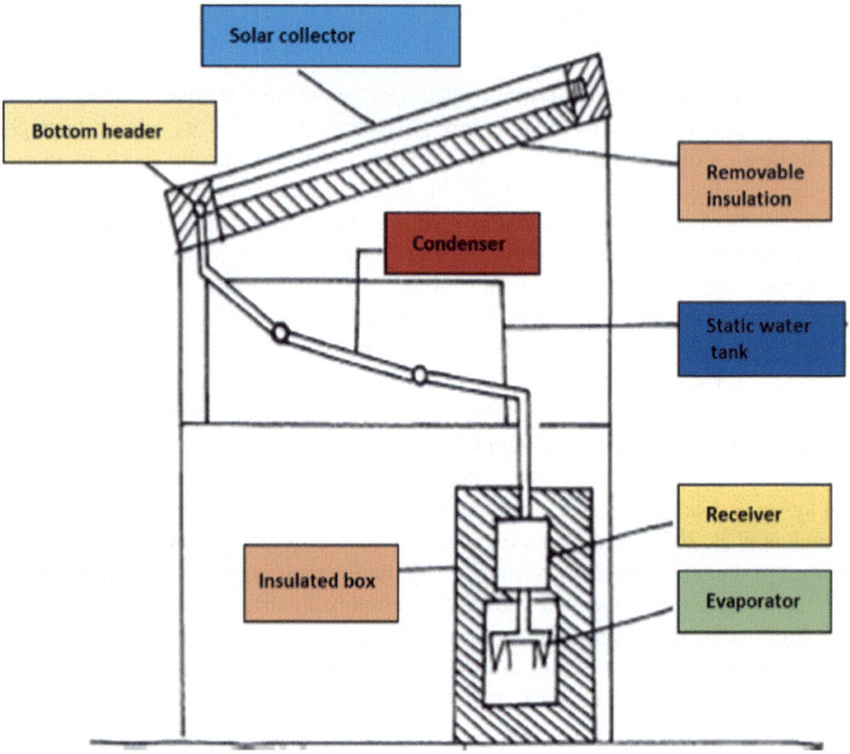

Fig. 2.5 Schematic diagrams of two types of intermittent-type solar adsorption cooling systems (Isaac and Adebayo 2010)

condense in the evaporator. At night, when the collector cools to ambient temperature, the adsorbent material re-absorbs the refrigerant which then vaporizes in the evaporator, extracting heat from the water. Effective cooling of the collector at night is crucial, highlighting the need for a design that allows the collector lid to open at night and close during the day.

2.6.2 Applications in Buildings

Furthermore, waste heat adsorption chillers are extensively researched and utilized with solar adsorption chillers. Adsorbents are essential for optimizing the effectiveness of refrigerant adsorption and desorption processes. These materials are commonly used in various shapes such as spherical pellets, rods, moldings, or monoliths, with hydrodynamic diameters ranging from 0.5 to 10 mm. The adsorbents must exhibit exceptional resistance to abrasion, exceptional thermal stability, and possess micropores with small diameters. These characteristics enhance the surface area of the absorbent, hence increasing their capacity for adsorption. In addition, it is important for adsorbents to possess a clearly defined macropore structure in order to enhance the efficient movement of gaseous vapors.

(a) **Oxygen-containing compounds**: These are hydrophilic or polar adsorbents, such as silica gel and hydrophilic zeolites.
(b) **Carbon-based compounds**: These are hydrophobic or non-polar adsorbents, such as activated carbon.
(c) **Polymer-based compounds**: These adsorbents feature polar or non-polar functional groups within a porous polymer matrix.

i. **Silica Gel**

Silica gel is a chemically inert, non-toxic, polar, and stable form of SiO_2 that remains unchanged at temperatures below 400 °C, as seen in Fig. 2.6a. The synthesis of this compound involves the reaction of sodium silicate with sulphuric acid, followed by subsequent post-treatment procedures including ageing and pickling (Islam et al. 2020). These processes result in a variety of pore size distributions on its surface. Silica gel is widely used to remove moisture from process air like oxygen and natural gas and to adsorb polar compounds from natural gas. In adsorption cooling systems silica gel is often combined with refrigerants like water or methanol to create an effective adsorbent-refrigerant mixture. This combination is extensively used in various applications.

ii. **Zeolites**

Zeolites can be either naturally occurring or synthetically produced. These aluminum silicates form a regular crystal lattice and release water at elevated temperatures, as shown in Fig. 2.6b. These materials are polar by nature. Zeolites are synthesized hydrothermally from sodium alumina silicate in an autoclave and subsequently undergo ion exchange with cations such as Na^+, Li^+, Ca^{2+}, and K^+ (Collins et al. 2020). The channels within the zeolite structure typically range from 2 to 9 Å in diameter. Following synthesis, the microcrystals are dried and pelletized with a binder to form macropores, then thermally activated at 650 °C. Zeolites are often paired with water to create an adsorbent-refrigerant system, which is widely utilized in adsorption cooling applications.

iii. **Activated Carbon**

Activated carbon is a porous solid with a disordered structure made up of microcrystallites that resemble graphite as shown in Fig. 2.6c. While it is non-polar and cost-effective, its main drawback is combustibility. It is produced from carbon-rich sources such as coal, peat, wood, or nutshells like coconut through a process involving carbonization and activation (Gayathiri et al. 2022). During activation, carbonized particles are treated with an agent like high-temperature steam to enhance adsorption capabilities. Due to its large micropore and mesopore volumes, activated carbon has a vast surface area, making it highly effective for adsorbing organic molecules and

Fig. 2.6 **a** Silica gel product, **b** zeolite, **c** activated carbon and **d** calcium chloride

non-polar chemicals. It is widely used in adsorption cooling systems, where it is combined with refrigerants like methanol, ammonia, and ethanol to create efficient adsorbent-refrigerant pairs.

iv. **Calcium Chloride**

Calcium chloride is a chemical compound consisting of calcium and chlorine. It is illustrated in Fig. 2.6d. It has a strong ability to dissolve in water and is deliquescent, which means it can absorb moisture from its surroundings. At room temperature, this substance exhibits the characteristics of a conventional ionic halide due to its solid state. Calcium chloride is commonly used as brine in refrigeration plants, for ice and dust control on roadways, and in cement making (Boldyryev et al. 2021). While it can be directly obtained from limestone, large quantities are also produced as a byproduct of the Solvay process. Due to its hygroscopic nature, calcium chloride must be stored in tightly sealed containers to prevent moisture absorption. In adsorption cooling systems, calcium chloride is used as an effective adsorbent due to its strong moisture absorption capacity, which is critical in maintaining efficient cooling cycles. It often pairs with water as the refrigerant, enhancing the ability of the system to operate efficiently at low temperatures, thereby contributing to the overall performance and energy efficiency of the cooling process.

2.7 Desiccant Cooling

Desiccant cooling is a novel way of air conditioning that provides a clear alternative to traditional cooling techniques. Desiccant cooling differs from conventional systems by prioritizing the reduction of air humidity instead of relying solely on electrical energy to lower air temperature, resulting in a more pleasant indoor atmosphere (Shakir 2024a). This strategy exploits the principle that air with reduced relative humidity creates a sensation of coolness for people, even if the real temperature remains the same. Desiccant cooling can improve comfort and energy efficiency by reducing the reliance on temperature-based cooling alone. This technology is especially beneficial in areas with elevated humidity levels, where conventional air conditioning systems face difficulties in maintaining pleasant conditions without excessive energy usage.

The increasing fascination in sustainable and energy-efficient technology has thrust desiccant cooling into the limelight. The capacity to harness low-quality heat sources, such as solar thermal energy, waste heat, and natural gas establishes it as an eco-friendly choice. By integrating desiccant materials that absorb moisture from incoming air, these systems decrease the burden on traditional cooling units. This will result in reducing energy usage and operational costs. Moreover, desiccant cooling systems can greatly enhance indoor air quality by decreasing humidity levels, so alleviating concerns associated with mould proliferation and other humidity-related difficulties. Desiccant cooling systems are becoming increasingly popular as a feasible and sustainable choice for modern air conditioning demands due to the growing need for green building solutions and energy efficiency.

2.7.1 System Description and Working Mechanism

Desiccant cooling functions in a distinct manner compared to conventional cooling technologies, as it does not immediately lower the temperature of the air. On the contrary, it decreases the level of moisture in the air resulting in a feeling of coolness (Liang et al. 2022). It frequently employs other cooling techniques such as evaporative cooling to produce a decrease in temperature. A desiccant cooling system consists of a dehumidifier that directs incoming air through a desiccant substance to eliminate moisture, as well as a cooling unit. Reducing humidity via desiccant cooling can be regarded an effective cooling approach since lower relative humidity at the same air temperature creates a cooler sensation compared to higher humidity.

The cooling unit often utilizes an evaporative cooling system, often employing an indirect cooler. The use of desiccant-prepared dry air greatly improves the effectiveness of the evaporative cooling process. Once the desiccant material collects moisture, it reaches its saturation point and needs to be regenerated. Therefore, a regenerator is a crucial element of the desiccant cooling system. The concept of this system is depicted in Fig. 2.7. In desiccant cooling systems, the desiccant dehumidifier serves as the primary unit among the three main components. Based on the phase of the desiccant material used, these systems are classified into solar desiccant cooling, which utilizes a solid desiccant, and liquid desiccant cooling, which employs a liquid desiccant.

(a) **Solid Desiccant Cooling: Two-Wheel Desiccant System (TWDS)**

Solid desiccant cooling functions as an open, heat-driven cycle, in contrast to traditional air conditioning systems that depend on electrical energy for the cooling process. The two-wheel desiccant system (TWDS) utilizes both a desiccant wheel and a thermal wheel to effectively accomplish cooling and dehumidification (Verma et al. 2022). Common desiccants such as silica gel, zeolite, and charcoal are frequently used in the system to remove moisture from the air. As a thermodynamic cycle, it has the capability to harness sustainable energy sources such as natural gas, heated water, waste heat, or solar thermal energy to elevate the air temperature to the required level for reactivation. The low humidity levels of the system yield ancillary advantages such as diminished corrosion and microbial proliferation. When appropriately planned, dimensioned, and supervised, this technology is exceedingly eco-friendly.

i. **Principle of Two-Wheel Desiccant System (TWDS)**

Technically, a revolving wheel incorporates the desiccant in a dry desiccant system. The desiccant alternatively passes through the incoming process air as the wheel spins, where it absorbs moisture, and a regeneration zone where it dries out and releases moisture (Goodarzia et al. 2017). This cycle never ends. During the procedure, around three-fourths of the desiccant wheel are exposed to the incoming air. The desiccant is heated using an indirect-fired water or steam coil or a direct-fired gas burner in order to regenerate it. The "process" and "reactivation" streams of air, which are held apart by seals that make contact with face and rim of the wheel, revolve

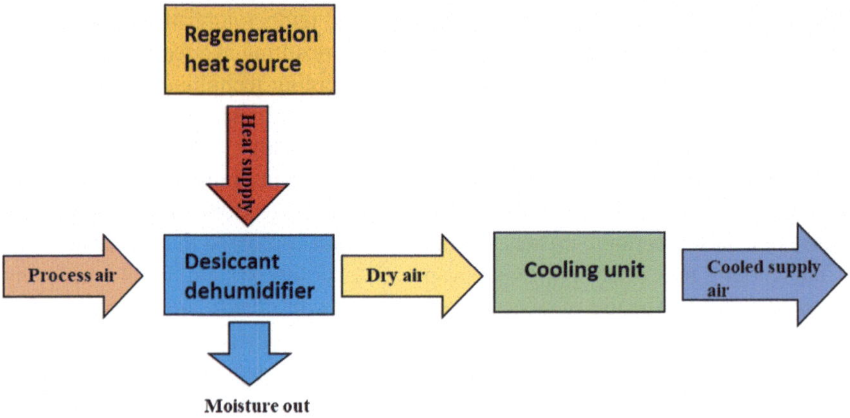

Fig. 2.7 Concept of desiccant cooling (Rambhad et al. 2016)

slowly as the desiccant wheel rotates between them. About eight to ten revolutions per hour is the sluggish speed at which the wheel revolves.

As the process air flows through the desiccant wheel, it releases its moisture to the desiccant material. The process of extracting moisture is exothermic, meaning it releases heat as the moisture changes from vapor to liquid when it comes into contact with the desiccant. As a result, the dry process air typically needs to be cooled using a heat exchanger or cooling coil once it exits the wheel. Once the desiccant wheel has absorbed moisture from the process air, it becomes fully saturated and rotates at a leisurely pace into the reactivation air stream. The reactivated air which is hot, flows through the wheel, raising the temperature of the honeycomb structure. The increase in temperature raises the vapor pressure of the moisture inside the wheel, leading the moisture to migrate to the surface and enter the reactivation air stream. The air which contains a high amount of moisture, is subsequently expelled from the structure.

Although the process air is significantly dehumidified, its temperature is slightly increased due to the latent heat released during moisture adsorption. As a result, this dry process air cannot be directly introduced into a building. To extract the heat from the dry process air, another heat exchanger known as a rotary thermal wheel is required, as shown in Fig. 2.8. The rotary heat exchanger, which recovers heat rather than moisture, resembles the desiccant wheel in both appearance and design. Various thermal energy sources typically at 180 °F (82.2 °C) or higher, can be used to dry and regenerate the desiccant. These include electric-resistance heaters, solar hot water coils, heat reclaim coils, hot water or steam from boilers, or natural gas burners. In most commercial applications, either direct- or indirect-fired natural gas burners are commonly used. Regardless of the system configuration, supplemental cooling is often necessary for most commercial building applications. This can be achieved through methods such as conventional vapor compression (direct expansion), chilled water coils, direct evaporative cooling pads, and indirect evaporative cooling coils. Figure 2.9 outlines the four possible operating modes: recirculation, pure ventilation, makeup, and mixed.

Process and regeneration air can be sourced from outdoor air and/or exhaust air for regeneration, and outdoor air and/or return air for the process. These air streams can be used in pure or mixed forms. In the recirculation mode, return air from the conditioned space serves as the process air source. While outdoor air is used for regeneration as shown in Fig. 2.9a. In pure ventilation mode, outdoor air is used for process air, supply air and exhaust air are utilized for regeneration (Fig. 2.9b). In makeup mode, outdoor air is used for both process air and regeneration air (Fig. 2.9c). In mixed mode, the sources for process and regeneration air can combine elements of the other modes (Fig. 2.9d). The selection of the appropriate mode depends on specific site conditions, application requirements, and the economic feasibility of the hybrid desiccant system.

(b) **Liquid Desiccant Cooling**

Solid desiccant cooling units are compact and less susceptible to corrosion and carry-over. However, they require high regeneration temperatures for the desiccant wheel

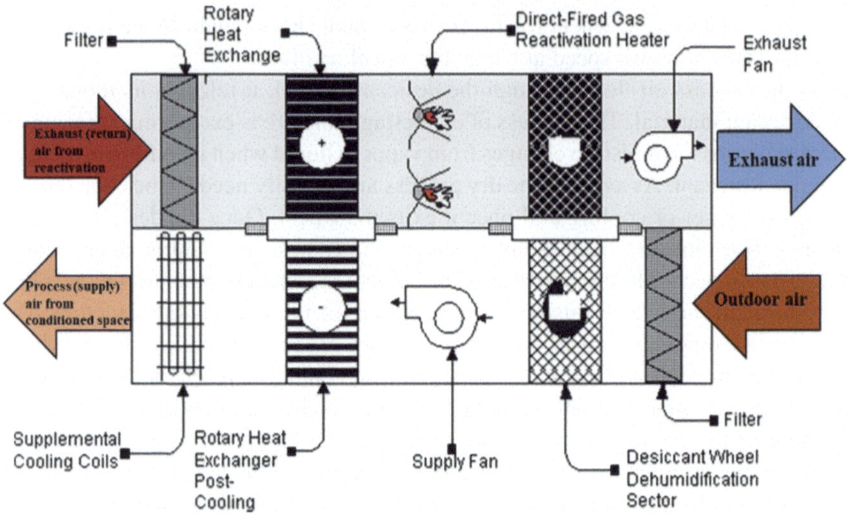

Fig. 2.8 Schematic of the two-wheel desiccant system (Bellia et al. 2000)

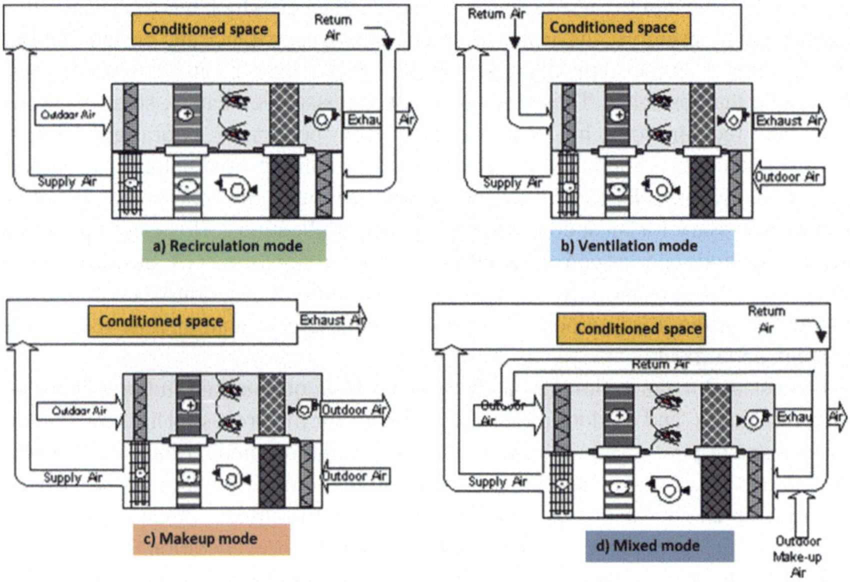

Fig. 2.9 Various operating modes for the hybrid cooling system (Bellia et al. 2000)

often reaching 120 °C or higher, and encounter significant air resistance against the two wheels. Liquid desiccant cooling units can mitigate these issues. The flexibility of liquid desiccants allows for their removal and regeneration, followed by cooling in a heat exchanger, which results in a lower pressure drop on the air side (Guan et al. 2022). As a result, research into liquid desiccant cooling systems has been gaining momentum. Commonly utilized liquid desiccants include aqueous solutions containing hygroscopic salts such as lithium chloride (LiCl), lithium bromide (LiBr), and calcium chloride ($CaCl_2$).

2.7.2 *Applications in Buildings*

Desiccant cooling systems are versatile in their application and can be broadly categorized into two types: liquid desiccant cooling systems and solid (dry wheel) desiccant systems. Liquid desiccant systems are typically used in large-scale industrial applications. Particularly in environments where controlling humidity is crucial to preventing the growth of microorganisms such as food processing facilities or pharmaceutical plants (Mahmood et al. 2020). These systems are effective in improving air quality by maintaining low humidity levels which can help reduce microbial contamination. In contrast, solid desiccant systems are more commonly found in commercial and light industrial applications including office buildings, retail spaces, and warehouses, where maintaining indoor comfort and reducing energy consumption are key priorities (Abd-Elhady et al. 2022). Both systems offer significant benefits in terms of energy savings and operational efficiency, especially when integrated with renewable energy sources like solar thermal energy or waste heat. One of the primary advantages of desiccant cooling systems is their ability to reduce energy consumption by utilizing low-grade thermal energy sources. Liquid desiccant systems are particularly advantageous in industrial settings where waste heat from processes can be repurposed to regenerate the desiccant material. This capability to utilize excess or waste heat makes liquid desiccant systems a highly efficient and eco-friendly option for industries with high energy demands. These systems can also be combined with other technologies such as evaporative cooling, further reducing the need for traditional air conditioning. In industrial applications, desiccant cooling systems are often preferred for controlling humidity in production environments where precise conditions are required for sensitive equipment and products.

Solid desiccant systems are highly effective in commercial buildings and light industrial applications because of their compact size and flexibility. These systems are commonly employed in office buildings, hotels, and shopping malls, where maintaining a comfortable indoor climate is essential. By removing moisture from the air and lowering humidity levels, solid desiccant systems can create a more pleasant indoor environment without relying solely on cooling air temperature. This approach helps reduce the energy consumption typically associated with conventional air conditioning systems, making solid desiccant systems an ideal solution for commercial spaces where energy efficiency and comfort are both critical (Ding et al.

2022). Looking at specific manufacturers, Table 2.3 highlights several companies that produce a wide range of desiccant cooling equipment. This diversity in the desiccant cooling market demonstrates the adaptability of these systems to various scales and requirements. When it comes to suggestions for the improvement of desiccant cooling systems, one key area of focus should be further integration with renewable energy sources. While desiccant systems can already harness low-grade heat from solar thermal energy and waste heat, increased research and development into optimizing these systems with renewable energy could further reduce operational costs and improve their environmental impact. For example, incorporating solar thermal panels or geothermal heat into the regeneration process of liquid desiccant systems could significantly lower the overall energy demand, making them even more sustainable and viable for a wider range of applications.

Another recommendation is to continue exploring new desiccant materials that can enhance system performance. Materials such as advanced silica gels, zeolites, and hybrid desiccant materials could improve the moisture absorption capacity, regeneration efficiency and overall performance of both liquid and solid desiccant systems. Additionally, modifying system designs to reduce air resistance and improve airflow could enhance the efficiency of solid desiccant systems, making them more effective in commercial and industrial environments. With ongoing innovation and adoption of desiccant cooling technologies, these systems are poised to play a significant role in the future of energy-efficient and sustainable building climate control.

Table 2.3 Some desiccant system manufactures

Company	Equipment type	Description
AIL Research Inc.	Desiccant—liquid	6000 cfm packaged liquid-desiccant AC
Bry-Air	Desiccant—wheel	600–10,000 CFM Brypac, 600–100,000 CFM VFB
Concepts and Designs	Desiccant—wheel	700–40,000 CFM
Drykor	Thermal air conditioner—liquid desiccant system	Dehumidification and cooling in one unit 10 tons cooling, 2500 CFM air
Kathabar Systems Division	Desiccant—liquid	Desic. Dehumid. to 7500 CFM Liquid Desic. to 84,000 CFM
Munters Corporation	Humidity control	Desiccants and other humidity control systems 300–80,000 CFM
Novelaire	Desiccant—wheel	BETA testing a residential product
Seasons4	Desiccant—wheel	4000–50,000 CFM
SEMCO	Desiccant—wheel	Desiccant systems, recovery wheels

2.8 Heat Recovery

Waste heat is generated from a variety of sources such as residential, commercial, and industrial sectors. This section specifically addresses the recuperation of waste heat from residential structures. In these conditions, significant quantities of heat are frequently dissipated by expelled hot air and water. For example, the thermal energy generated by common household devices, heating systems, and water heaters usually disperses into the surroundings, resulting in a substantial waste of usable energy. By incorporating effective waste heat recovery systems, the energy that is currently being wasted may be harnessed and reused, resulting in improved energy efficiency for residential buildings.

Furthermore, there has been a growing interest in the possibility of harnessing heat from Building-Integrated Photovoltaics (BIPV). BIPV systems, which incorporate solar panels into the architectural framework, concurrently generate energy and provide heat (Kuhn et al. 2021). Unmanaged, this heat might diminish the efficacy of the solar cells. Nevertheless, with the utilization of heat recovery methods, the thermal energy produced by BIPV systems can be captured and utilized for residential heating and hot water generation, hence enhancing the overall energy efficiency of the structure. These integrated techniques not only save energy but also encourage sustainable building practices by decreasing reliance on external energy sources.

2.8.1 System Description and Working Mechanism

Air-to-air energy recovery refers to the transfer of heat energy from one air stream (exhaust air) to another (incoming fresh air) within a ventilation system of the building. This technology is applicable for both heating and cooling purposes (Li et al. 2022). By utilizing heat exchanger technology, air-to-air energy recovery reduces the load on heating and cooling systems by reclaiming heat that would otherwise be lost to the atmosphere. In buildings, heat exchangers are integrated between the supply and exhaust ventilation ducts to enable this heat recovery process, making it a common practice in building ventilation and heating systems (Mansur et al. 2021). Riffat and Gillott have developed an innovative mechanical ventilation heat recovery heat pump system that employs revolving heat exchangers to both move air and transfer heat. This system captures low-grade heat from exhaust air, upgrades it through a heat pump, and uses it to heat incoming fresh air (Mansur et al. 2021; Gillott 2000). There are various designs of heat recovery devices for air conditioning. Four common types of heat exchanger technology qualify for Enhanced Capital Allowances such as:

(a) **Plate Heat Exchangers**

These systems are commonly known as heat recovery ventilators (HRV) or recuperators. They generally consist of a cubical plate heat exchanger with alternating channels for supply (inbound) and exhaust (outbound) airflows, which are separated by thin metal or plastic plates. HRV are designed to capture the heat energy from exhaust air and transfer it to the incoming fresh air as it enters the building.

(b) **Thermal Wheel**

The rotary regenerator commonly referred to as a heat wheel is a rotating heat exchanger that facilitates gas-to-gas energy transfer from an exhaust stream to a supply stream. It features a circular heat transfer medium that rotates slowly within an airtight enclosure. The exhaust air flows over one portion of the wheel, while the supply air flows in the opposite direction over the other portion. As the wheel turns, it transfers energy between the two air streams. Some designs are capable of recovering both latent and sensible energy.

(c) **Run-Around Coils**

The run-around heat recovery system features a simple piping loop. This loop connects a finned-tube coil located in the exhaust plenum to a finned-tube coil situated in the make-up air plenum (Mahmoud et al. 2022). The warm exhaust air heats the circulating fluid, which subsequently transfers this heat to the incoming cool make-up air. This system is cost-effective for long-distance applications, customizable to specific requirements, and capable of integrating multiple heat sources and applications. Additionally, it offers flexibility in construction materials to match both the heat source and sink, and is easy to control. However, the system typically exhibits lower temperature efficiency, operates within a limited temperature range, and contains moving parts that may require maintenance.

(d) **Heat Pipe Heat Recovery System**

A heat pipe is a sealed self-contained system for liquid evaporation and condensation, typically designed in a cylindrical form. Heat is transferred from the hot incoming gas to the evaporator section of the heat pipe, increasing the local vapor pressure (Jouhara et al. 2021). The evaporated fluid then flows along the pipe to the condenser section, where the vapor pressure is lower. Here, the fluid condenses and releases heat to the cooler gas. The condensed liquid returns to the evaporator through a combination of capillary action and gravity.

(e) **Heat Recovery Water Heating from Grey Water**

Domestic grey water, which primarily includes waste bath water and kitchen dishwater, can serve as a valuable energy source. Drainwater heat recovery (DHR) devices capture heat from grey water and use it to preheat cold water that supplies showers, water heaters, or dedicated storage units (Ravichandran et al. 2021). By recovering this energy, DHR systems reduce the overall energy required for heating water and effectively increase the capacity of water heaters. The design of heat exchangers used in DHR systems varies in terms of pipe sizing, orientation of the drain line (horizontal, vertical, or otherwise), and overall configuration. These systems also differ in cost and the extent of energy savings they achieve.

(f) **Heat Recovery for Sorption Cooling and Dehumidification**

Sorption cooling systems which include both absorption and adsorption technologies, rely on a heat source to operate effectively. This heat source can be provided by domestic waste heat, such as flue gas exhaust from industrial processes or heating systems. By recovering this otherwise wasted heat, sorption cooling systems can achieve greater energy efficiency. In addition to providing cooling, these systems can also be used for dehumidification, making them highly versatile for both residential and industrial applications. Utilizing waste heat in this manner reduces the reliance on external energy sources, enhancing overall system sustainability and reducing operational costs.

(g) **Waste Heat Recovery from BIPV**

Building-integrated photovoltaic (BIPV) systems are designed to produce electricity by converting solar energy. However, in the process of generating electricity, these systems also produce significant amounts of heat which is typically wasted. This excess heat can negatively impact the efficiency of electricity generation, as higher temperatures reduce the performance of photovoltaic cells. By incorporating a heat recovery unit into the BIPV system, the waste heat can be captured and repurposed for other applications such as heating water or air to enable cogeneration. This not only improves the overall energy efficiency of the system but also maximizes the utilization of solar energy by providing both electricity and thermal energy from the same system.

2.8.2 Applications in Buildings

Heat recovery systems have a broad range of applications across residential, commercial, and industrial sectors. In residential buildings, these systems can be used to reclaim waste heat from air ventilation systems, hot water plumbing, and grey water. Then it can then be repurposed to preheat fresh air or water to reduce the overall energy demand. Industrial applications such as those in manufacturing plants, data centers, and power stations, benefit significantly from heat recovery technologies (Brough and Jouhara 2020). These systems capture waste heat from machinery and processes, which can then be used for space heating, power generation, or even for re-heating in industrial processes. In commercial buildings, heat recovery units are integrated into heating, ventilation, and air conditioning (HVAC) systems to reduce energy consumption for air conditioning and hot water production (Shakir et al. 2017a; Asasian-Kolur et al. 2022). Furthermore, integrating heat recovery with Building-Integrated Photovoltaic (BIPV) systems not only captures excess heat generated during electricity production but also improves the efficiency of both heating and electricity generation.

To maximize the efficiency of heat recovery systems, it is essential to adopt advanced technologies tailored to specific applications and conditions. For instance, combining heat recovery units with modern HVAC systems or thermal storage devices can significantly enhance energy savings in commercial and residential buildings (Abdelaziz et al. 2021). In industrial settings, integrating heat recovery systems with renewable energy sources such as solar or geothermal can optimize both economic and environmental outcomes. Additionally, systems designed to recover heat from low-grade heat sources such as grey water or exhaust air can be further improved by employing high-efficiency heat exchangers for better control. Proper maintenance and regular inspection of heat recovery systems will ensure peak performance and extend the lifespan of the equipment. In addition, ongoing research into emerging materials and technologies such as nanofluids and phase-change materials has the potential to further improve the thermal efficiency of future heat recovery systems.

2.9 Heat Pipe

Heat pipes are highly efficient thermal devices designed to transfer heat from one location to another with minimal temperature difference. They consist of a sealed tube that contains a working fluid, allowing heat to be absorbed at one end and released at the other (Weragoda et al. 2023). This straightforward design enables rapid heat transfer without requiring external power sources. Heat pipes are widely utilized in various fields due to their ability to handle significant amounts of heat while maintaining consistent performance, making them crucial in applications such as electronics cooling, spacecraft thermal management, and renewable energy systems.

One of the key advantages of heat pipes is their ability to function across a wide range of temperatures, making them versatile for use in both industrial and domestic environments. Their efficiency and simplicity make heat pipes a reliable option for systems that require effective heat management (Abdelkareem et al. 2022). Whether in high-temperature applications like industrial furnaces or low-temperature uses in refrigeration systems, heat pipes play an essential role in ensuring optimal thermal performance. Their adaptability, coupled with their ability to maintain heat transfer efficiency over long periods, makes them indispensable in modern technology and engineering.

2.9.1 System Description and Working Mechanism

A heat pipe is a highly efficient device specifically designed to facilitate quick heat transfer from one point to another. Often referred to as "heat superconductors," heat pipes are renowned for their ability to transport thermal energy with minimal losses, making them essential in applications where precise temperature regulation is critical. The concept of heat pipes was first introduced by R. S. Gaugler in 1942 (Wahlquist et al. 2023). However, their unique properties were not fully recognized until innovation by G.M. Grover in 1962, which sparked significant technological advancements and widespread use in industries such as aerospace, electronics, and energy systems (Grover et al. 1964).

A typical heat pipe consists of a hermetically sealed container usually constructed from materials like aluminum or copper, with interior surfaces coated in a capillary wicking material. This wick material often made of metal mesh, sintered powder, or grooved channels. It plays a critical role in the heat transfer process by facilitating the capillary action necessary to return condensed fluid back to the heat source. Heat pipes function through the continuous phase transition of a working fluid, typically water or ammonia, which evaporates at the heated end (evaporator) and releases the absorbed heat as it condenses at the cooler end (condenser) (Maghrabie et al. 2022). This process allows heat pipes to transfer heat even in challenging conditions such as when operating against gravity. The efficiency of the heat pipe is largely determined by the design and material of the wick structure. Also it can be determined based on the selection of the working fluid, which varies depending on the specific application, operating temperature range, and environmental conditions.

(a) **Container**

The primary role of container is to isolate the working fluid from the external environment. Therefore, it must be leak-proof, maintain the pressure differential across its walls, and facilitate the transfer of heat to and from the working fluid. Heat pipes typically operate under reduced internal pressure. Many of these factors are self-explanatory. For example, a high strength-to-weight ratio is particularly important for spacecraft applications. The container material must be non-porous to prevent vapor diffusion. Additionally, high thermal conductivity is necessary to minimize the temperature gradient between the heat source and the wick. The choice of container material is influenced by several critical factors:

i. Compatibility with both the working fluid and the external environment
ii. Strength-to-weight ratio
iii. Thermal conductivity
iv. Ease of fabrication, including weldability, machinability, and ductility
v. Porosity
vi. Wettability.

(b) **Working Fluid**

The initial factor in selecting an appropriate working fluid is the operating vapor temperature range. Within this temperature range, multiple potential working fluids may be considered, and various characteristics must be evaluated to identify the most suitable fluid for the application. Additionally, the choice of working fluid must account for thermodynamic factors that influence heat flow limitations within the heat pipe, such as viscous, sonic, capillary, entrainment, and nucleate boiling thresholds. These considerations are crucial for ensuring the heat pipe functions efficiently in its intended application. The key requirements include:

i. Excellent thermal stability
ii. Adequate wettability of wick and container materials
iii. Vapor pressure that remains within acceptable limits across the operating temperature range
iv. High latent heat
v. Superior thermal conductivity
vi. Low viscosities in both liquid and vapor states
vii. High surface tension
viii. Suitable freezing or pour point.

In designing a heat pipe, high surface tension is essential to allow the pipe to function against gravity and generate a significant capillary driving force. The effective surface tension can be substantially reduced by the accumulation of contaminants at the liquid/vapor, liquid/liquid, or solid interfaces. Along with high surface tension, the working fluid must adequately wet the wick and container materials, meaning the contact angle should be zero or very small. The vapor pressure across the operating temperature range must be sufficiently high to avoid excessive vapor velocities. It also can create large temperature gradients, entrain the returning condensate in the countercurrent flow, or cause flow instabilities associated with compressibility. A high latent heat of vaporization is beneficial for transferring large amounts of heat with minimal fluid flow, thereby maintaining low pressure drops within the heat pipe. Ideally, the working fluid should have high thermal conductivity to minimize radial temperature gradients and reduce the risk of nucleate boiling at the wick or wall surfaces. Choosing fluids with low vapor and liquid viscosities will minimize resistance to fluid flow.

(c) **Wick or Capillary Structure**

The wick or capillary structure in a heat pipe is a porous material made from materials such as steel, aluminum, nickel, or copper, and it comes in a variety of pore sizes. These structures are commonly fabricated using metal foams, with felts being more frequently used. By adjusting the pressure on the felt during assembly, different pore sizes can be created. Additionally, incorporating removable metal mandrels allows for the molding of an arterial structure within the felt. The wick plays a critical role in a heat pipe by generating capillary pressure, which drives the movement of the working fluid from the condenser to the evaporator. Additionally, it must effectively distribute the liquid within the evaporator section to regions where heat absorption is most needed. These two functions often require wicks with different designs. The choice of wick material and structure is influenced by several factors, many of which are closely related to the specific properties of the working fluid used in the system.

Fibrous materials like ceramics have been extensively used as wick structures in heat pipes due to their fine pore sizes. However, ceramic fibers have a notable limitation where they lack rigidity and typically require continuous support from metal mesh (Jouhara et al. 2021). While ceramic fibers may be chemically compatible with various working fluids, the metal mesh used for support can introduce complications. Recently, carbon fibers have emerged as a promising alternative for wick materials. Carbon fiber filaments are equipped with numerous fine longitudinal grooves, offering high capillary pressures, and they are chemically stable (Zhang et al. 2021a). Heat pipes that utilize carbon fiber wicks have demonstrated enhanced heat transport efficiency. The maximum capillary head produced by a wick increases as pore size decreases, while wick permeability improves with larger pore sizes. Another critical characteristic to optimize is the thickness of the wick. Increasing the thickness enhances the heat transport capacity. Additionally, the overall thermal resistance at the evaporator depends on the thermal conductivity of the working fluid within the wick. Essential properties of the wick also include compatibility with the working fluid and good wettability.

(d) **Working Principle**

Within the container, a liquid under its own pressure enters the pores of the capillary material, wetting all internal surfaces. When heat is applied at any point along the surface of the heat pipe, the liquid at that location boils and transitions into a vapor state, absorbing the latent heat of vaporization. This vapor, now at a higher pressure, moves through the sealed container to a cooler region, where it condenses and releases the latent heat of vaporization, thus transferring heat from the input to the output end of the heat pipe. Heat pipes are extensively used in various fields such as construction, industry, and personal computing, leading to a wide range of shapes. Based on their structures, particularly the cross-sectional geometrical shape of the pipe and wick, heat pipes can be categorized into several types: tubular, annular, baffled, bent tubular, flexible tubular, variable conductance, diode, flat plate, loop, rotating, micro-heat pipes, and others. Below are some notable applications of heat pipes in building construction:

i. **Tubular Heat Pipe**

Tubular heat pipes are the most traditional and widely used type of heat pipe, dominating the market. These heat pipes are the simplest and most popular, commonly employed in various applications to transfer heat energy from one location to another. They can also function as heat spreaders to achieve a uniform temperature distribution, essential for isothermal conditions. Although predominantly utilized in cooling applications, tubular heat pipes are also very effective in heating applications. This conventional type of heat pipe is extensively documented in books and publications.

ii. **Flat Plate Heat Pipe**

Flat plate heat pipes are commonly used to cool printed circuit boards or to achieve heat leveling, creating an isothermal plane. These this heat pipes have been the subject of extensive research and are also applied in buildings for both cooling and heating purposes.

iii. **Loop Heat Pipe**

A loop heat pipe (LHP) is a self-regulating heat transmission system that utilizes capillary forces produced in a small-pore wick to circulate the working fluid through evaporation and condensation within a sealed loop. A basic LHP consists of an evaporator, a reservoir, a condenser, and two smooth-wall pipes: one for conveying vapor to the condenser and the other for returning liquid to the evaporator, as depicted in Fig. 2.10. The capillary forces occurring at the interface between the vapor and liquid within the wick structure, referred to as the "primary wick," generate the required pressure for the circulation of fluid. A "secondary wick" establishes a connection between the reservoir and the evaporator, guaranteeing a constant movement of liquid from the reservoir to the inner surface of the primary wick. This connection effectively prevents the evaporator from running out of liquid.

iv. **Micro-Heat Pipes (MHP)**

A micro heat pipe is defined as being so small that the average curvature of the liquid–vapor interface becomes comparable to the inverse of the hydraulic radius of the flow channel. An extensive analytical and experimental research has been conducted to identify the key factors influencing the performance of these devices (Zhang et al. 2023b). Micro heat pipes have been studied, modeled, and fabricated in various sizes, with hydraulic diameters ranging from 1 mm to 30 μm and lengths between 60 and 10 mm. Unlike conventional heat pipes, micro heat pipes generally do not include a wick structure to help transport condensate back to the evaporator. Instead, they rely on capillary forces generated at the sharp edges of the pipe cross section.

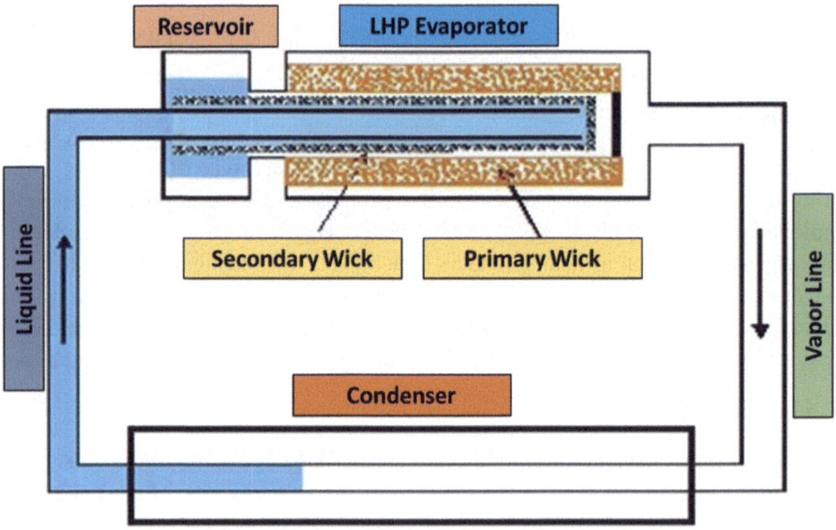

Fig. 2.10 Schematic of Single Loop Heat Pipe (Tang et al. 2018)

2.9.2 Applications in Buildings

Heat pipes are widely used in buildings for various applications, including heat pipe heat exchangers, solar heat pipe collectors, and heat pipe panel coolers.

(a) **Heat Pipe Heat Exchangers (HPHE)**

Heat pipe heat exchangers (HPHE) are mainly employed in waste heat recovery systems, where the reclaimed heat can be harnessed for room heating, steam production, or warming up combustion air in fossil-fired power plants and process furnaces (Brough et al. 2020). HPHE are classified according to the phases of the fluids they involve, including air-to-air, air-to-liquid, and liquid-to-liquid. A standard air-to-air heat pipe heat exchanger is composed of several tubular, finned heat pipes that are organized in a staggered pattern and rely on gravity for operation. The arrangement is customized to suit the unique requirements of the application. The HPHE contains a splitter plate that divides the high-temperature and low-temperature fluids. Heat pipes are used to connect the two fluid streams. The HPHE operates by utilizing the evaporator section of the heat pipes in the high-temperature fluid flow and the condenser section in the low-temperature fluid flow (Mohsen et al. 2021). The heat transmission in a HPHE can be improved by incorporating exterior fins, which augment the surface area available for convective heat transfer. One significant benefit of HPHE is its capacity to function without any cross-contamination occurring between the two gas streams. Air-to-liquid heat pipe heat exchangers are similar to air-to-air units, with the main difference being the addition of a liquid/water tank at the condensation section to preheat the liquid or water.

(b) **Solar Heat Pipe Collector**

Gravity-assisted heat pipes can be effectively used in flat plate solar collectors, providing an efficient method for harnessing solar energy. These heat pipes enhance thermal performance by using gravity to drive the circulation of the working fluid, ensuring consistent heat transfer even under varying solar conditions. A compact, integral heat exchanger within the system reduces the amount of water in direct contact with the collector. This design allows the collector to respond rapidly to fluctuations in solar radiation intensity, improving its overall efficiency. The heat pipes also help to prevent overheating by distributing excess heat evenly throughout the system.

(c) **Heat Pipe Panel Cooler**

In modern buildings, electrical and electronic control panels, such as photovoltaic (PV) control units are typically housed in sealed, airtight enclosures to protect sensitive and expensive components from dust, moisture, and other environmental contaminants. However, this enclosure can trap heat generated by the components inside leading to overheating which can result in equipment failure, reduced efficiency, or even permanent damage to the systems (Yuan et al. 2023). To address this challenge, heat pipes are integrated into the panel designs, equipped with both internal and external heat sinks. These heat pipes work by transferring heat away from the internal components to the external environment, effectively cooling the system without disrupting the controlled environment inside the panel. This allows for continuous operation while maintaining the integrity of the enclosed system.

2.10 Novel Thermal Insulation Materials

Thermal insulation pertains to the utilization of materials and methods to decrease the speed at which heat is transferred. Heat transfer takes place through conduction, convection, and radiation. Therefore, thermal insulation seeks to hinder the movement of heat through these three mechanisms (Shakir et al. 2023a). Solid foams and fibrous materials are employed to decrease the transmission of heat through conduction and convection, while heat mirror coatings are utilized to minimize the transmission of heat by radiation. Evacuated glazing is employed to reduce both conduction and convection heat transmission. Thermal insulation materials are characterized by their low thermal conductivity, also known as the k-value, which quantifies the ability of material to conduct heat (Shakir et al. 2023b). The k-value is the paramount characteristic to consider while evaluating thermal efficiency and energy preservation.

Insulation materials vary widely, ranging from traditional organic foams that work well in low-temperature situations to cutting-edge inorganic materials specifically created for high-temperature settings such as microporous ceramics and recently discovered aerogels. Common types of organic foams include polystyrene (in the forms of EPS and XPS), polyurethane (PU), polyisocyanurate (PIR), phenolic foam, expanded nitrile rubber, and polyethylene foam (Kumar et al. 2020). In addition to mineral wools such as glass wool, the main types of inorganic insulation materials include calcium silicate, cellular glass, ceramic fiber, perlite, vermiculite, microporous silica, and aerogels (Mawardi et al. 2024). Moreover, natural waste fibers have gained interest as sustainable options for thermal insulation boards, as shown in Fig. 2.11. These boards provide favorable thermal conductivity values, making them suitable for various applications. A summary of thermal insulation boards made from natural waste fibers with their respective thermal conductivity values is listed in Table 2.4.

Along with natural fibers, several other notable insulation materials stand out:

(a) **Microporous Aerogel**

Microporous aerogel excels in space efficiency, weight reduction, and insulation quality. However, its high cost limits its commercial use. It is particularly suitable for applications requiring minimal space and weight, such as window glazing.

(b) **Polyisocyanurate Insulation**

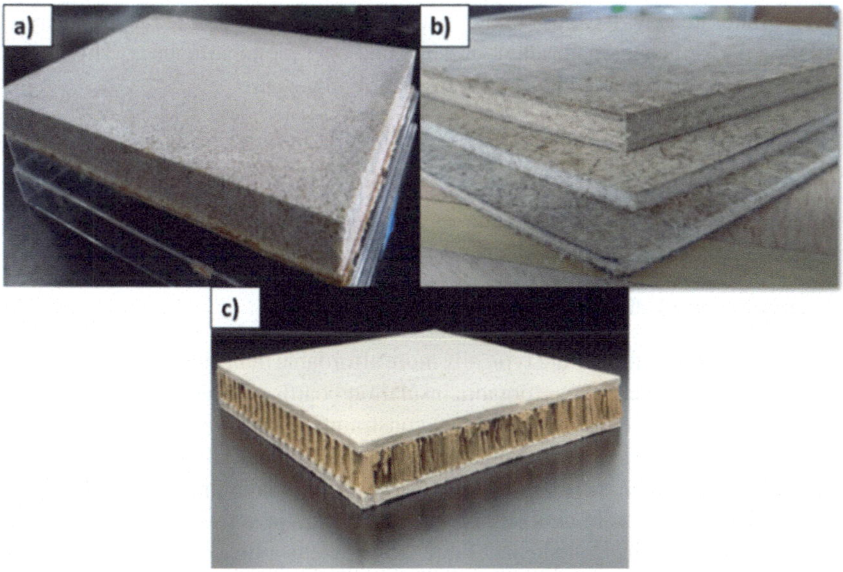

Fig. 2.11 Figures of several common thermal insulation boards. **a** Mycelium thermal insulation board (Shakir et al. 2020a), **b** fibrous thermal insulation boards (Shakir et al. 2023c), and **c** honeycomb sandwich thermal insulation boards (Shahbazi et al. 2022)

Table 2.4 List of thermal insulation boards made of natural waste fibers and their thermal conductivity values

Composite insulation panel made of natural waste fibers	Density (kg/m^3)	Thermal conductivity value (W/(m K))	References
Binderless rubberwood fiber	150–450	0.048–0.055	Cetiner and Shea (2018)
SMS fiber	190–200	0.06–0.07	Schritt et al. (2021)
Binderless coconut husk fiber	250–350	0.046–0.068	Sharma et al. (2021)
Sugarcane fiber	250–350	0.049–0.055	Jose et al. (2020) Sharma et al. (2021)
Empty fruit bunch fiber	652–816	0.174–0.187	Buasri and Surin (2021)
Pineapple leaves fiber	178–232	0.039	Cascone and Cascone (2019)
Rice husk fiber	154–168	0.08	Ali et al. (2020)
Hemp fiber	369–478	0.090–0.0108	Ramlee et al. (2020)
Wheat husk fiber	448–450	0.071–0.1	Muthuraj et al. (2019) Hysek et al. (2018)
Corn stalk fiber	280–540	0.075–0.11	Philip and Rakendu (2020)
Sunflower fiber	358–687	0.078–0.099	Ramlee et al. (2020)

Polyisocyanurate is widely accepted due to its balance of economic factors, thermal insulation performance, material processing, and operational efficiency. This makes it one of the most favoured choices for insulation materials.

(c) **Conventional Insulations (EPS, PU, Phenolic Foams)**

Expanded polystyrene (EPS), polyurethane (PU), and phenolic foams offer effective insulation properties and are the second-best choices after polyisocyanurate, providing viable secondary options for thermal insulation.

(d) **Organic Insulation Materials**

Organic insulation materials are typically more affordable than inorganic alternatives. Due to their cost-effectiveness, organic insulation boards or blankets are commonly used in building envelopes, where budget constraints are a priority.

There is a wide variety of trademarks for insulation materials currently available on the market. A trademark refers to any sign that differentiates the goods and services of one trader from those of another. This sign can encompass words, logos, images, or a combination of these elements. Table 2.5 provides a list of various trademarks for thermal insulation in the United Kingdom.

2.10.1 System Description and Working Mechanism

Novel thermal insulation materials are designed to minimize heat transfer by addressing the three main mechanisms such as conduction, convection, and radiation. These materials achieve thermal insulation through various means, depending on their composition and structure. Many novel insulation materials, such as microporous ceramics and aerogels, have highly porous structures (Sun et al. 2024). These pores are typically filled with air or other gases, which impede the transfer of heat through conduction and convection. Microporous ceramics use a network of nanosized pores that trap air, significantly reducing the pathways for heat to travel (Zhao et al. 2021). Aerogels known for their extremely low density, have an intricate internal structure that further hinders the movement of heat, making them highly effective insulators. Similarly, fibrous materials, such as glass wool or organic waste fibers, slow heat transfer by trapping air within their fibers.

Heat mirror coatings and other reflective surfaces are often applied to insulation materials to reduce heat transfer through radiation. These coatings reflect infrared radiation, preventing the material from absorbing or emitting heat. In evacuated glazing, a vacuum is created between layers of glass to reduce both conduction and convection, while a reflective layer helps block radiative heat transfer (Di et al. 2020). These innovative combinations of materials and techniques result in highly efficient thermal insulation solutions. The k-value (thermal conductivity) is the most critical factor to evaluate the performance of these materials. Lower k-values indicate higher resistance to heat flow, meaning the material is better at insulating. Novel

Table 2.5 Trademarks of common thermal insulation materials in buildings

Type of insulation material	Trade marks	Producers and sources
Polystyren e		http://www.styrotech.uk.com/
Polyuretha nePenolic	**Kingspan Kooltherm®**	http://www.insulation.kingspan.com
Fiberglass	Fiberfrax® Blanket,Insulfrax® S Blanket	http://www.unifrax.com
Microporous ceramic	Min-K®	http://www.thermalceramic.com
Microporous ceramic	Fiberfrax® Blanket	http://www.unifrax.com
Microporous ceramic	BTU-BLOCK	http://www.thermalcerамics.com
Aerogel		http://www.aerogel.com/
Microporous silica	Microtherm®	http://www.microtherm.uk.com
Microporous silica	**Mintherm®**	http://www.armilcfs.com/

materials like aerogels and microporous ceramics boast exceptionally low k-values, making them highly effective for both low- and high-temperature applications. This superior thermal resistance makes these materials ideal for use in environments where maintaining temperature control is crucial such as in industrial equipment or building insulation. By optimizing the structure of the material and integrating features like reflective coatings, manufacturers can create insulation materials that provide better energy efficiency and thermal performance in a variety of environments.

2.10.2 Applications in Buildings

Novel thermal insulation materials play a pivotal role in improving the energy efficiency of buildings, where heat regulation is critical for reducing energy consumption. These materials are extensively used in walls, roofs, floors, and foundations to create effective thermal barriers that prevent heat loss in winter and heat gain in summer. By reducing the need for artificial heating and cooling, these materials help lower energy bills and contribute to a more sustainable environment. In modern building design, materials like aerogels and microporous ceramics have gained prominence for their superior insulating properties. Aerogels are increasingly used in high-performance building applications due to their extremely low thermal conductivity and lightweight structure.

One notable application is in aerogel window glazing as shown in Fig. 2.12a. Aerogels provide excellent insulation while maintaining transparency. This makes them ideal for windows, skylights, and other architectural elements where natural light is needed, but thermal performance must not be compromised (Buratti et al. 2021). This technology helps retain indoor temperatures without sacrificing daylight, thereby improving the overall energy efficiency of the building. In addition to aerogel-glazed windows, other super-insulating glazing options include vacuum insulation panels (VIP), as shown in Fig. 2.12b (Katsura et al. 2020). VIP are highly efficient insulators that work by creating a vacuum-sealed environment to drastically reduce heat transfer through conduction and convection. These panels are typically coated with low-emissivity (low-e) coatings and filled with inert gases such as krypton or argon, further enhancing their insulating properties. In residential and commercial settings, traditional insulation materials like expanded polystyrene (EPS), polyurethane (PU), and mineral wool are still widely used. However, there is growing interest in adopting more advanced materials such as aerogels and microporous insulation for high-end construction projects. These materials while more expensive, offer long-term energy savings and superior performance in both new builds and retrofitting older structures.

As energy efficiency regulations become more stringent, further development in the cost-effectiveness of advanced insulation materials such as aerogels will be essential. Research should focus on reducing the production costs of these materials to make them more accessible for everyday residential and commercial building projects. Additionally, hybrid materials that combine the insulation properties of aerogels with more affordable materials like organic fibers could strike a balance between performance and cost. Moreover, integrating smart insulation solutions that adapt to changing environmental conditions could enhance building performance. Combining aerogel technology with smart windows, which can adjust their opacity based on sunlight exposure. This may offer further energy savings in buildings by optimizing natural light and temperature regulation.

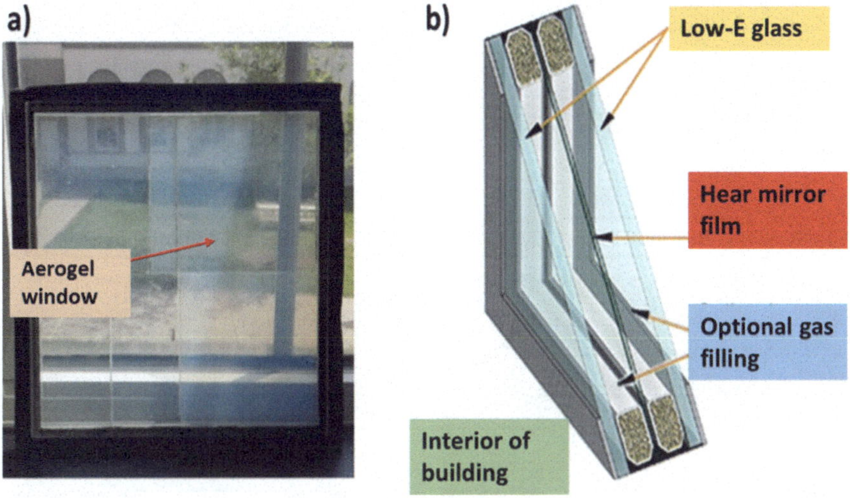

Fig. 2.12 **a** Aerogel window glaze (Carroll et al. 2022) and **b** vacuum insulation panel (Mempouo et al. 2010)

2.11 Biomass Energy Systems

Biomass encompasses live and recently deceased organic matter such as plants and animals that can be utilized as fuel or in industrial applications. Biomass typically refers to plant material cultivated specifically for the manufacture of biofuels (Binhweel 2024; Senusi 2024). However, it can also encompass plant or animal matter utilized in the creation of textiles, chemicals, or heat. Moreover, biomass might consist of organic waste that is capable of being decomposed and used as a source of energy through combustion. Biofuels include bioethanol, biobutanol, biodiesel, and biogas. Among these, biodiesel and biobutanol stand out as biofuels that can be directly utilized in petroleum-based engines without the need for significant modifications (Ahmad et al. 2024; Senusi et al. 2024). The utilization of biomass as an energy supply is a longstanding tradition. With the expansion of the population and the advancement of industry, there was a surge in the need for energy which prompted the investigation of alternate energy sources.

Biomass is classified as a renewable energy source due to its ability to be replenished by the continuous growth of plants and crops, as well as the perpetual generation of organic waste. However, burning biomass is not the only way to extract its energy. As previously mentioned, biomass can be transformed into alternative energy sources such as methane gas or transportation fuels like ethanol and biodiesel. Methane, the primary component of natural gas is released during the decomposition of organic matter, including agricultural waste and human excrement, commonly referred to as landfill gas or biogas (Larson et al. 2021). Additionally, crops like corn and sugarcane can undergo fermentation to produce ethanol for transportation fuel, while biodiesel can be made from waste materials such as vegetable oils and animal fats.

Fig. 2.13 The wastes from agricultural production (Shakir et al. 2024c)

Classification of Biomass Biomass sources can be categorized into two main groups: plant-based and animal-based. Vegetal biomass mostly consists of agriculture and forestry leftovers. Agricultural activities generate a significant amount of trash, including several types of husks (rice, olive, chestnut, and camellia), corncobs, corn stalks, cotton stalks, and different varieties of straw (wheat, barley, and rice), as well as coconut shells and bagasse, as depicted in Fig. 2.13. Meanwhile, animal excrement is a significant form of biomass, with dairy cow waste serving as a prominent example.

2.11.1 System Description and Working Mechanism

Biomass technologies refer to the various methods used to convert organic materials, such as plants and waste, into energy or other valuable products. Four key methods are particularly important: biomass gasification, pyrolysis (bio-oil), municipal waste processing, and anaerobic digestion (bio-ethanol, biogas). These processes enable the transformation of biomass into bioenergy, making them essential for sustainable energy production. There are two main strategies for harnessing biomass such as recycling organic waste and cultivating energy-rich crops and trees. Biomass can be processed through one of four major treatments: thermochemical, chemical, physical, or biological processes (Shakir et al. 2024d). Thermochemical processes like gasification and pyrolysis involve the decomposition of biomass at high temperatures to produce syngas, bio-oil, and biochar. Chemical processes such as transesterification and acid hydrolysis, convert biomass into biofuels like biodiesel and bioethanol by employing chemical reactions. These methods allow for precise control, making them highly effective for producing targeted biofuels.

On the other hand, physical processes including size reduction and palletization, involve the mechanical treatment of biomass to make it more manageable for transport or more efficient in combustion and conversion processes. Last but not least, biological processes such as anaerobic digestion and fermentation, rely on microbial activity to break down organic material and produce biogas or bioethanol. These low-temperature processes are particularly suited for wet biomass, such as food waste or agricultural residues. Biological methods are eco-friendly and are often used to generate renewable energy in the form of methane-rich biogas which can be used for heating or electricity generation. Together, these methods form the foundation of biomass technologies, offering sustainable pathways for renewable energy production.

(a) **Biomass Gasification and Pyrolysis**

Gasification is a series of reactions in which solid fuel combines with high-temperature steam and air or oxygen to produce a fuel in the form of gas. Gasifiers of different kinds function over a temperature range of a few hundred degrees to over 1000 °C and at pressures ranging from near atmospheric levels to 30 atmospheres. The primary element of a gasifier is a reactor in which wood chips or cotton stalks are combusted with a regulated quantity of air to generate flammable gas (Umar et al. 2021). This gas is comprised of a blend of carbon monoxide, hydrogen, methane, carbon dioxide, and nitrogen. The composition of these gases is contingent upon the processing parameters and the choice between air or oxygen as the reactant. Biomass gasification has gained considerable attention and has been intensively researched in recent years, mostly because of its sustainability and ability to recycle waste.

Meanwhile, bio-oil can be produced by employing quick pyrolysis to leverage a bubbling fluidized bed reactor and carefully control reactive conditions. Pyrolysis is a basic and perhaps the most ancient technique for transforming a fuel into a higher quality one (Wang et al. 2022b). In traditional pyrolysis, the original material is heated without air at temperatures ranging from 300 to 500 °C, causing the release of volatile components (Shakir et al. 2020b). The resultant residue generally referred to as charcoal, which has an energy density that is around two times more than that of the initial substance and combusts at a considerably higher temperature. During the 1980s, scientists made the finding that the production of pyrolysis liquid may be enhanced by employing fast pyrolysis. This procedure entails the swift application of heat to biomass feedstock such as hardwoods, softwoods, grasses, and agricultural wastes, which have a moisture content of less than 10% and a particle size of 1–2 mm. The subsequent vapors are promptly condensed. The complete reaction, starting from injection and ending with quenching, takes place over a mere two-second timeframe, resulting in the formation of a dark brown liquid often known as bio-oil.

(b) **Transesterification and Acid Hydrolysis**

Among the most impactful chemical processes for biomass conversion are transesterification and acid hydrolysis. Transesterification is critical in the production of biodiesel from vegetable oils or animal fats. In this process, triglycerides present in fats or oils react with alcohol (typically methanol or ethanol) in the presence of a catalyst like sodium hydroxide or potassium hydroxide (Murugesan and Kannan 2023). This reaction yields biodiesel and glycerin as by-products. Transesterification is widely used due to its efficiency in producing a renewable fuel that can serve as a substitute for conventional diesel. The process is scalable and can be adapted to various biomass feedstocks, making it a versatile solution in the biodiesel industry. Similarly, acid hydrolysis plays an essential role in bioethanol production by breaking down complex carbohydrates such as cellulose and hemicellulose found in plant materials into simple sugars (Ahmad et al. 2022). These sugars can then undergo fermentation to produce ethanol. Acid hydrolysis is particularly valuable in converting lignocellulose biomass (such as wood, straw, or agricultural residues) into fermentable sugars, opening pathways for creating biofuels from non-food biomass sources, further enhancing sustainability in biofuel production.

(c) **Size Reduction and Palletization**

Physical processes such as size reduction and palletization are essential for preparing biomass for efficient conversion. Size reduction involves the mechanical breakdown of biomass into smaller particles, increasing the surface area available for further chemical or thermochemical processes like gasification, pyrolysis, or fermentation. Common methods for size reduction include grinding, shredding, or milling, which transform large biomass feedstocks into finer materials that can be handled more easily. This process is crucial for improving the overall efficiency of subsequent conversion processes, ensuring better heat transfer or chemical reactions. Palletization compresses finely ground biomass into dense, uniform pellets. These pellets are more compact, making them easier to store, transport, and burn efficiently in boilers or stoves. Palletization not only improves the energy density of biomass but also offers logistical advantages as pellets are less bulky and more manageable than raw biomass. As such, palletization is widely used in both domestic heating systems and industrial bioenergy applications.

(d) **Municipal Solid Waste (MSM)**

Municipal solid waste (MSW) usually known as garbage is an additional kind of biomass. MSW comprises organic trash derived from plant or animal sources, such as food remnants, grass trimmings, and foliage. In contrast, materials such as glass, plastic, and metals are excluded from the category of biomass due to their origin from non-renewable sources. MSW can be utilized as an energy source through two methods: incineration in waste-to-energy plants or the extraction of biogas from landfills. Waste-to-energy plants incinerate garbage to generate steam, which can be utilized for heating structures or producing electricity (He et al. 2023). Waste incineration is generally more favorable than landfill disposal, as it has the potential to decrease waste volume and provide energy.

Meanwhile, biomass undergoes decomposition in landfills resulting in the release of methane gas which is also referred to as biogas or landfill gas. Certain landfills are equipped with methane collection devices allowing for the utilization of this gas as a fuel. Furthermore, biogas is harnessed by dairy farmers through the utilization of digesters, which are containers designed to store manure and other waste materials from barns. Biogas also known as marsh gas or methane is a combustible gas generated by microorganisms that ferment organic substances under certain circumstances of temperature, moisture, and acidity in the process of anaerobic digestion. The resultant of this procedure, a degraded slurry abundant in ammonia can serve as fertilizer. The utilization of biogas plants has resulted in many advantages such as energy conservation and enhancements in sanitation, agricultural methods, and ecological conditions.

(e) **Biofuel**

Biofuel is a term used to describe liquid or gaseous fuels that are primarily made from biomass and are used in the transportation industry. Biofuels are widely acknowledged for their myriad advantages, including their sustainability, ability to reduce greenhouse gas emissions, support for regional development, enhancement of social and agricultural structures, and contribution to increased energy security.

i. **Bioethanol**

Bioethanol is a term that refers to the production of ethanol from biological sources.

Carbohydrates, such as hemicelluloses and cellulose present in plant materials, can undergo hydrolysis to be transformed into sugars (Shakir et al. 2017b). Fermentation is an anaerobic biological process that converts carbohydrates into alcohol utilizing microbes, usually yeast. Ethanol also referred to as bioethanol is produced from sustainable sources such as wheat, sugar beet, corn, straw, and wood. Bioethanol can serve as an addition or replacement for gasoline. Moreover, it is economically viable to transform wood, straw, and even household garbage into bioethanol.

ii. **Biodiesel**

Biodiesel often known as vegetable oil methyl esters which is a feasible substitute for conventional diesel fuels. Biodiesel derived from vegetable oil through the process of transesterification is compatible with diesel engines. Biodiesel has gained fresh attention because of its renewable nature and lower pollution levels in comparison to traditional diesel, which is a finite fossil fuel. Biodiesel is technologically comparable to traditional petroleum diesel fuel and provides several technical benefits.

2.11.2 Application in Buildings

Biomass energy systems have a wide range of practical applications across various sectors, particularly in agriculture, waste management, and energy production. One of the most significant applications is biogas production through anaerobic digestion, which is particularly beneficial in rural and agricultural settings. Organic waste from farms such as manure, crop residues, and other biodegradable materials can be processed in biogas plants to generate biogas. This biogas can be used as a renewable energy source for electricity generation, heating, or as a substitute for natural gas. Additionally, the anaerobic digestion process produces a nutrient-rich byproduct known as digestate, which can be used as a high-quality organic fertilizer. This dual benefit helps farmers manage waste sustainably while simultaneously lowering energy costs and improving soil fertility. The integration of biogas plants in agricultural operations supports sustainable farming practices by reducing the environmental impact of waste and providing a closed-loop system for energy and nutrient recovery.

In the waste management sector, biomass technologies are increasingly employed to convert municipal solid waste (MSW) into energy through processes such as waste-to-energy (WTE). WTE plants incinerate organic waste to produce steam, which is then used to generate electricity or provide district heating for urban areas. This not only significantly reduces the volume of waste sent to landfills but also offers a renewable energy solution that can help reduce cities' reliance on fossil fuels. Moreover, WTE plants are designed to capture harmful emissions during combustion, minimizing their environmental impact. Additionally, the implementation of landfill gas recovery systems, which capture methane emissions from decomposing organic waste in landfills, presents an opportunity to harness biogas for energy generation. By capturing and using methane a potent greenhouse gas for energy, cities can further mitigate their greenhouse gas emissions while producing a renewable energy source.

Looking ahead, there are several areas where biomass energy systems could be further improved and optimized. Research and development in improving the efficiency of anaerobic digestion and expanding its use to a broader range of organic feedstocks could increase the potential for biogas production. Furthermore, enhancing waste-to-energy technologies by making them more efficient and economically viable could boost their adoption in urban areas. Another suggestion is to explore advanced biomass conversion technologies such as gasification and pyrolysis, which can produce higher-quality biofuels and chemicals, adding value to the waste conversion process. The integration of biomass technologies with other renewable energy sources, such as solar and wind, could also create hybrid systems that maximize energy production while maintaining sustainability. These advancements would ensure that biomass energy systems continue to play a critical role in the global transition to renewable energy and circular economy models.

2.12 Energy-Efficient Boiler

A boiler is an enclosed container specifically intended to generate heat by applying pressure on water or other fluids. After being heated, the fluid is circulated for a variety of process or heating purposes. Boilers are commonly constructed using materials such as copper, steel, or cast iron. They can be heated through the process of fuel combustion, using sources such as biomass, coal, oil, or natural gas. Electric boilers, on the other hand, utilize resistance or immersion heating elements (Riet and Verhaert 2020). Energy-efficient boilers, like condensing boilers, improve efficiency by recovering heat from flue gases, reducing energy waste. Generally, boilers can be classified into the following types based on their construction.

(a) **Fire-Tube Boilers**

Fire-tube boilers are designed with the heat source typically combustion gases, flowing through tubes, while the water to be heated surrounds the tubes. This configuration is commonly used in steam locomotives and industrial applications, where the high-temperature gases produced by fuel combustion pass through long, linear tubes to transfer heat to the surrounding water. The simplicity of fire-tube boilers makes them robust and cost-effective for a variety of applications, though they tend to be larger and slower to respond to changes in steam demand compared to other boiler types. Despite their widespread use in earlier steam engines, modern fire-tube boilers are less commonly used in residential settings, as other boiler designs offer higher efficiency for home heating.

(b) **Water-Tube Boilers**

Water-tube boilers function by circulating water through narrow tubes that are surrounded by the heat source. These tubes are often coiled or bent to increase the surface area for more efficient heat transfer. Water-tube boilers are known for their ability to withstand high pressure and produce steam more rapidly than fire-tube designs. This makes them ideal for large-scale industrial applications, power plants, and situations requiring high-pressure steam. Their compact design and ability to achieve high thermal efficiency make water-tube boilers a preferred choice in modern industrial settings. However, they are generally more complex and expensive to manufacture than fire-tube boilers.

(c) **Primitive Boilers**

Primitive boilers often referred to as pot boilers or early steam boilers, lack the advanced design features of modern fire-tube or water-tube systems. In this simple design, the fire directly heats the water container with no tubes or internal structures to enhance heat transfer. While primitive boilers were widely used in early steam engines and heating systems, they are highly inefficient due to the minimal surface area for heat exchange. The absence of heat recovery systems and their direct heating method results in significant energy losses, making them largely obsolete today. Modern boilers have evolved to include sophisticated designs that optimize heat transfer and fuel efficiency.

The objective is to optimize the heat transmission from the source to the water. Hydronic boilers are frequently utilized in residential settings to provide central heating. These systems function by raising the temperature of water or another liquid to a certain level, which is then distributed by radiators, baseboard heaters, or floors (Rizi and Heidarinejad 2022). Hydronic systems commonly utilize natural gas as the main heat source in urban areas with access to piped gas. The fluid flows within a sealed system propelled by a motorized pump. Steam boilers employ comparable technology to hot-water boilers, however without the need for circulating pumps. Contemporary systems commonly use condensing boilers to enhance efficiency, rendering them appropriate for single-family residences as well.

2.12.1 System Description and Working Mechanism

Energy-efficient boilers use various mechanisms to enhance heat transfer and reduce energy loss. For example, condensing boiler is a high-efficiency, modern boiler that includes an additional heat exchanger. This extra heat exchanger enables the hot exhaust gases to transfer much of their sensible heat energy to preheat the water in the boiler system. When operating at peak efficiency, the water vapor generated during combustion condenses back into liquid, releasing the latent heat of vaporization. This process produces condensate, which is usually acidic and needs to be drained away. The condensing boiler achieves a fuel-to-heat conversion rate of over 88%, surpassing the 72% efficiency of a new conventional boiler. This makes the condensing boiler very efficient and reduces energy wastage to a minimum (Men et al. 2021). Condensing boilers are offered in standard and combination configurations. In a traditional boiler, combustion occurs in a designated chamber enclosed by a water jacket, resulting in the heating of the water that subsequently flows through the heating system. Regrettably, a significant portion of the generated heat dissipates into the surrounding atmosphere or is expelled through the chimney. Condensing boilers utilize an extra heat exchanger to lower the temperature of the flue gases, allowing the water vapor generated during combustion to condense. This process releases latent heat that would otherwise be wasted. As a result, condensing boilers have been the predominant choice in the market and have been subject to substantial research.

i. **Coke Dry Quenching (CDQ) Boiler**

In addition, there is a developing field of study and retrofitting focused on the Coke Dry Quenching (CDQ) boiler. A CDQ boiler is a type of waste heat boiler that utilizes waste heat from the system to generate steam. This steam is then used to spin a turbine, which in turn produces electric power (Gangadharan et al. 2023). In the steelmaking business, it is necessary to rapidly cool down the extremely hot coke from coke ovens in order to prevent it from reacting with oxygen and forming ash, before it is transported to the blast furnace. The process of quenching is used to rapidly reduce the coke from a temperature of around 1100–200 °C in order to prevent combustion and gasification. This procedure is largely accomplished through the utilization of either a CDQ system or a coke wet quenching (CWQ) system. Recent studies highlight the exegetic and ecological advantages of CDQ over CWQ. Due to the numerous advantages, many steel plants are adopting CDQ systems to replace CWQ. Currently, numerous Chinese steel companies are importing CDQ boilers from Japan.

i. Environmental Benefits: CDQ is a clean process that significantly reduces environmental emissions and simplifies the permitting process.
ii. Energy Production: It produces high-pressure steam suitable for power generation.
iii. Energy Efficiency: CDQ systems have low energy consumption and result in energy savings in the blast furnace process due to lower coke moisture content.
iv. Coke Quality: CDQ provides improved and more consistent coke quality.
v. Reliability: CDQ systems are more reliable and require less maintenance.
vi. Cost Efficiency: They are more cost-effective compared to wet quenching systems.

ii. **Biomass-Fired Boiler**

Biomass-fired refers to boilers that utilize biomass as their primary fuel source. For residential use, solid biomass fuels can include agricultural leftovers such as rice husks, olive husks, chestnut husks, camellia husks, corncobs, corn stalks, cotton stalks, different types of straw like wheat straw, barley straw, and rice straw, coconut shells, and bagasse (Shakir et al. 2024b). Wood is commonly considered the superior kind of solid fuel. Wood-fired boilers can be classified into many categories based on the form and size of the wood fuel.

i. **Log-Fired Central Heating Boilers**

Log-burning boilers clearly distinguish between two combustion processes. Wood is heated in the primary combustion chamber, causing the release of non-combustible wood gases by oxidation on the left side. Subsequently, an induced-draft fan is utilized to pull these gases into a secondary combustion chamber, which is equipped with either ceramic material or high-temperature-resistant steel (Szubel et al. 2024). This secondary chamber harnesses the wood gases generated in the primary combustion chamber. The wood gas is combined with additional oxygen in a vortex chamber and undergoes burning as it is pulled past a combustion plate. The high-temperature gases produced from the combustion process in both stages are guided through heat exchangers that are built into the boiler. In this process, the heat is transferred to the circulating water in the heating system, resulting in a condensing boiler. Consequently, the flue gases undergo a cooling process and reach a temperature that is lower than 200 °C.

ii. **Central Pellet Boilers**

Central pellet boilers are an increasingly popular choice for biomass heating due to their efficiency, sustainability, and automation. These boilers burn wood pellets, which are compacted biomass materials derived from wood waste, such as sawdust and wood chips. The boilers are designed to automatically load pellets into the combustion chamber, minimizing the need for manual input and ensuring a consistent fuel supply (Zima and Ojczyk 2022). There are three primary methods for loading wood pellet boilers: the bottom-feed, retort system, and top-feed systems. The bottom-feed system introduces pellets from beneath the combustion chamber, allowing ash to naturally settle and reducing the need for frequent cleaning. The retort system features a burner where pellets are continuously fed to create a controlled and efficient combustion process. In contrast, the top-feed system loads pellets from above, dropping them directly into the combustion zone. This design helps prevent clogging or jamming, especially with low-quality pellets. The choice of loading method depends on the boiler manufacturer and the specific needs of the installation, such as fuel quality and desired maintenance intervals. These boilers are commonly used in residential and commercial settings due to their ability to provide reliable and clean heating solutions.

iii. **Woodchip Boilers**

Woodchip boilers are a versatile and efficient biomass heating solution, similar in design and operation to wood pellet boilers. These systems utilize woodchips as a primary fuel source, making them an attractive option for both residential and commercial applications where wood waste is readily available. The automatically fed mechanism ensures a steady supply of fuel, reducing the need for manual intervention while maximizing combustion efficiency (Sousa et al. 2024). Like pellet boilers, woodchip boilers are equipped with advanced combustion controls and feed systems, which enable precise fuel management, ensuring optimal burn conditions. This results in reduced emissions and enhanced energy efficiency. Woodchips, which are generally more affordable than pellets as it can be sourced from forestry residues, sawmills, or agricultural waste, making woodchip boilers not only cost-effective but also environmentally sustainable. These boilers are particularly well-suited for larger heating installations, such as district heating systems, schools, or industrial plants, where high-capacity heating is required.

iv. **Combined Wood Pellet-Solar Thermal System**

In regions with high levels of solar energy, combining a wood pellet boiler with a solar thermal system creates a highly efficient, integrated heating solution. This hybrid configuration represents the cutting edge of renewable heating technology, offering the flexibility of multiple energy sources (Henning 2004). During sunny periods, the solar thermal system captures solar energy to heat water, reducing the demand on the pellet boiler. When solar energy is insufficient, the pellet boiler takes over to ensure a continuous supply of heat. This approach significantly reduces fuel consumption, lowers emissions, and provides a sustainable solution for both space heating and domestic hot water production. This advanced setup is ideal for residential buildings, eco-homes, or small commercial spaces looking to maximize renewable energy utilization while maintaining reliable heating year-round.

2.12.2 Applications in Buildings

Energy-efficient boilers are widely used in both residential and industrial sectors, playing a crucial role in reducing energy consumption and promoting sustainability.

In residential applications, condensing boilers are particularly favoured due to their ability to capture waste heat from exhaust gases, thereby significantly improving efficiency. These systems utilize natural gas or propane and are highly effective in reducing household energy consumption while providing reliable and consistent heating. Moreover, they can be combined with solar thermal systems to further enhance energy efficiency, making them an ideal choice for eco-friendly homes. In urban settings where piped natural gas is available, hydronic systems are often employed for central heating, circulating heated water through radiators, baseboard heaters, or underfloor systems, ensuring even and effective distribution of heat throughout the building.

In industrial settings, biomass-fired boilers have gained popularity for their ability to utilize renewable fuel sources such as wood pellets, agricultural residues, and other organic materials. These systems are particularly useful in industries like agriculture and forestry, where biomass waste is readily available, providing a sustainable alternative to fossil fuels. Biomass boilers not only reduce dependence on non-renewable resources but also help lower greenhouse gas emissions, making them an attractive option for industries aiming to improve their environmental footprint. Additionally, the emerging Coke Dry Quenching (CDQ) process in the steel industry is an innovative technology that allows for waste heat recovery, improving energy efficiency and reducing emissions during coke production. In the UK, boilers are a staple in nearly every household due to the cold and damp winters. Gas-fired boilers are the most prevalent heating solution, and the boiler market in the UK is robust, with numerous manufacturers catering to the demand. Table 2.6 provides a comprehensive list of leading boiler manufacturers in the UK, illustrating the diversity and competitiveness of the market.

Future advancements in boiler technology should focus on increasing the adoption of biomass-fueled systems and improving the durability of condensing boilers, especially in environments with high humidity or when using corrosive fuels. Additionally, integrating boilers with renewable energy systems such as solar or geothermal

Table 2.6 Leading boiler manufacturers and types of boilers available in the UK

Type of boiler	Manufacturer of the boilers
Wall Mounted Combi's	Alpha, Ariston, Baxi Glow-Worm, Ideal, Potterton, Vaillant, Vokera, Worcester
Floor standing Combi's	Vokera, Worcester
Condensing Boilers	Baxi, Glow-Worm, Ideal, Potterton, Vaillant, Vokera, Worcester, Keston
System Boilers	Alpha, Ariston, Baxi, Glow-Worm, Ideal, Potterton, Vaillant, Vokera, Worcester
Oil Combi Boilers	Boulter, Grant, Worcester
Oil Condensing System Boilers	Grant,
Oil External Boilers	Boulter, Grant
Oil Wall Mounted Combi Boilers	HRM

sources will further promote energy efficiency and sustainability across multiple sectors. Modern control systems that offer real-time monitoring and adjustments can also optimize boiler performance, reducing fuel consumption, improving operational reliability, and minimizing maintenance costs. By continuing to innovate in boiler technology, we can ensure that these systems remain a cornerstone of sustainable energy solutions for both residential and industrial applications.

2.13 Evaporative Cooler

Evaporative coolers are energy-efficient devices that utilize the natural process of water evaporation to cool air. Unlike conventional air conditioning systems, which rely on refrigerants and compressors, evaporative coolers pass air over water-soaked pads or media (Yang et al. 2021). As air flows through these damp surfaces, the water evaporates, absorbing heat and reducing the air temperature. This process mirrors the cooling sensation felt when moisture evaporates from the skin making evaporative coolers a simple yet effective cooling method. One of the key advantages of evaporative coolers is their energy efficiency. They consume far less electricity than traditional air conditioners because they do not require the energy-intensive use of compressors or refrigerants. This makes them an eco-friendly option, relying solely on water and airflow.

However, their performance is highly dependent on the surrounding climate. Evaporative coolers work best in hot, dry environments where low humidity allows for more efficient evaporation. As a result, they are particularly popular in arid regions and are often used in agricultural, industrial, and commercial settings (Kapilan et al. 2023). In recent years, advancements in evaporative cooler design have made them more suitable for residential use as well. Many modern units are equipped with air filtration systems to remove airborne particles, enhancing indoor air quality. Additionally, newer models are designed to work in conjunction with traditional HVAC systems, offering hybrid solutions for cooling homes and reducing energy consumption. With a growing emphasis on sustainability, evaporative coolers represent a practical, low-energy alternative to conventional air conditioning, particularly in regions where water availability and dry climates make them a viable option.

2.13.1 System Description and Working Mechanism

Evaporative coolers are devices that lower the temperature of air by utilizing the process of water evaporation. When water evaporates, it absorbs heat from the surrounding air, cooling both the air and the remaining water. This results in a noticeable drop in temperature, a process that is especially effective in dry climates (Sajjad et al. 2021). Cooled air absorbs moisture from water vapor, hence increasing its humidity. This method is known as direct evaporative cooling (DEC). In DEC systems, while the air becomes cooler, it also becomes more humid, which can lead to discomfort in areas that already have high humidity. An alternative approach called indirect evaporative cooling (IEC) addresses the humidity issue by passing the cooled, moist air through the walls of a heat exchanger (Sajjad et al. 2021). This process cools the adjacent air without adding moisture. This method produces cooler, drier air, making it better suited for indoor environments like homes and offices. Since IEC systems keep the cooled air separate from the humid air where the evaporation takes place, they avoid the moisture-related drawbacks of DEC systems, making them a more comfortable choice in regions with higher humidity levels.

Evaporative coolers are energy-efficient compared to traditional air conditioning systems because they use less electricity to operate. However, they work best in dry, arid climates where the humidity is low. In areas with high humidity, the performance of DEC systems decreases because the air is already near saturation, reducing the cooling effect of evaporation. While evaporative cooling is widely used in commercial and industrial settings such as in warehouses and agriculture, IEC systems are becoming more popular for residential use because they provide greater comfort and save energy. Additionally, modern evaporative coolers often include air filtration systems that can remove airborne particles, further improving indoor air quality. Both DEC and IEC systems can be enhanced through the use of specific materials and mechanisms. For example, many modern systems utilize absorbent paper media that draws water from a reservoir to optimize the cooling process. Systems that employ water-spraying pumps on intermittent cycles can further increase efficiency by ensuring that only the right amount of water is used, avoiding waste. Additionally, integrating dehumidification units at the air intake can improve the cooling efficiency in areas with moderate humidity levels.

2.13.2 Applications in Buildings

Evaporative coolers have broad applications in both residential and industrial sectors, particularly in regions with hot, dry climates. Direct Evaporative Cooling (DEC) is frequently used in large industrial spaces such as warehouses, greenhouses, and factories, where increased humidity may not be an issue. It is also used in agricultural settings to maintain optimal growing and storage conditions for crops and livestock. The low operational cost and energy efficiency of evaporative coolers make them an attractive option for industries that require high cooling capacities without significant energy consumption as shown in Fig. 2.14. While refrigeration and cold storage facilities are effective, but they are often energy-intensive and require substantial investment. As a result, many small-scale farmers particularly in tropical regions, find these options inaccessible, making it difficult to maintain optimal growing and storage conditions for their crops. Fresh horticultural products continue to metabolize after harvest, and without proper storage, their rate of deterioration accelerates due to factors like heat and humidity. Evaporative cooling which are low-cost, energy-efficient, and made from locally available materials, offer a viable solution to this problem. These chambers reduce temperature and increase relative humidity, helping to preserve fresh produce on farms, especially in rural areas where expensive refrigeration is not an option.

Indirect Evaporative Cooling (IEC) is also gaining popularity in residential settings, particularly in areas with moderate humidity. These systems provide cooling without raising indoor humidity, which makes them more suitable for homes and offices. In hot and dry regions, combining IEC coolers with solar-powered systems can significantly reduce energy consumption while ensuring a comfortable indoor climate. Furthermore, these systems can be integrated with air filtration units, which improve indoor air quality by removing dust and other airborne particles. This approach making them a healthier alternative to traditional cooling systems. In some areas that have hot and dry climate like Central Iran, evaporative coolers are the dominant cooling technology due to their affordability and effectiveness in arid environments (Naderi et al. 2020). About 9 million evaporative coolers, with 130,000 units being sold in the region during the months of April and May in 2006. Nevertheless, if the moist paper media utilized in these coolers is not adequately dried after usage, it can develop mould. Hence, it is imperative to explore water-absorbent polymers that exhibit resistance to degradation. In the UK, despite its cooler climate, evaporative coolers have found use in industrial settings where manufacturing processes generate high temperatures, making efficient and cost-effective cooling solutions essential. Government-backed initiatives program have further incentivized businesses to adopt evaporative coolers, making them a sustainable choice for industries looking to reduce their carbon footprint. For example, the UK government endorses the utilization of renewable energy, and evaporative cooling is qualified for

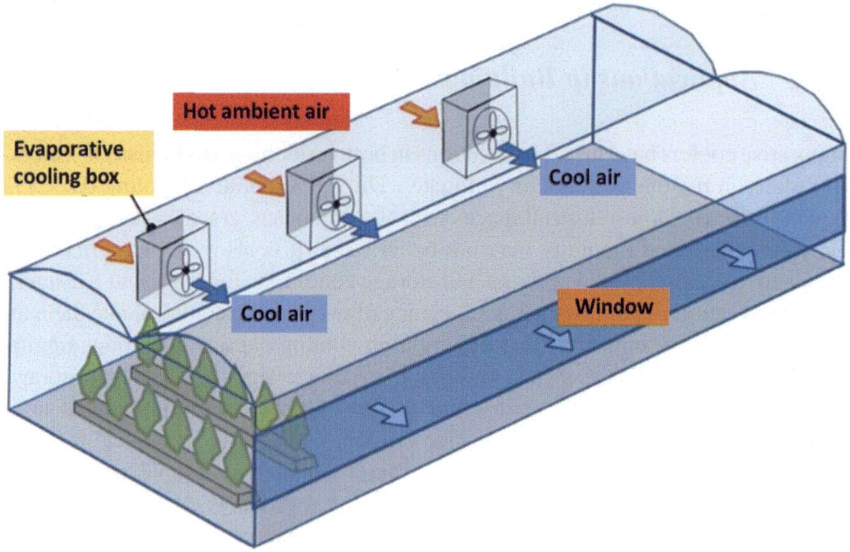

Fig. 2.14 Schematic of direct evaporative cooling system in farmhouse (Tawalbeh et al. 2023)

the Carbon Trust's "Energy Efficiency Loans" program. This program enables small or medium-sized firms in England or Wales to obtain unsecured and interest-free loans ranging from £5000 to £100,000 for project funding. Looking ahead, the integration of evaporative coolers with renewable energy systems such as solar power, could provide additional environmental and economic benefits, further expanding their applicability across different regions.

2.14 Phase Change Material (PCM) Thermal Storage

A phase change material (PCM) is a substance capable of storing and releasing large amounts of latent heat during its transition between solid and liquid states as shown in Fig. 2.15. This phase transition occurs at a constant temperature allowing PCM to maintain stable temperatures over extended periods. This mechanism making them ideal for thermal regulation applications (Zhang et al. 2021b). This unique property allows PCM to provide effective thermal regulation by maintaining temperatures at specific set points for extended periods. Importantly, certain PCM can undergo these phase transitions repeatedly without any significant change in their physical or chemical properties, ensuring long-term reliability. PCM are available in various

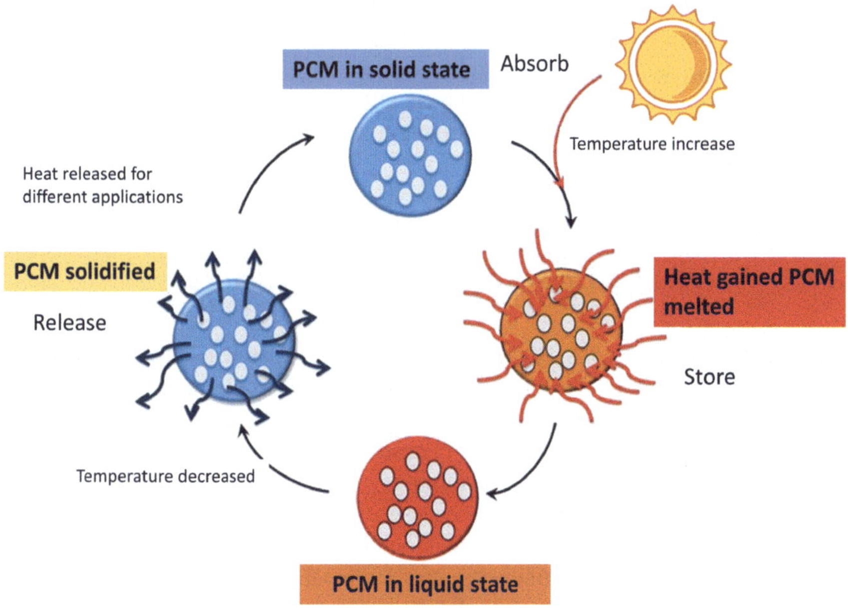

Fig. 2.15 Representative diagram of melting/solidification cycle of phase change material (PCM) (Tony 2021)

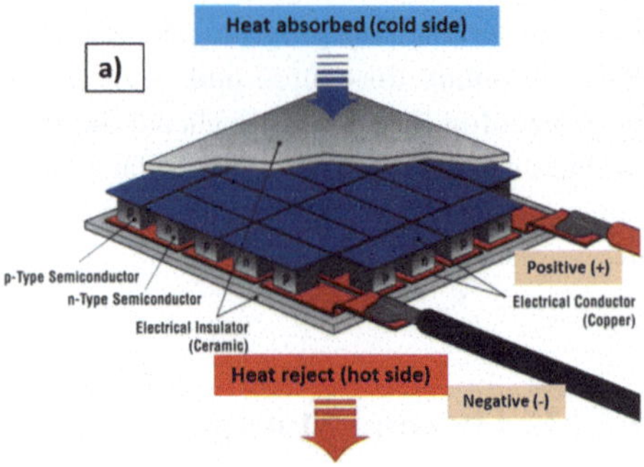

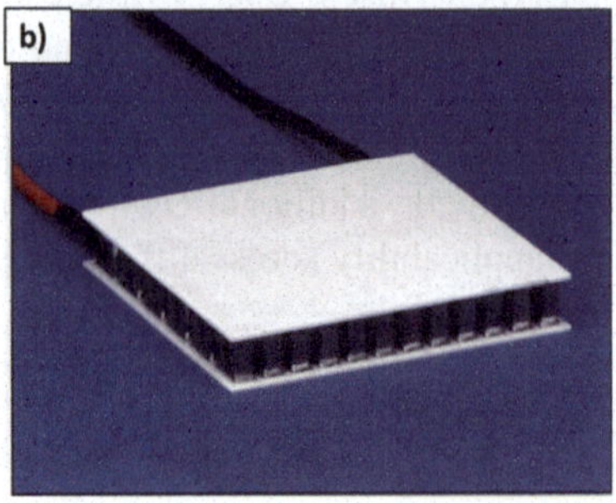

Fig. 2.16 Thermoelectric module, **a** Operating principle (Kumar et al. 2019) and **b** Photograph of a thermoelectric module (Goncalves et al. 2010)

types, including organic, inorganic, and eutectic blends, each with different melting points, making them suitable for a broad range of thermal storage applications.

The variety of PCM includes organic compounds like paraffin waxes and fatty acids, which are known for their non-corrosive and chemically stable characteristics. In contrast, inorganic materials such as hydrated salts, offer higher energy storage densities but can be more corrosive and complex to handle. Eutectic mixtures combine the properties of organic and inorganic substances to optimize performance across different temperature ranges. These materials have gained significant attention in applications such as building energy management, where they are integrated into walls, floors, and ceilings to regulate indoor temperatures. PCM are also used in industries requiring temperature-sensitive environments such as pharmaceuticals and food storage, offering an energy-efficient solution for maintaining consistent thermal conditions.

2.14.1 System Description and Working Mechanism

Phase Change Material (PCM) thermal storage systems operate by utilizing the latent heat absorbed or released during the phase transition of a material between its solid and liquid states. The essence of this mechanism lies in the fact that when a PCM undergoes a phase change, it absorbs or releases large quantities of heat at a constant temperature, making it ideal for temperature regulation in various applications. This process is particularly advantageous because it allows for the storage and release of thermal energy over long periods without significant fluctuations in temperature, enhancing the efficiency of energy use. A typical PCM thermal storage system comprises a container or thermal module housing the PCM, designed to facilitate efficient heat exchange between the PCM and its surrounding environment (Khudhair and Farid 2021). During the phase transition, as the PCM absorbs heat, it melts, storing the thermal energy in the form of latent heat. When the surrounding environment cools, the PCM solidifies, releasing the stored heat to maintain a stable temperature. This cyclical process of melting and solidifying can be repeated numerous times, making PCM systems highly reliable and durable.

The working mechanism is generally classified into two stages: the charging and discharging phases. During the charging phase, heat is absorbed by the PCM causing it to transition from solid to liquid, thereby storing energy. In contrast, during the discharging phase, the environment cools down causing the PCM to solidify and release the previously absorbed heat. The selection of the PCM material is crucial in determining the effectiveness of the system, as it must match the temperature range required for the specific application (Zsembinszki et al. 2020). PCM can be integrated into various systems, including heat exchangers, wall structures, ceiling tiles, or other forms of thermal reservoirs, to enhance their thermal regulation capabilities. In heat exchangers, the PCM is typically enclosed within a capsule or container that allows for direct contact with the heat transfer fluid (Togun et al. 2024). As the fluid flows over the PCM, thermal energy is transferred to or from the material, facilitating the melting or solidification process. In building materials, PCM are embedded in panels or structures to regulate indoor temperatures by absorbing excess heat during the day and releasing it during cooler nights.

The efficiency of a PCM system is heavily influenced by the type of material used. Organic PCM such as paraffin and fatty acids are chemically stable, non-toxic, and non-corrosive, making them easy to integrate into various systems. However, they generally have lower thermal conductivity, requiring enhanced heat transfer mechanisms to optimize performance. Inorganic PCM such as hydrated salts, offer higher thermal storage capacities and conductivity but may present challenges such as corrosiveness and phase segregation, requiring careful system design. Eutectic PCM combine organic and inorganic properties, offering a balanced solution for systems that require precise temperature control over specific ranges. Thermal management is further enhanced through the use of encapsulation techniques, where PCM are encapsulated in small, micro, or macro capsules. This encapsulation prevents material leakage during phase transitions and increases surface area for improved heat transfer. Additionally, advanced PCM systems may incorporate heat transfer fluids or solid conductive materials within the storage module to improve the overall efficiency of the energy exchange process.

(a) **Organic PCM**

Organic Phase Change Materials (PCM) consist of aliphatic chemicals and other organic substances such as paraffin wax and fatty acids. These materials are typically costly, possess a moderate latent heat per unit volume, and exhibit poor density. A multitude of organic phase change materials (PCM) are flammable and display a diverse array of melting temperatures. Paraffins are very suitable phase change material (PCM) options for domestic heating because of their broad temperature range and diverse structures, enabling the selection of certain paraffins for specific temperature ranges. These organic resources are both cost-effective and abundant. These compounds have a greater storage capacity than others and can freeze without the need for supercooling. Paraffins offer a significant benefit in terms of their melting points, as certain varieties have melting points ranging from 22 to 24 °C. This characteristic makes them very suitable for household heating purposes (Bharathiraja et al. 2024). Fatty acids are being organic compounds, exhibit melting ranges and heat of fusion values that are comparable to those of paraffin substances, including coco fatty acid and other similar varieties. However, in contrast to paraffins, there is a limited selection of acceptable PCM that possess melting temperatures that are optimal for heating purposes. Only a limited number of fatty acids exhibit melting and freezing temperatures within the range of 20–30 °C.

(b) **Inorganic PCM**

Hydrated salts provide the basis of inorganic phase change materials (PCM). These materials have undergone extensive examination for their ability to store latent heat, including multiple experiments and smaller-scale studies (Li et al. 2021). Their melting points, which span from 0 to 120 °C, render them well-suited for a variety of thermal storage applications. Salt hydrates provide a variety of materials for heating homes, with melting temperatures ranging from 18.5 to 36.4 °C. Hydrated salts can separate into layers due to their various hydrates and a dehydrated form, resulting in the solid phase settling at the bottom of the container, a phenomenon referred to as decomposition. In addition, they have a tendency to undergo sub-cooling, which is why performance-enhancing chemicals are employed to postpone degradation. An additional risk is the corrosion of metal containers caused by salts. Salt hydrates can be used effectively and adaptably for storing heat energy in household heating applications when the appropriate design and material choices are employed.

(c) **Eutectic PCM**

Eutectic phase change materials (PCM) are a unique class of materials composed of a combination of two or more components that, when mixed, solidify or melt at a lower temperature than any of the individual components melting points. These materials exhibit excellent temperature control and are particularly valued for their precise and predictable phase change behavior. Eutectic PCM can be composed of organic-organic, inorganic-inorganic, or organic–inorganic mixtures, offering versatility in their use (Sun et al. 2023). Mixtures of inorganic salts or organic compounds can be tailored to achieve desired phase change temperatures. This will make them suitable for applications requiring precise thermal management such as refrigeration, air conditioning, or even solar energy storage systems.

The primary advantage of eutectic PCM is their ability to maintain a consistent phase change temperature, ensuring that heat absorption and release occur at a predictable and constant rate. This consistency makes eutectic materials highly desirable for systems where temperature control is critical, such as in building materials for passive heating or cooling. They are also valuable in cold storage and medication transport, where maintaining a constant temperature is vital. Additionally, eutectic mixtures have been shown to have high latent heat storage capacities, allowing them to store large amounts of energy in a relatively small volume, thus improving the overall energy efficiency of thermal storage systems. However, the formulation of eutectic mixtures requires careful design to prevent phase segregation and ensure long-term stability during repeated cycles of melting and solidification.

(d) **Molecular Alloy as Phase Change Material (MAPCM)**

Molecular alloys possess desirable characteristics for thermal protection and energy storage, such as a high heat of melting and suitable melting temperatures for diverse applications. These alloys have the ability to modify their melting range at various temperatures by adding different molecular components (Usman et al. 2022). Molecular alloys are solid solutions that can be either binary or multicomponent. They can be composed of organic compounds or comprise aluminum and silicon (Al-Si) compositions. These alloys exhibit significant latent heat and possess high thermal conductivity. MAPCM are well-suited for use in high-temperature applications such as solar power systems and residential heaters.

2.14.2 Applications in Buildings

PCM thermal storage systems provide an effective solution for managing temperature fluctuations by leveraging the latent heat exchange properties of phase change materials. These systems can be integrated into a wide range of applications, from residential building materials to industrial heat exchangers, offering energy-efficient temperature regulation. The versatility and repeatability of PCM systems along with their ability to store large quantities of heat at consistent temperatures, make them indispensable in modern energy management strategies. Thermal energy storage utilizes both latent heat and sensible heat. Cement board can serve as a thermal storage material in structures. PCM on the other hand, utilize the concept of latent heat to efficiently store energy in a compact space by undergoing significant heat release and absorption during phase transitions. The heat storage capacity of concrete is 0.784 kWh/m^3, but a typical PCM may store between 38 and 105 kWh/m^3. PCM only utilize solid–liquid phase transitions, while liquid–gas phase transitions are not feasible for thermal storage due to the substantial quantities or elevated pressures necessary in the gas phase. Consequently, there has been significant research and practical use of solid–liquid PCM in the fields of building construction and solar energy systems.

Phase Change Materials are generally characterized by their low heat conductivity and exhibit optimal performance when confined within compact containers partitioned into cells. PCM can be packaged in different forms, such as enclosed PCM, microencapsulated PCM, slurries, and shape-stabilized PCM. The conventional approach entails the storage of organic or inorganic PCM within containers. Several PCM products used in building applications are described below:

(a) **Latent Heat Fiber Boards**

Latent heat fiber boards are an innovative type of thermal energy storage boards that include PCM into a secondary supporting framework, such as natural wood fiber boards, gypsum boards, or high-density polyethylene. Thermal energy storage applications utilize a bound PCM that undergoes melting and solidification to store and release latent heat, while ensuring that it remains contained within the board without any leakage. The heat storage capacity of the bound PCM is dependent on the porosity of the supporting material, as it governs the quantity of PCM that can be incorporated into the capillary structure.

(b) **Microencapsulated PCM and PCM Microcapsule Slurry**

PCM fiber boards, wallboards, and panels contain paraffin waxes that are exposed and can undergo oxidation, resulting in the production of hazardous chemicals such as ketones and aldehydes. To prevent exposure, it is necessary to have a storage system that is totally sealed. In recent times, there has been a growing interest in the utilization of microencapsulated Phase Change Materials for energy storage applications (Rathore et al. 2020). Microencapsulation is the process of enclosing fine particles, including liquids and solids, into capsules that vary in size from less than 1 μm to more than 300 μm. The walls of the capsule can be composed of a variety of materials, such as both natural and synthetic polymers. Microencapsulated paraffin wax has several advantages, including less reactivity with the surroundings, enhanced surface area for heat transfer, and safeguarding of the core substance during phase transitions (Chang et al. 2022). Two common ways of microencapsulation include complicated coacervation and spray drying. By combining microencapsulated phase change materials, a fluid, a slurry of PCM microcapsules is formed. This slurry possesses a substantial heat capacity and can be easily transported. Microencapsulated phase change materials and slurry containing PCM microcapsules are widely utilized in developing energy-efficient systems and for storing solar energy.

(c) **Building Floorboard Air-Conditioning Heating**

Heating and cooling are crucial for ensuring a pleasant and comfortable indoor atmosphere. Utilizing floorboard air-conditioning, provides a superior and more even distribution of temperature compared to conventional approaches. Maintaining low temperatures for the head helps to preserve the freshness of the brain, while keeping high temperatures for the feet ensures that the body stays warm, resulting in a more effective and healthier heating solution. Underfloor heating effectively combats the growth of mildew, rot, and fungus, while also eradicating dust mites in mattresses and carpets by applying heat to the floorboards or carpet. This method can utilize shape-stabilized PCM or PCM slurries (Sun et al. 2020). To comply with building requirements, it is essential to manage the combustion hazard of PCM in floorboard PCM electric heating systems.

2.15 Thermoelectric Heating and Cooling

Thermoelectric heating and cooling are based on the Peltier effect, where an electric current flows through the junction of two dissimilar materials, causing one side to absorb heat (cooling) while the other releases heat (heating). This technology offers an efficient, solid-state solution for temperature control, making it an ideal choice for compact and portable applications (Salah and Abuhelwa 2020). Thermoelectric devices, also known as thermoelectric coolers (TEC) or heaters, consist of semiconductor elements that can quickly switch between heating and cooling modes with the simple reversal of electrical current. Due to the absence of moving parts or refrigerants, these devices are highly reliable, durable, and environmentally friendly, contributing to their increasing adoption in industries such as consumer electronics, automotive, and medical devices.

Thermoelectric systems are particularly advantageous for precise temperature control in small spaces, where traditional cooling and heating methods may not be practical. In applications like electronics cooling, thermoelectric modules are used to manage the heat generated by high-performance processors, improving overall efficiency and lifespan (Baru and Bhatia 2020). Additionally, thermoelectric technology has found use in climate-controlled seats in automobiles, where individual cooling and heating adjustments can be made for optimal comfort. Although current thermoelectric systems are most effective for small-scale applications, ongoing advancements in material science and thermoelectric materials have the potential to increase the efficiency and scalability of these systems, paving the way for broader usage in residential and industrial heating and cooling applications.

2.15.1 System Description and Working Mechanism

When two metals or semiconductors that are not alike are joined and their junctions are at different temperatures, five occurrences happen at the same time. The following effects are the Joule effect, the Fourier effect, the Seebeck effect, the Peltier effect, and the Thomson effect. All of these consequences are permanent and cannot be reversed. The Peltier effect, which was first observed by Peltier in 1834, occurs when an electric current passes through a circuit containing two dissimilar semiconductors, one being p-type and the other n-type. This leads to the heating of one contact region and the cooling of the other, depending on the direction of current flow which showcases two PN thermoelements.

Bismuth telluride (Bi_2Te_3) is a semiconductor that exhibits a more pronounced Peltier effect. The application of a DC voltage enables the transmission of heat from one junction to another. In order to enhance heat transfer, it is possible to connect several n-type and p-type semiconductor thermoelements in series using copper strips. These thermoelements can then be positioned between two ceramic plates that are electrically insulating but thermally conductive. The configuration shown in Fig. 2.16a, b represents a thermoelectric module that enables the transfer of heat without the need for any mechanical components. By reversing the current direction, the heat transfer direction is reversed, resulting in the hot side becoming cold and the cold side being hot.

Commercial single-stage thermoelectric modules may attain a maximum temperature difference (Th-Tc) of 77 °C or a maximum heat transfer rate of 125 W. The dimensions of these modules vary from around 62 × 62 × 4.6 mm to about 9.4 × 9.4 × 4.7 mm for the CP series. A thermoelectric cooler often referred to as a Peltier cooler or thermoelectric device, can be created by connecting many thermoelectric modules either in series or in parallel. By combining multiple thermoelectric modules, a cascade cooler can be created, resulting in lower refrigeration temperatures and bigger temperature differences (ΔT) of up to 131 °C. A thermoelectric device necessitates the presence of heat sinks on both the hot and cold sides.

Ensuring optimal power input for a thermoelectric (TE) module requires a crucial connection of the module and a careful selection of the power supply. A thermoelectric cooling module is a semiconductor device with low impedance that functions as a resistive load to its power source. Due to the characteristics of bismuth telluride, these modules exhibit a positive resistance temperature coefficient of approximately 0.5 percent per degree Celsius, as determined by the average module temperature. As the TE device functions, the temperature of the module rises, leading to an increase in its resistance. The performance of a TE module is determined by its maximum current (I_{max}), maximum voltage (V_{max}), and maximum cooling load (P_{max}), as specified by manufacturers. The parameters are displayed in Table 2.7.

2.15.2 Applications in Buildings

Thermoelectric devices have been widely adopted due to their numerous advantages. Some of these advantages include:

(a) Reversible characteristics: When the direction of current flow is reversed, the cold junction becomes the hot junction and vice versa.
(b) Solid-state design: With no moving parts, compressed gas, or chemicals, these devices are low noise and environmentally friendly.
(c) High reliability: These devices have a lifespan of approximately 250,000 h, or 28.5 years, at room temperature.
(d) Compact design: Their small volume and lightweight nature make them easy to incorporate into various systems.
(e) Simple configuration: Their straightforward design makes them easy to use and integrate.
(f) Fast start-up and easy control: Refrigeration velocity and temperature can be easily adjusted via the working electric current.
(g) High efficiency at low cooling power: These devices perform well even with low cooling power requirements.

Table 2.7 Definitions about Q_{max}, I_{max}, V_{max} and DT_{max}

Parameters	Physical definitions
Qmax (Watts)	The amount of heat (Watts) that a TE cooler can pump at a zero degree temperature difference and at a hot-side temperature of 300 K
Imax (Amps)	The current at which the maximum DT is produced when the hot-side of the thermoelectric module is held at 300 K. Generally, it is not a good to operate a TE cooler at Imax because the amount of input power increases significantly without a significant change in delta T. 70–80% of Imax is usually an optimal operating condition
Vmax (Volts)	The voltage that is produced at DT_{max} when Imax is applied and the hot-side temperature of the thermoelectric module is 300 K
DTmax (°C or K)	The maximum obtainable temperature difference between the cold and hot side of the thermoelectric elements within a module when Imax is applied and there is no heat load applied to the module. This parameter is measured with the hot-side of an element at a temperature of 300 K. In reality, it is virtually impossible to remove all sources of heat in order to achieve the true DT_{max}. Therefore, the number only serves as a standardized indicator of the cooling capability of a thermoelectric module. DTmax is most important thermoelectric cooler/device specifications
Optimum current (Amps)	The specific level of electrical current that will produce the greatest heat absorption by the cold side of a thermoelectric module. At the optimum current, a thermoelectric module will be capable of pumping the maximum quantity of heat; maximum temperature differential (DTmax) typically occurs at a somewhat lower current level

(h) Steady temperature maintenance: Thermoelectric coolers (TEC) maintain enclosures at a steady temperature by operating with a proportional control system rather than an on–off system.

However, thermoelectric devices also possess many disadvantages such as a low coefficient of performance (COP), limited power output, and often a substantial electric current (DC) demand in comparison to traditional vapor compression air conditioners. Furthermore, they necessitate the use of two heat sinks. If the heat sink on the hot side is not functioning well, a substantial quantity of heat will be transferred back to the cold side. The coefficient of performance (COP) of a thermoelectric cooler equipped with air-cooled heat sinks varies between 0.38 and 0.45. Nevertheless, this value increases by nearly two times when a water-cooled heat sink is utilized on the hot side. Notwithstanding these disadvantages, thermoelectric gadgets are extensively utilized in diverse applications. They function as refrigerators, air conditioners, and cooler/warmer boxes. For example, a 48-L thermoelectric minibar operates without the need for fans since it utilizes a heat pipe for both the hot and cold heat exchangers. Thermoelectric air conditioners have minimal power consumption, with certain models are capable of producing up to 1200 BTU/h or 352 W. This renders them appropriate for cooling confined areas like electrical enclosures. Cooler/warmer boxes are well-suited for activities such as picnicking, camping, and touring. Additionally, they are valuable for the preparation of infant milk and juice since they have the capability to rapidly heat and chill.

In thermoelectric refrigerators, the continuous transfer of heat from the cold side to the hot side will be disrupted if the discharge of heat from the hot side is not efficient. Heat flow can potentially be reversed (Kaiprath et al. 2023). In the absence of power supply to the TE modules, heat from the hot side will rapidly dissipate due to the low thermal insulation properties of the TE modules. Thermal diode heat pipes are a highly effective option. The purpose of these pipes is to guarantee that the TE modules are completely contained within the enclosure and to transport heat to the outside using heat pipes, which take up less space on the enclosure. A commercially available heat pipe cooling equipment utilizes a heat pipe radiator and blower on the hot side.

Further investigations for thermoelectric cooling and heating in domestic applications are suggested as follows:

(a) The hot side heat sink of the TE device may use a water-cooled heat sink.
(b) The optimum power input and module connection in a TE device should be investigated. Allowing the lead ends of modules to be outside the base plate of the heat sink would enable modifications to the connection method for modules such as in series in parallel or hybrid configurations.
(c) Improving TE materials would lead to more efficient TE refrigerators and TE air conditioners.
(d) Further tests on refrigerators and air conditioners powered by PV may reduce greenhouse gas emissions.

References

Abdelaziz GB et al (2021) Energy saving via Heat Pipe Heat Exchanger in air conditioning applications "experimental study and economic analysis." J Build Eng 35:102053

Abd-Elhady MM et al (2022) Solid desiccant-based dehumidification systems: a critical review on configurations, techniques, and current trends. Int J Refrig 133:337–352

Abdelkareem MA et al (2022) Heat pipe-based waste heat recovery systems: background and applications. Therm Sci Eng Progr 29:101221

Abdukarimov B, O'tbosarov S, Abdurazakov A (2021) Investigation of the use of new solar air heaters for drying agricultural products. In: E3S web of conferences. EDP Sciences

Ahmad A, Rita N, Noorjannah L (2022) Effect of acid hydrolysis on bioethanol production from oil palm fruit bunches. Mater Today: Proc 63:S276–S281

Ahmad MI et al (2024) Fueling the future: extraction and production technological development of marine macroalgae-seaweed biodiesel. Multidisciplinary applications of marine resources: a step towards green and sustainable future. Springer, pp 285–302

Ahmadi A et al (2021) Recent residential applications of low-temperature solar collector. J Clean Prod 279:123549

Ahmed SF et al (2021) Recent progress in solar water heaters and solar collectors: a comprehensive review. Therm Sci Eng Progr 25:100981

Ali M et al (2020) Thermal and acoustic characteristics of novel thermal insulating materials made of *Eucalyptus globulus* leaves and wheat straw fibers. J Build Eng 32:101452

Almutairi K et al (2022) An updated review on solar air heating systems. Sustain Energy Technol Assess 53:102573
Alqaed S et al (2022) Using nanoparticles in solar collector to enhance solar-assisted hot process stream usefulness. Sustain Energy Technol Assess 52:101992
Ameri M, Farzan H, Nobari M (2021) Evaluation of different glazing materials, strategies, and configurations in flat plate collectors using glass and acrylic covers: an experimental assessment. Iranica J Energy Environ 12(4):297–306
Amirgaliyev Y et al (2021) Experimental comparative analysis of operating characteristics of double circuit flat-plate solar collector with thermosiphon circulation and flat solar collector with chemical coating. Therm Sci 26:173–173
Anupam Rao Y, Somwanshi A (2022) A comprehensive review on integrated collector-storage solar water heaters. Mater Today Proc 63:15–26
Asasian-Kolur N et al (2022) Membrane-based enthalpy exchangers for coincident sensible and latent heat recovery. Energy Convers Manage 253:115144
Assad MEH et al (2022) Geothermal heat pumps: principles and applications. In: 2022 advances in science and engineering technology international conferences (ASET). IEEE
Bakthavatchalam B, Habib K (2021) Thermoelectric air conditioners for tropical countries. In: Sulaiman SA (ed) Clean energy opportunities in tropical countries. Springer Singapore, Singapore, pp 289–307
Baroutaji A, Arun A, Robinson J, Wilberforce T, Abdelkareem MA, Olabi AG (2021) PEMFC Poly-generation systems: developments, merits, and challenges. Sustain, https://doi.org/10.3390/su132111696
Baru S, Bhatia S (2020) A review on thermoelectric cooling technology and its applications. In: IOP conference series: materials science and engineering. IOP Publishing
Bellia L et al (2000) Air conditioning systems with desiccant wheel for Italian climates. Int J Archit Sci 1(4):193–213
Bharathiraja R et al (2024) Thermal characteristics enhancement of Paraffin Wax Phase Change Material (PCM) for thermal storage applications. Renew Energy 222:119986
Bilardo M et al (2020) Design and experimental analysis of an Integral Collector Storage (ICS) prototype for DHW production. Appl Energy 259:114104
Binhweel F et al (2024) Kinetics, thermodynamics, and optimization analyses of lipid extraction from discarded beef tallow for bioenergy production. In: Separation science and technology, pp 1–20
Boldyryev S, Shamraev AA, Shamraeva EO (2021) The retrofit of the calcium chloride production by Pinch approach and process modifications. Appl Therm Eng 189:116775
Bouhdjar A, Semai H, Amari A (2021) New technique to evaluate the overall heat loss coefficient for a flat plate solar collector. J Energy Technol 14(1):11–25
Brough D, Jouhara H (2020) The aluminium industry: A review on state-of-the-art technologies, environmental impacts and possibilities for waste heat recovery. Int J Thermofluids 1:100007
Brough D et al (2020) An experimental study and computational validation of waste heat recovery from a lab scale ceramic kiln using a vertical multi-pass heat pipe heat exchanger. Energy 208:118325
Buasri K, Surin P (2021) Effect of adhesive on physical and thermal properties roofing material from oil palm empty fruit bunch fibers use as building material
Buratti C et al (2021) Aerogel glazing systems for building applications: a review. Energy Build 231:110587
Caprì A, Frazzica A, Calabrese L (2020) Recent developments in coating technologies for adsorption heat pumps: a review. Coatings 10(9):855
Carroll P, Chesser M, Lyons P (2020) Air source heat pumps field studies: a systematic literature review. Renew Sustain Energy Rev 134:110275
Carroll MK et al (2022) Aesthetic aerogel window design for sustainable buildings. Sustainability 14(5):2887

Cascone S, Cascone S, Vitale M (2019) Building insulating materials from agricultural by-products: a review, pp 309–318
Cetiner I, Shea AD (2018) Wood waste as an alternative thermal insulation for buildings. Energy Build 168:374–384
Chang Z et al (2022) Review on the preparation and performance of paraffin-based phase change microcapsules for heat storage. J Energy Storage 46:103840
Chauhan VK et al (2021) A comprehensive review of direct solar desalination techniques and its advancements. J Clean Prod 284:124719
Chauhan P, Kaushik S, Tyagi S (2022) A review on thermal performance enhancement of green cooling system using different adsorbent/refrigerant pairs. Energy Convers Manage X 14:100225
Chinnasamy S, Arunachalam A (2023) Experimental investigation on direct expansion solar-air source heat pump for water heating application. Renew Energy 202:222–233
Collins F et al (2020) A critical review of waste resources, synthesis, and applications for Zeolite LTA. Micropor Mesopor Mater 291:109667
de Sousa MA et al (2024) CRFD modeling of high-temperature reciprocating grate degradation in a 15 t/h eucalyptus wood chip boiler. Renew Energy 230:120858
Di X, Chen J, Zheng S (2020) Residual gas analysis in vacuum insulation panel (VIP) with glass fiber core and investigation of getter for VIP. Build Environ 186:107337
Ding Z et al (2022) On-site measurement and simulation investigation on condensation dehumidification and desiccant dehumidification in Hong Kong. Energy Build 254:111560
Esmaeilpour M, Gholami Korzani M, Kohl T (2021) Performance analyses of deep closed-loop U-shaped heat exchanger system with a long horizontal extension. In: Proceedings of the 46th workshop on geothermal reservoir engineering, Stanford, CA, USA
Evangelisti L, Vollaro RDL, Asdrubali F (2019) Latest advances on solar thermal collectors: a comprehensive review. Renew Sustain Energy Rev 114:109318
Fallah A et al (2021) Globally scalable geothermal energy production through managed pressure operation control of deep closed-loop well systems. Energy Convers Manage 236:114056
Fort K (2013) Hypocaust and murocaust storage. Solar air systems. Routledge, pp 175–184
Fudholi A, Sopian K (2019) A review of solar air flat plate collector for drying application. Renew Sustain Energy Rev 102:333–345
Gangadharan S et al (2023) Coke dry quenching and implications on coke quality and blast furnace performance. Ironmaking Steelmaking 50(1):75–83
Gayathiri M et al (2022) Activated carbon from biomass waste precursors: factors affecting production and adsorption mechanism. Chemosphere 294:133764
Gillott MC (2000) A novel mechanical ventilation heat recovery/heat pump system. University of Nottingham
Goncalves L et al (2010) Heat-pipe assisted thermoelectric generators for exhaust gas applications. In: ASME international mechanical engineering congress and exposition
Goodarzia G et al (2017) Energy saving opportunities in air drying process in high-pressure compressors. Energy Procedia 110:428–433
Govindasamy K et al (2024) Performance analysis of evacuated tubes with thermosyphon heat pipe solar collector integrated with compound parabolic concentrator under different operating conditions. Energy Explor Exploit 42(1):231–249
Grover G, Cotter T, Erickson G (1964) Structures of very high thermal conductance. J Appl Phys 35(6):1990–1991
Guan B et al (2022) Review of internally cooled liquid desiccant air dehumidification: materials, components, systems, and performances. Build Environ 211:108747
He D et al (2023) Thermal separation of heavy metals from municipal solid waste incineration fly ash: a review. Chem Eng J 467:143344
Henning A (2004) Social anthropological and interdisciplinary research on the conversion of electrically heated single family houses to heating by combined pellet–solar systems. Biomass Bioenerg 27(6):547–555

Hu Y et al (2020) Experimental and numerical study of a PCM solar air heat exchanger and its ventilation preheating effectiveness. Renewable Energy 145:106–115

Humia GM et al (2021) Experimental study and simulation model of a direct expansion solar assisted heat pump to CO_2 for water heating: Inventory, coefficient of performance and total equivalent warming impact. Sol Energy 230:278–297

Hysek S et al (2018) Effect of wheat husk surface pre-treatment on the properties of husk-based composite materials. Ind Crops Prod 125:105–113

Isaac O, Adebayo J (2010) Solar adsorption technologies for ice-making and recent developments in solar technologies: a review. Int J Adv Eng Technol 1:284–303

Islam MA, Pal A, Saha BB (2020) Experimental study on thermophysical and porous properties of silica gels. Int J Refrig 110:277–285

Jesper M et al (2021) Large-scale heat pumps: Uptake and performance modelling of market-available devices. Renew Sustain Energy Rev 137:110646

Joao S et al (2022) The utilisation of geothermal spring water for Tilapia aquaculture to promote food security and skills development at the Brandvlei Correctional Services Centre. Water Research Commission

Johansson P (2021) Heat pumps in Sweden—a historical review. Energy 229:120683

Jose J et al (2020) Eco-friendly thermal insulation material from cellulose nanofibre. J Appl Polym Sci 137(2):48272

Jouhara H et al (2021) Investigation on a full-scale heat pipe heat exchanger in the ceramics industry for waste heat recovery. Energy 223:120037

Kaiprath J (2023) A review on solar photovoltaic-powered thermoelectric coolers, performance enhancements, and recent advances. Int J Air-Condition Refriger 31(1):6

Kalaiarasi G et al (2020) Experimental analysis and comparison of flat plate solar air heater with and without integrated sensible heat storage. Renew Energy 150:255–265

Kalogirou SA (2004) Solar thermal collectors and applications. Prog Energy Combust Sci 30(3):231–295

Kapilan N, Isloor AM, Karinka S (2023) A comprehensive review on evaporative cooling systems. Results Eng 18:101059

Karthick L et al (2022) Energy performance of a compression refrigeration cycle using environment-friendly refrigerants. Mater Today: Proc 66:1519–1525

Katsura T et al (2020) Thermal performance analysis of a new structured-core translucent vacuum insulation panel in comparison to vacuum glazing: experimental and theoretically validated analyses. Sol Energy 199:326–346

Khudhair AM, Farid M (2021) A review on energy conservation in building applications with thermal storage by latent heat using phase change materials. In: Thermal energy storage with phase change materials, pp 162–175

Koholé YW, Tchuen G (2020) Experimental and numerical investigation of a thermosyphon solar water heater. Int J Ambient Energy 41(4):384–394

Krarti M (2022) Energy performance of control strategies for smart glazed windows applied to office buildings. J Build Eng 45:103462

Kuhn TE et al (2021) Review of technological design options for building integrated photovoltaics (BIPV). Energy Build 231:110381

Kumar PM, Jagadeesh Babu V, Subramanian A, Bandla A, Thakor N, Ramakrishna S, Wei H (2019) The design of a thermoelectric generator and its medical applications. Designs 4(1):8

Kumar D et al (2020) Comparative analysis of building insulation material properties and performance. Renew Sustain Energy Rev 131:110038

Kumar KR, Chaitanya NK, Kumar NS (2021a) Solar thermal energy technologies and its applications for process heating and power generation—a review. J Clean Prod 282:125296

Kumar A, Said Z, Bellos E (2021b) An up-to-date review on evacuated tube solar collectors. J Therm Anal Calorim 145(6):2873–2889

Larson SL et al (2021) Amendment for increased methane production rate in municipal solid waste landfill gas collection systems. Sci Total Environ 772:145574

Lazzarin R (2020) Heat pumps and solar energy: A review with some insights in the future. Int J Refrig 116:146–160

Li S et al (2017) Comparative study on the performance of a new solar air collector with different surface shapes. Appl Therm Eng 114:639–644

Li X et al (2020) A new energy saving ventilation system assisted by transpired solar air collectors for primary and secondary school classrooms in winter. Build Environ 177:106895

Li Y et al (2021) Review on tailored phase change behavior of hydrated salt as phase change materials for energy storage. Mater Today Energy 22:100866

Li J et al (2022) A review of air-to-air membrane energy recovery technology for building ventilation. Energy Build 265:112097

Liang J-D et al (2022) Performance investigation of a hybrid ground-assisted desiccant cooling system. Energy Convers Manage 265:115765

Lima AA et al (2020) Absorption refrigeration systems based on ammonia as refrigerant using different absorbents: review and applications. Energies 14(1):48

Lin T, Quan X, Cheng P (2022) Experimental investigation of superlong two-phase closed thermosyphons for geothermal utilization. Int J Therm Sci 171:107199

Liu B, Zhang X, Ji J (2021) Review on solar collector systems integrated with phase-change material thermal storage technology and their residential applications. Int J Energy Res 45(6):8347–8369

Ma Q, Ahmadi A, Cabassud C (2020) Optimization and design of a novel small-scale integrated vacuum membrane distillation-solar flat-plate collector module with heat recovery strategy through heat pumps. Desalination 478:114285

Maghrabie HM et al (2022) Numerical simulation of heat pipes in different applications. Int J Thermofluids 16:100199

Mahmood MH, Sultan M, Miyazaki T (2020) Experimental evaluation of desiccant dehumidification and air-conditioning system for energy-efficient storage of dried fruits. Build Serv Eng Res Technol 41(4):454–465

Mahmoud M et al (2022) Flow rate optimization in run-around heat recovery systems. Appl Therm Eng 200:117599

Mansur FZ et al (2021) Thermal performance of a fixed-plate air-to-air energy recovery system for building application in hot and humid environment. Int J Energy Res 45(6):8900–8918

Mawardi I et al (2024) Eco-friendly production and performance evaluation of water-resistant cellulose nanofiber bioaerogels. Polym Eng Sci 64(2):733–748

Mempouo B, Cooper E, Riffat SB (2010) Novel window technologies and the Code for Sustainable Homes in the UK. Int J Low-Carbon Technol https://doi.org/10.1093/ijlct/ctq013

Men Y, Liu X, Zhang T (2021) A review of boiler waste heat recovery technologies in the medium-low temperature range. Energy 237:121560

Mohsen OA, Muhammed MA, Hasan BO (2021) Heat transfer enhancement in a double pipe heat exchanger using different fin geometries in turbulent flow. Iranian J Sci Technol Trans Mechan Eng 45(2):461–471

Mota-Babiloni A et al (2020) Assessment of the utilization of equivalent warming impact metrics in refrigeration, air conditioning and heat pump systems. Renew Sustain Energy Rev 129:109929

Mufti IR (1971) Geothermal aspects of radioactive waste disposal into the subsurface. J Geophys Res 76(35):8568–8585

Muhammad A et al (2021) Thermo-economic analysis of integrated gasification combined cycle co-generation system with carbon capture and integrated with absorption refrigeration system. Energy Convers Manage 248:114782

Muneer T, Maubleu S, Asif M (2006) Prospects of solar water heating for textile industry in Pakistan. Renew Sustain Energy Rev 10(1):1–23

Murugesan E, Kannan GR (2023) An experimental study on synthesis of ternary biodiesel through potassium hydroxide catalyst transesterification. Environ Prog Sustain Energy 42(1):e13958

Muthuraj R et al (2019) Sustainable thermal insulation biocomposites from rice husk, wheat husk, wood fibers and textile waste fibers: Elaboration and performances evaluation. Ind Crops Prod 135:238–245

Naderi E et al (2020) Simulation-based performance analysis of residential direct evaporative coolers in four climate regions of Iran. J Build Eng 32:101514
Nihadh S, Alkaff SA (2021) New design arrangement for the PV roof-integrated system
Paneri A, Wong IL, Burek S (2019) Transparent insulation materials: an overview on past, present and future developments. Sol Energy 184:59–83
Paraschiv S, Bărbuţă-Mişu N, Paraschiv LS (2020) Technical and economic analysis of a solar air heating system integration in a residential building wall to increase energy efficiency by solar heat gain and thermal insulation. Energy Rep 6:459–474
Patel AK (2023) Efficiency enhancement of solar water heaters through innovative design. Int J Sci Res Archive 10(1):289–303
Philip S, Rakendu R (2020) Thermal insulation materials based on water hyacinth for application in sustainable buildings. Mater Today Proc 33:3803–3809
Pomianowski MZ et al (2020) Sustainable and energy-efficient domestic hot water systems: a review. Renew Sustain Energy Rev 128:109900
Rambhad KS, Walke PV, Tidke DJ (2016) Solid desiccant dehumidification and regeneration methods—a review. Renew Sustain Energy Rev 59:73–83
Ramlee NA, Naveen J, Jawaid M (2020) Potential of oil palm empty fruit bunch (OPEFB) and sugarcane bagasse fibers for thermal insulation application—a review. Constr Build Mater 271:121519
Rao YA, Somwanshi A (2022) A comprehensive review on integrated collector-storage solar water heaters. Mater Today: Proc 63:15–26
Rathore PKS, Shukla SK, Gupta NK (2020) Potential of microencapsulated PCM for energy savings in buildings: a critical review. Sustain Cities Soc 53:101884
Ravichandran A et al (2021) An assessment of the influence of local conditions on the economic and environmental sustainability of drain water heat recovery systems. J Clean Prod 279:123589
Rizi BS, Heidarinejad M (2022) Analysis of hydronic heating and cooling systems in commercial buildings using CBECS microdata. J Archit Eng 28(3):04022016
Sajjad U et al (2021) A review of recent advances in indirect evaporative cooling technology. Int Commun Heat Mass Transfer 122:105140
Salah WA, Abuhelwa M (2020) Review of thermoelectric cooling devices recent applications. J Eng Sci Technol 15(1):455–476
Sarbu I, Sarbu I (2021) Solar heating and cooling systems. Springer
Sarbu I, Sarbu I (2021) Heat pumps for sustainable heating and cooling. In: Advances in building services engineering: studies, researches and applications, pp 447–557
Sathkumara TM, Waidyasekara AS, Victar HC (2023) The feasibility of transparent solar panels for high-rise building façade in Sri Lanka. Constr Innov
Saxena A, El-Sebaii A (2015) A thermodynamic review of solar air heaters. Renew Sustain Energy Rev 43:863–890
Schritt H, Vidi S, Pleissner D (2021) Spent mushroom substrate and sawdust to produce mycelium-based thermal insulation composites. J Clean Prod 313:127910
Senusi W et al (2024) Biodiesel production and characteristics from waste frying oils: sources, challenges, and circular economic perspective. Environ Sci Pollut Res, pp 1–20
Senusi W et al (2024) Comparative assessment for biodiesel production from low-cost feedstocks of third oil generation. Renew Energy 121369
Sezen K et al (2021) Effects of ambient conditions on solar assisted heat pump systems: a review. Sci Total Environ 778:146362
Shahbazi S et al (2022) Cementitious insulated drywall panels reinforced with Kraft-paper honeycomb structures. Buildings 12(8):1261
Shakir MA, Yhaya MF, Ahmad MI (2017b) Physical effects of pre-hydrolysis and alkali treatment of empty fruit bunch (EFB) fibers. Int J Chem Eng Appl 8(6):387–391
Shakir MA et al (2020a) Preparation and characterization of mycelium as a bio-matrix in fabrication of bio-composite. J Adv Res Fluid Mechan Therm Sci 65(2):253–263

Shakir MA et al (2020b) Structural, thermal, and mechanical properties of spent mushroom substrate (sms) and rubberwood sawdust (rws) binderless particleboards. J Adv Res Fluid Mechan Therm Sci 75(2):113–124

Shakir MA et al (2023a) Sandwich composite panel from spent mushroom substrate fiber and empty fruit bunch fiber for potential green thermal insulation. Buildings 13(1):224

Shakir MA et al (2023b) Effect of fibrillation degree of empty fruit bunch fiber on sandwich composite incorporate with spent mushroom substrate. J Ind Text 53:15280837231153660

Shakir MA et al (2023c) Review on the influencing factors towards improving properties of composite insulation panel made of natural waste fibers for building application. J Ind Text 53:1–33

Shakir MA et al (2023d) From waste to wealth: converting rubber wood sawdust into green mycelium-based composite. Biomass Convers Biorefin 1–19

Shakir MA et al (2024a) Development of natural waste fiber desiccant paper coated with biopolymeric material for energy efficient dehumidification. Clean Technol Environ Policy 1–13

Shakir MA et al (2024b) Processing of waste biomass: production of composite thermal insulation panels from empty fruit bunch and spent mushroom substrates. Waste and Biomass Valorization 1–18

Shakir MA, Ahmad MI (2024c) Bioproduct advances: insight into failure factors in mycelium composite fabrication. Biofuels, Bioproducts and Biorefining

Shakir MA et al (2024d) Biomass to biofuel: palm kernel shells as catalyst supports for enhanced biodiesel production. Biofuels, Bioproducts and Biorefining

Shakir M, Yhaya MF, Ahmad MI (2017) The effect of crosslinking fibers with polyvinyl alcohol using citric acid. Imperial J Interdisc Res (IJIR) 3(4)

Sharma A, Singh PK, Sharma VK (2021) Analysis on WH cement composite thermal insulation material for increasing efficiency of building. Mater Today Proc 45:3036–3041

Sheikholeslami M, Farshad SA (2021) Investigation of solar collector system with turbulator considering hybrid nanoparticles. Renew Energy 171:1128–1158

Shemelin V, Matuska T (2022) Unglazed solar thermal collector for building facades. Energy Rep 8:605–617

Singh AP, Kumar A, Singh O (2021) Effect of integrating high flow naturally driven dual solar air heaters with Trombe wall. Energy Convers Manage 249:114861

Sinha S (2020) An energy efficient solar-assisted advanced adsorption chiller system. University of Sydney

Soni P et al (2021) Review on improvement of adsorption refrigeration systems performance using composite adsorbent: current state of art. In: Energy sources, Part A: recovery, utilization, and environmental effects, pp 1–25

Su X et al (2022) Energy performance of a reversible window integrated with photovoltaic blinds in Harbin. Build Environ 213:108861

Sun W et al (2020) Experimental investigation on the thermal performance of double-layer PCM radiant floor system containing two types of inorganic composite PCMs. Energy Build 211:109806

Sun M et al (2023) A review on thermal energy storage with eutectic phase change materials: fundamentals and applications. J Energy Storage 68:107713

Sun M et al (2024) A strategy to fabricate hierarchical microporous architecture of polyimide nanofibrous aerogels with efficient electromagnetic wave absorption and thermal insulation. Compos A Appl Sci Manuf 177:107940

Szubel M, Papis-Frączek K, Podlasek S (2024) Impact of the air supply system configuration on the straw combustion in small scale batch-boiler-experimental and numerical studies. Renew Energy 220:119540

Tabarhoseini SM, Sheikholeslami M, Said Z (2022) Recent advances on the evacuated tube solar collector scrutinizing latest innovations in thermal performance improvement involving economic and environmental analysis. Sol Energy Mater Sol Cells 241:111733

Tang H et al (2018) Review of applications and developments of ultra-thin micro heat pipes for electronic cooling. Appl Energy 223:383–400

Tawalbeh M et al (2023) Selection criteria of cooling technologies for sustainable greenhouses: a comprehensive review. Therm Sci Eng Progress 38:101666

Tian X et al (2023) Comparison analysis of the glazed and unglazed curved water-based PV/T roofs in the non-heating season. Renewable Energy 205:899–917

Togun H et al (2024) A critical review on phase change materials (PCM) based heat exchanger: different hybrid techniques for the enhancement. J Energy Storage 79:109840

Tony MA (2021) Recent frontiers in solar energy storage via nanoparticles enhanced phase change materials: succinct review on basics, applications, and their environmental aspects. Energy Storage 3(4):e238

Tuncer AD, Amini A, Khanlari A (2023) Experimental and transient CFD analysis of parallel-flow solar air collectors with paraffin-filled recyclable aluminum cans as latent heat energy storage unit. J Energy Storage 70:108009

Tyagi V, Kaushik S, Tyagi S (2012) Advancement in solar photovoltaic/thermal (PV/T) hybrid collector technology. Renew Sustain Energy Rev 16(3):1383–1398

Umar HA et al (2021) Syngas production from gasification and co-gasification of oil palm trunk and frond using a down-draft gasifier. Int J Energy Res 45(5):8103–8115

Usman A et al (2022) Emerging solid-to-solid phase-change materials for thermal-energy harvesting, storage, and utilization. Adv Mater 34(41):2202457

Van Riet F, Verhaert I (2020) Hydronic configurations of hybrid heat production systems in buildings: general design methodology and case studies. Appl Therm Eng 164:114454

Van Toan N, Tuoi TTK, Ono T (2020) Thermoelectric generators for heat harvesting: from material synthesis to device fabrication. Energy Convers Manage 225:113442

Verma AK et al (2022) Mathematical investigation of two-stage dehumidification from single rotary dehumidifier with and without precooling. Numer Heat Transf Part a: Appl 82(3):82–108

Violante AC et al (2021) Preliminary study of a closed loop vertical ground source heat pump system for an experimental pilot plant (Rome, Italy). Renew Energy 176:415–422

Wahlquist S et al (2023) A critical review of heat pipe experiments in nuclear energy applications. Nucl Sci Eng 197(5):719–752

Wang X et al (2020) Synergetic process of condensing heat exchanger and absorption heat pump for waste heat and water recovery from flue gas. Appl Energy 261:114401

Wang X et al (2022a) A critical review on phase change materials (PCM) for sustainable and energy efficient building: design, characteristic, performance and application. Energy Build 260:111923

Wang Y et al (2022b) Catalytic pyrolysis of lignocellulosic biomass for bio-oil production: a review. Chemosphere 297:134181

Weragoda DM et al (2023) A comprehensive review on heat pipe based battery thermal management systems. Appl Therm Eng 224:120070

Williamson KH et al (2001) Geothermal power technology. Proc IEEE 89(12):1783–1792

Wu D, Hu B, Wang R (2021) Vapor compression heat pumps with pure Low-GWP refrigerants. Renew Sustain Energy Rev 138:110571

Xiao B et al (2020) Comparison and analysis on air-to-air and air-to-water heat pump heating systems. Renew Energy 146:1888–1896

Xu W, Guo H, Ma C (2022) An active solar water wall for passive solar greenhouse heating. Appl Energy 308:118270

Yadav JS, Rao CM, Kumar VV (2020) Performance of vapour absorption refrigeration system using waste heat source. i-Manager's J Mechan Eng 10(3):45

Yang H et al (2021) Research development of indirect evaporative cooling technology: an updated review. Renew Sustain Energy Rev 145:111082

Yildiz B et al (2021) Analysis of electricity consumption and thermal storage of domestic electric water heating systems to utilize excess PV generation. Energy 235:121325

Yuan T et al (2023) The review of the application of the heat pipe on enhancing performance of the air-conditioning system in buildings. Processes 11(11):3081

Zhang Y et al (2021a) Visualization study on start-up characteristics of a loop heat pipe with a carbon fiber capillary wick. Int J Heat Mass Transf 169:120940

Zhang S et al (2021b) A review of phase change heat transfer in shape-stabilized phase change materials (ss-PCMs) based on porous supports for thermal energy storage. Renew Sustain Energy Rev 135:110127

Zhang S et al (2023a) Analysis of a solar-assisted heat pump system with hybrid energy storage for space heating. Appl Therm Eng 231:120884

Zhang Y et al (2023b) Performance prediction on a novel dual-mode heat pump with a hybrid photovoltaic/micro-channel heat pipe/fin heat exchanger. Energy Convers Manage 293:117505

Zhang D et al (2024) Thermal performance of two evacuated tube solar collectors with flat heat pipes. Appl Therm Eng 241:122366

Zhang J et al (2020) Review on advanced technology for waste heat recovery of absorption heat pump. Power Gener Technol 41(3):269

Zhao K et al (2021) Formation of ultra-high temperature ceramic hollow microspheres as promising lightweight thermal insulation materials via a molten salt-assisted template method. ACS Appl Mater Interfaces 13(31):37388–37397

Zhou Y et al (2020) Passive and active phase change materials integrated building energy systems with advanced machine-learning based climate-adaptive designs, intelligent operations, uncertainty-based analysis and optimisations: a state-of-the-art review. Renew Sustain Energy Rev 130:109889

Zima W, Ojczyk G (2022) Combustion of wood pellets in a low-power multi-fuel automatically stoked heating boiler. Arch Thermodyn 43(1)

Zotos M (2022) Aventicum: protection of roman monuments in Switzerland. In: Protection of historical constructions: proceedings of PROHITECH 2021. Springer

Zsembinszki G, Fernández AG, Cabeza LF (2020) Selection of the appropriate phase change material for two innovative compact energy storage systems in residential buildings. Appl Sci 10(6):2116

Chapter 3
Sustainable Power Generation and Application in Energy-Efficient Buildings

Contents

Abstract In the pursuit of sustainable development, the integration of renewable energy technologies in building designs has become a key strategy for reducing energy consumption and environmental impact. This chapter explores the integration of sustainable power generation technologies in energy-efficient buildings, focusing on enhancing energy efficiency through advanced electrical systems and renewable energy sources. It delves into the role of smart grids, energy management systems, and dynamic pricing models in optimizing energy consumption, supported by predictive algorithms and real-time monitoring. The chapter also examines the integration of renewable energy sources, such as solar photovoltaics (PV), wind turbines, and geothermal heat pumps, highlighting their contribution to energy independence and grid reliability. Building-integrated photovoltaics (BIPV) and microgrids are emphasized for their ability to provide on-site power generation and contribute to the resilience and sustainability of urban infrastructure. Additionally, the chapter discusses innovative applications like solar-powered smart windows, thin-film solar cells, and photovoltaic thermal collectors (PV/T), showcasing their potential to

S. Riffat et al., *Sustainable Energy Technologies and Low Carbon Buildings*, Lecture Notes in Energy 45, https://doi.org/10.1007/978-3-031-78853-6_3

revolutionize energy efficiency in modern buildings. By combining technological advancements with sustainable architectural practices, this chapter outlines a path toward reducing energy consumption, lowering operational costs, and mitigating environmental impacts in the built environment.

3.1 Overview of Electrical Application in Energy-Efficient Buildings

Energy-efficient buildings leverage advanced electrical systems to minimize energy consumption while maximizing performance and comfort. The integration of smart grids and energy management systems allows buildings to monitor and optimize their energy usage in real-time. In particular, smart grids provide enhanced control over electricity flows, ensuring that both energy generation and consumption are balanced efficiently. These systems incorporate dynamic pricing models that allow consumers to adjust their energy usage based on the fluctuating costs of electricity, leading to more efficient demand-side management. Additionally, predictive algorithms are increasingly being integrated into smart grids to forecast energy demand based on historical data, weather patterns, and occupancy trends, enabling more precise energy distribution and reducing the need for excess capacity during peak hours. Smart grids enable two-way communication between buildings and energy suppliers, facilitating demand response strategies where energy use can be adjusted during peak times to prevent overloading the grid (Hua et al. 2022). This interactivity helps optimize energy distribution, lowers operational costs, and ensures a more reliable supply.

By incorporating sensors, automated controls, and data-driven algorithms, these systems can adjust lighting, heating, ventilation, and air conditioning (HVAC) based on occupancy and environmental conditions (Sendrayaperumal et al. 2021). For example, occupancy sensors can detect when rooms or areas are unoccupied and automatically reduce or turn off HVAC systems, thereby minimizing unnecessary energy use. Similarly, temperature sensors and automated shading systems can work in tandem to regulate indoor climate by blocking excessive sunlight during the hottest parts of the day, reducing the need for air conditioning. Furthermore, machine learning algorithms can be employed to continuously refine these systems operations by learning from user behaviour and environmental changes, resulting in a more adaptive and efficient energy system over time. This not only reduces energy wastage but also enhances the overall efficiency and sustainability of the building.

In addition to automation, the adoption of energy-efficient appliances and electrical components plays a vital role in reducing the power demand of buildings. Energy-efficient lighting systems, such as LED technology, consume significantly less electricity compared to traditional incandescent or fluorescent lighting. LED lights are more durable and have a longer operational life, often lasting up to 50,000 h compared to 1000 h for incandescent bulbs. This reduces the need for frequent replacements, which further contributes to the sustainability of the building. Moreover, smart lighting systems can adjust brightness levels based on natural light availability, further minimizing energy use. Daylight harvesting technology, which adjusts indoor lighting based on available natural light can cut lighting energy use by up to 60% in some cases. Moreover, smart lighting systems can be programmed to dim or switch off in areas not in use, further increasing efficiency.

Furthermore, the use of high-efficiency motors and transformers in HVAC systems helps in cutting down electricity consumption, which is one of the major contributors to energy use in buildings (Alavy et al. 2020). For instance, advanced HVAC systems can incorporate heat recovery ventilators (HRV) that capture heat from outgoing air and transfer it to incoming fresh air, reducing the energy needed to heat or cool indoor spaces. Additionally, geothermal heat pumps can be integrated into HVAC systems to harness stable underground temperatures for heating and cooling, significantly lowering the energy demand for climate control. Variable speed drives (VSD) are often integrated with these motors to optimize their operation by adjusting motor speed according to load demand, leading to significant energy savings. VSD technology can reduce energy consumption by 20–30% in HVAC systems compared to traditional fixed-speed systems. By deploying such technologies, energy-efficient buildings achieve substantial reductions in operational energy consumption, leading to lower utility costs and a smaller carbon footprint.

Renewable energy integration is another critical factor in the electrical systems of energy-efficient buildings. By incorporating on-site power generation technologies such as solar photovoltaic (PV) panels or wind turbines, buildings can reduce their dependence on conventional power grids (Singh et al. 2021). In addition, building-integrated photovoltaics (BIPV) are gaining popularity as they allow solar panels to be seamlessly integrated into the architecture of buildings, such as in facades or rooftops. BIPV systems not only generate clean energy but also contribute to the aesthetics and functionality of modern architecture by replacing traditional building materials. Wind turbines for residential use are less common than solar panels, but they are being designed to be smaller and more efficient, making them suitable for buildings in areas with enough wind. Incorporating renewable energy also provides financial benefits, as surplus energy generated can be sold back to the grid in regions with net metering policies. This not only enhances the energy independence but also serves as a potential revenue stream. Additionally, advancements in energy storage solutions such as lithium-ion batteries, enable buildings to store excess renewable energy, ensuring a steady power supply even when solar or wind resources are unavailable. Recent advances in energy storage technology like flow batteries and solid-state batteries are improving the reliability and capacity of on-site storage systems, leading to greater energy independence.

In combination with energy storage systems, these renewable sources can supply clean, reliable electricity, further enhancing the energy of the building to be independence and sustainable. Some energy-efficient buildings employ microgrids, a localized energy systems that operate independently or in conjunction with the main power grid. Microgrids enhance resilience by ensuring a consistent energy supply during power outages or natural disasters. Moreover, microgrids can integrate multiple forms of renewable energy, such as solar, wind, and even bioenergy, allowing for greater flexibility and stability in energy generation and consumption. Microgrids also allow for the decentralization of energy, reducing the risk of large-scale power failures and enhancing the security of energy supply in critical facilities such as hospitals and data centres. This approach aligns with global initiatives to shift toward cleaner energy and reduce greenhouse gas emissions, making energy-efficient buildings a cornerstone of sustainable urban development. Energy-efficient buildings contribute to the economic growth of green industries and drive innovation in building design and construction. They also promote healthier indoor environments by improving air quality and reducing exposure to pollutants through better ventilation systems. These innovations not only support climate goals but also create healthier living and working environments, as energy-efficient buildings often incorporate better air quality control systems and reduce heat island effects in urban areas.

3.2 Solar Cells

A solar cell is a device engineered to convert sunlight into electrical energy through a process known as the photovoltaic effect. It can also function under artificial light sources, such as incandescent bulbs though its efficiency is optimized for sunlight. The fundamental role of a solar cell is to generate charge carriers, which are electrons and holes, and guide them towards conductive contacts to produce electrical current (Al-Ezzi and Ansari 2022). This process occurs in most solar cells, which are primarily made from a silicon-based p–n junction. When photons from sunlight strike the semiconductor material, they energize electrons, creating electron–hole pairs. The electric field present at the junction then separates these charges, directing electrons to the n-region and holes to the p-region (Tang et al. 2023). This movement creates a flow of electricity when the cell is connected to an external circuit, allowing the energy to be harnessed for various applications.

The efficiency of a single solar cell is limited by its surface area and its ability to capture and convert sunlight into usable electricity. Currently, the efficiency of commercial solar cells ranges from about 15–22%, though more advanced designs in research laboratories have achieved efficiencies over 40% (Benda and Černá 2020). The efficiency of a cell is also affected by factors like temperature, shading, and dust accumulation which can reduce its overall performance. To address this limitation, individual cells are grouped together to form modules, which are then connected to create panels and larger arrays capable of producing substantial amounts of power. By linking multiple solar cells into larger modules, the output can be scaled up to meet the electricity demands of different applications, from small residential systems to large-scale solar farms. Residential systems typically have capacities between 3 and 10 kilowatts (kW), whereas utility-scale solar farms can generate hundreds of megawatts (MW). This assembly technique increases the overall energy output and is commonly used in residential, commercial, and industrial settings to provide renewable energy solutions.

Solar panels have long been the exclusive power source for satellites and other space-based technologies due to their reliability and ability to operate without direct fuel input. Satellites use solar panels to power their instruments and communication systems, relying on sunlight for long-term missions in orbit. Solar cells in space are more efficient because they are not affected by atmospheric interference or weather conditions, allowing for continuous and uninterrupted energy generation. However, despite their success in space, the widespread adoption of solar panels on Earth has been hindered by the relatively high production and installation costs, limiting their use to specialized or high-value applications. The production of traditional silicon solar cells requires high-purity silicon, which is energy-intensive to refine. The cost of manufacturing, combined with installation and maintenance expenses, has traditionally made solar power less accessible for widespread use.

Recent advancements in photovoltaic technology have been pivotal in reducing the cost of solar power, bringing it closer to parity with traditional energy sources. The development of thin-film solar cells has helped reduce costs by using less material than traditional silicon wafers. Thin-film technology also allows for flexible and lightweight panels that can be applied in a wider range of settings, including curved surfaces and portable devices. Innovations in materials such as thin-film solar cells and multi-junction designs as shown in Fig. 3.1, are improving both efficiency and affordability. Multi-junction solar cells use multiple layers of semiconductors, each optimized to capture different parts of the solar spectrum. This allows them to convert a greater portion of sunlight into electricity compared to conventional single-junction cells. These cells are particularly useful in concentrated solar power (CSP) systems, where mirrors or lenses focus sunlight onto the cells, increasing their efficiency even further.

Furthermore, the integration of solar energy systems with energy storage solutions, like batteries, enhances their practicality for consistent energy supply. Energy storage is critical for overcoming one of the key limitations of solar power. Since solar panels

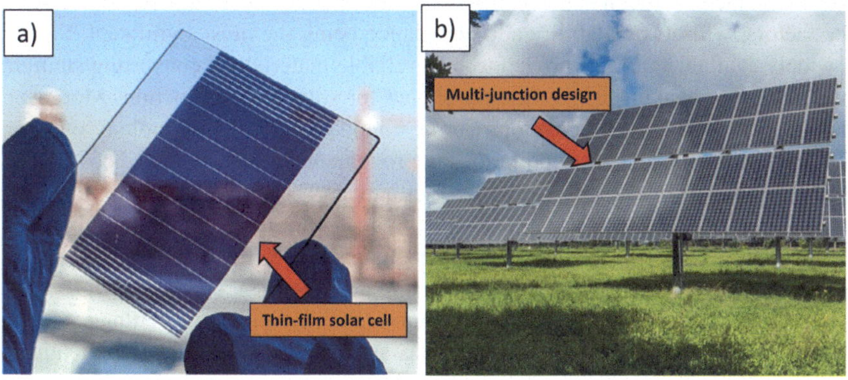

Fig. 3.1 **a** Photograph of thin-film solar cell (Machín and Márquez 2024) and **b** multi-junction design solar cell

only generate electricity during the day, storage systems such as lithium-ion batteries allow excess energy produced during peak sunlight hours to be stored for use at night or during cloudy periods. As global energy demands rise and environmental concerns grow, solar technology is increasingly seen as a key component in the transition to a sustainable and renewable energy future. In addition to lithium-ion batteries, newer storage technologies such as flow batteries and solid-state batteries are emerging as alternatives that offer longer lifespans, greater storage capacity, and faster charging times. This makes them ideal for larger solar installations, where consistent energy storage is essential.

Governments, industries, and research institutions are now investing heavily in solar infrastructure, positioning it as a competitive alternative to conventional fossil fuels. Many countries are offering incentives such as tax rebates, feed-in tariffs, and renewable energy credits to encourage the adoption of solar technology. These financial supports have made solar power increasingly attractive for homeowners, businesses, and utilities alike. Additionally, solar energy is now being integrated into national grids in ways that stabilize energy prices and reduce reliance on fossil fuels. Solar power is becoming an essential element of global energy strategies, especially as efforts to reduce carbon emissions intensify.

3.2.1 Classification of Solar Cells

Over 95% of solar cells produced annually rely on crystalline silicon wafers (Fraunhofer Institute for Solar Energy Systems 2019). Crystalline silicon has maintained its dominance in the photovoltaic market due to its favourable semiconductor properties, which include a high energy conversion efficiency and long-term stability. Silicon well-established supply chain, infrastructure, and widespread availability contribute to its position as the most reliable and scalable material for solar cell production. The dominance of silicon in the photovoltaic (PV) market can be attributed to several key factors, with its availability and abundance being the most significant. Silicon is widely available and has proven to be a reliable material for converting sunlight into electricity, making it the primary choice for solar cell production. Moreover, the extensive research and development in silicon-based solar technology over the past few decades have led to continuous improvements in efficiency and cost reductions, further solidifying its position in the market. These solar cells are generally categorized into three primary types, depending on the form of silicon employed: monocrystalline, polycrystalline, and amorphous silicon. Each type offers unique advantages, depending on efficiency, cost, and design needs, making them suitable for different applications in residential, commercial, and industrial settings.

Each type of silicon-based solar cell has specific advantages and applications, depending on factors such as performance, cost, and design requirements. Monocrystalline solar cells are often used in residential and commercial settings where high efficiency and space-saving design are crucial. Polycrystalline cells, with their lower cost are widely adopted in large-scale solar farms and industrial applications. Amorphous silicon is favoured in portable solar devices, consumer electronics, and applications where flexibility and lightweight characteristics are essential. The wide range of silicon-based solar cells available today ensures that there is a suitable option for nearly every solar energy project, from small consumer electronics to massive utility-scale solar power plants.

(a) **Monocrystalline Silicon Solar Cells**

Monocrystalline solar cells are made from a single and pure crystal of silicon, which is produced using the Czochralski process (Hendawi 2021). This process involves melting high-purity silicon in a crucible and slowly pulling a single crystal seed from the molten silicon, creating a large cylindrical ingot. The uniform structure of this ingot is key to producing solar cells that are highly efficient, as the uninterrupted crystal lattice minimizes defects that could interfere with electron flow. Once the ingot is formed, it is sliced into thin wafers, which are then processed to form individual solar cells. This method produces high-quality wafers with minimal impurities, which is crucial for maximizing the efficiency of the cell.

As previously mentioned, monocrystalline cells are known for their high efficiency, typically ranging between 20 and 22% (Kumari et al. 2022). The high efficiency is due to the uniform crystal structure of the cell. This structure allows electrons to move more freely without being obstructed by grain boundaries or impurities, resulting in better energy conversion from sunlight. These cells also have a longer operational life, often performing well for 25 years or more with minimal degradation in efficiency. In fact, monocrystalline cells typically retain around 80% of their initial efficiency even after decades of use, making them one of the most reliable long-term investments in solar technology. A key advantage of monocrystalline solar cells is their performance under low-light conditions (Enaganti et al. 2020). These cells can generate electricity even when sunlight is weak such as during overcast days or early mornings, making them ideal for areas with varying sunlight availability. Their superior efficiency also means that fewer panels are needed to generate the same amount of electricity as other types of solar cells, which is particularly important in space-constrained installations. In urban environments where rooftop space is limited, the compact design of monocrystalline cells ensures that the available area is used efficiently allowing for maximum energy production.

However, the manufacturing process for monocrystalline cells is more complex and costly compared to other types of solar cells. The Czochralski process is energy-intensive, and the slicing of the cylindrical ingots into wafers results in material waste, as the round shape of the ingot does not fully align with the rectangular shape of most solar panels. This waste increases production costs and makes the process less efficient in terms of material usage. Despite these challenges, the high efficiency and durability of monocrystalline cells often justify their higher upfront cost. In the long term, their greater energy production and longevity can result in lower overall costs, especially in applications where maximizing power output is critical. Monocrystalline cells are also aesthetically appealing often featuring a sleek, uniform black appearance. This makes them a popular choice for residential installations where visual design is a consideration, especially in homes where the aesthetics of the solar panels are important to homeowners. Their design is optimized for energy conversion, with a detailed internal structure that ensures maximum capture of sunlight. The architecture of these cells is designed to minimize internal resistance and heat build-up, further enhancing their performance in both hot and cool climates.

Additionally, monocrystalline cells are more resistant to high temperatures than other types of solar cells. While all solar cells lose efficiency as temperatures rise, monocrystalline cells tend to perform better than polycrystalline or amorphous cells in hot climates (Sun et al. 2022). This temperature resilience makes them particularly suitable for regions with high solar irradiance and hot temperatures, where maintaining efficiency is crucial. These factors make monocrystalline solar cells an excellent choice for various applications, including residential rooftop systems, commercial installations, and off-grid solar setups. Their high efficiency and smaller footprint make them ideal for installations where space is limited, while their longevity and performance in low-light and high-temperature conditions make them suitable for a wide range of climates and environments. Although they come with a higher initial cost, their long-term energy production, reduced maintenance requirements, and higher energy output per square meter make them a cost-effective solution over the lifespan of the system.

(b) **Polycrystalline Silicon Solar Cells**

Polycrystalline solar cells also known as multicrystalline cells are made from silicon that has been melted and poured into moulds, allowing multiple silicon crystals to form. This process is simpler and faster than the Czochralski method used for monocrystalline cells. It involves cooling the molten silicon in a mould, which causes the formation of multiple crystal structures within the cell. As a result, polycrystalline cells are less expensive to produce because the manufacturing process is less resource-intensive, requiring less precision and time compared to monocrystalline production (Becker et al. 2013). The simplified production process not only reduces costs but also results in less waste, making polycrystalline cells a more affordable option for large-scale solar installations. Despite this, polycrystalline cells remain a popular choice for large-scale installations such as large-scale solar farms, industrial installations, and utility projects where vast areas of land are available due to their lower cost and sufficient performance levels, which typically range from 13 to 16% in efficiency (Benghanem et al. 2023). The main reason for this reduced efficiency is the presence of grain boundaries, where the individual crystals meet within the cell. These boundaries create disruptions in the electron flow, which causes internal resistance and reduces the overall performance of the cell. As a result, electrons cannot move as freely through the polycrystalline structure, leading to lower energy conversion rates compared to monocrystalline cells (Zeina and Almaz 2023).

In addition to their cost-effectiveness, polycrystalline cells have a distinctive appearance. Compared to monocrystalline cells which are typically black, polycrystalline cells are characterized by a blue hue, caused by the way light interacts with the silicon crystals. This blue coloration is one of the most recognizable features of polycrystalline solar cells, making them easy to distinguish from other types of cells. The

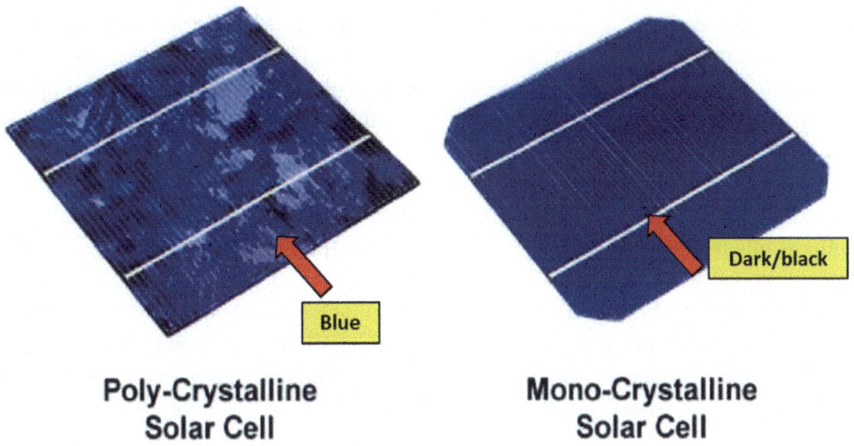

Fig. 3.2 Visual comparison of polycrystalline and monocrystalline silicon solar cells (Zeina and Almaz 2023)

variation in the crystal structure gives the cells a fragmented, reflective look, adding to their visual distinctiveness. This unique appearance is shown in Fig. 3.2, which illustrates the visual differences between polycrystalline and monocrystalline silicon solar cells. The latest laboratory-record efficiencies for monocrystalline and multicrystalline silicon solar cells have reached 26.7 and 24.4%, respectively (Sabatino et al. 2024). Although polycrystalline cells do not perform as well as monocrystalline cells in terms of efficiency, they remain a popular choice due to their lower production cost and widespread availability. They are particularly useful in applications where space is not a limiting factor, such as ground-mounted solar farms in rural areas or regions with abundant sunlight. Moreover, advances in technology continue to improve the performance of polycrystalline cells, narrowing the efficiency gap between them and monocrystalline options.

(c) **Amorphous Silicon Solar Cells**

Amorphous silicon (a-Si) solar cells are a type of thin-film solar technology that does not rely on a crystalline structure. Unlike monocrystalline, amorphous silicon consists of silicon atoms arranged in a random, non-crystalline form. This disordered structure allows the cells to be more flexible and lightweight, making them suitable for a variety of applications where traditional crystalline cells would not be practical. A thin layer of non-crystalline silicon is deposited onto a substrate such as glass, metal, or plastic, allowing for greater flexibility in applications (Dambhare et al. 2021). The substrates used can vary depending on the specific application, offering versatility in design and installation, which is a key advantage of amorphous silicon cells. These cells benefit from a simpler and less resource-intensive manufacturing process. Amorphous silicon can be deposited at lower temperatures compared to crystalline silicon, and the thin-film layers are applied in a vacuum chamber, which reduces production costs and energy consumption. In the process, hydrogen atoms are introduced to bond with the dangling silicon atoms in the material. This hydrogenation process improves electron movement within the disordered structure by reducing defects that would otherwise impede electron flow, thus enhancing the ability of the cell to generate electricity despite the lack of a regular atomic structure (Liu et al. 2022).

While amorphous silicon solar cells have efficiencies typically ranging from 6 percent to 10%, making them less efficient than crystalline-based cells, they absorb sunlight much more efficiently (Shah 2020). Amorphous silicon has a higher light absorption coefficient, which allows it to capture a broader range of the solar spectrum more effectively than crystalline silicon. In fact, amorphous silicon can absorb sunlight 40 times more efficiently than crystalline silicon. It means that even with a thinner layer of material, it can still capture a significant amount of solar energy. A film as thin as one micron can capture 90% of usable solar energy, making amorphous silicon cells ideal for applications with low power demands such as consumer electronics and portable devices. Their ability to function with such thin layers also makes them useful in innovative applications, including building-integrated photovoltaic (BIPV), where solar cells are embedded into windows, walls, or roofs without adding significant weight or altering the appearance of the structure.

In addition to being cheaper to produce, amorphous silicon cells perform better in low-light conditions compared to their crystalline counterparts. Their high absorption rate allows them to generate electricity in cloudy weather or shaded areas, making them suitable for environments where sunlight is not consistently strong. These cells are also less affected by high temperatures, which can degrade the performance of crystalline silicon cells. Amorphous silicon cells maintain relatively stable performance in hot climates, as their thin-film structure dissipates heat more efficiently, preventing excessive energy loss due to temperature increases. However, the lower efficiency of amorphous silicon cells requires a larger surface area to generate the same amount of power as crystalline cells, which limits their use in some larger-scale applications. For instance, to generate the same amount of electricity, an amorphous silicon array would need to cover more space compared to a monocrystalline array. This requirement makes amorphous silicon less suitable for installations where space is limited, such as urban rooftops or areas where maximizing power output per square meter is crucial.

Despite this limitation, the ability to produce a-Si cells at lower temperatures and deposit them on inexpensive substrates makes them a leading choice for thin-film photovoltaic applications. They are especially well-suited for niche markets, including portable solar chargers, flexible solar panels for backpacks and tents, and low-power devices like calculators and remote controls. Amorphous silicon unique characteristics continue to make it a viable option in the growing market for flexible and low-cost solar technologies. Ongoing advancements in thin-film technology and hybrid approaches, such as tandem cells that combine amorphous silicon with other materials are helping to improve efficiency and expand the range of potential applications for a-Si cells.

(d) **Polycrystalline-Compound Cell**

Polycrystalline-compound cells are recognized as part of the second generation of photovoltaic technology. These cells use thin-film layers of semiconductors such as gallium arsenide, cadmium telluride, and copper indium selenide to convert sunlight into electricity (Rahman 2022). While these cells generally exhibit lower efficiencies compared to traditional silicon-based solar cells, they offer significant reductions in manufacturing costs. The use of thin-film technology allows for less material to be used during production, which reduces the overall cost of manufacturing. This translates to a lower cost per watt of electricity produced, making polycrystalline-compound cells more affordable for large-scale solar installations (Efaz et al. 2021). One of the main advantages of polycrystalline-compound cells is their lightweight design. These cells are thinner and lighter than conventional silicon cells, which means they require less structural support for installations. This is especially beneficial for rooftop installations and other projects where reducing weight is a priority. Their lightweight nature also makes them suitable for use on flexible and lightweight materials, including textiles and other innovative applications in building-integrated photovoltaics (BIPV). In BIPV systems, these cells can be integrated directly into the structure of buildings, such as in windows, facades, or roofs, allowing for seamless solar energy generation without compromising the design of the building.

However, one challenge associated with polycrystalline-compound cells is the inclusion of heavy metals such as gallium, cadmium, and copper, which raise potential environmental and health concerns regarding safe disposal and manufacturing processes (Efaz et al. 2021). These materials can be harmful if not properly managed during production or at the end of the cell lifecycle, posing risks for both the environment and human health. The presence of heavy metals requires careful handling and disposal to minimize the potential for contamination. Despite these concerns, the materials used in polycrystalline-compound cells are generally stable and insoluble in water, which reduces some of the risks associated with their environmental impact. While there are still challenges related to the use of heavy metals, the stability of these compounds helps mitigate some of the dangers associated with water contamination or leaching into soil. As research continues to focus on improving both the performance and environmental safety of these cells, polycrystalline-compound technology remains a promising option for cost-effective solar energy generation. Advancements in recycling technologies and safer production methods are likely to address the environmental concerns in the future, making these cells an increasingly viable choice for renewable energy solutions.

(e) **Polymer Solar Cell**

Polymer solar cells also known as organic or transparent solar cells are part of the latest generation of photovoltaic modules and differ significantly from traditional silicon-based cells. Unlike conventional cells that rely on a rigid silicon-based p–n junction to separate photogenerated charge carriers, polymer solar cells use organic materials such as carbon-based polymers to convert sunlight into electricity (Jin et al. 2021). These materials are flexible and lightweight, which gives polymer solar cells unique properties that set them apart from traditional technologies. One of the key advantages of polymer solar cells is their versatility. They are lightweight which makes them easy to integrate into a variety of surfaces, including curved or flexible materials. This flexibility opens up new possibilities for solar applications in areas like portable electronics, wearable technology, and even fabrics. In addition, polymer solar cells are semi-transparent, which allows them to be used in innovative ways, such as in windows or glass facades, where they can generate electricity while still allowing light to pass through (Fan et al. 2020). This transparency makes them particularly attractive for building-integrated photovoltaics (BIPV), where aesthetics and functionality can be combined to create energy-generating architectural elements.

Polymer solar cells are also inexpensive to manufacture. The organic materials used in these cells can be processed at lower temperatures compared to silicon, and the production methods such as roll-to-roll printing, are cost-effective and scalable. This low-cost production capability makes polymer solar cells a promising option for large-scale solar energy deployment, especially in applications where affordability is a priority. Additionally, these cells are disposable, offering environmental benefits over traditional silicon-based cells, which can be more challenging to recycle (Sreekanth et al. 2023). Another important advantage of polymer solar cells is their potential for reduced environmental impact. Because they are made from organic materials, the environmental footprint of their production and disposal is smaller compared to traditional silicon cells. Their ability to be designed at the molecular level allows researchers to develop materials that are less harmful to the environment, and ongoing efforts are focused on making polymer solar cells fully biodegradable.

Despite these benefits, polymer solar cells face several challenges. One of the most significant limitations is their lower energy output. Currently, the efficiency of polymer solar cells is about one-quarter that of conventional silicon-based cells, which means that they generate less electricity for the same surface area. This lower efficiency limits their use in high-demand applications, where maximizing energy production is critical. Another challenge is their tendency to degrade over time due to exposure to environmental conditions such as UV light, moisture, and oxygen. This degradation leads to reduced efficiency and a shorter lifespan compared to more durable silicon-based cells. Ongoing research is focused on overcoming these issues. Scientists are working on developing new materials and protective coatings that can improve the stability and longevity of polymer solar cells. By enhancing their resistance to environmental stressors, these cells could become more competitive with traditional technologies in terms of both efficiency and durability. While polymer solar cells represent a promising and evolving technology, their efficiency and stability remain areas of active research. Efforts are being made to improve their performance, making them a viable alternative for low-cost, flexible, and sustainable solar energy solutions in the future.

3.2.2 *Solar Cell Application in Buildings*

Photovoltaic (PV) panels are increasingly being integrated directly into building structures, a concept known as Building Integrated Photovoltaics (BIPV). BIPV refers to the incorporation of solar technologies within the architectural design of a building, allowing the structure itself to act as a generator of renewable energy. Unlike traditional solar panel installations which are mounted on existing surfaces, BIPV systems are embedded into the building materials, making the solar panels part of the design from the outset (Kuhn et al. 2021). This approach has gained popularity as a key strategy in promoting sustainable architecture, enabling buildings to be more energy-efficient and environmentally friendly. BIPV systems can be incorporated into various parts of the building such as facades, roofs, or windows. For example, solar panels can be used as roofing materials, replacing conventional roof tiles, or they can be integrated into glass facades and windows, where semi-transparent solar cells allow light to enter the building while still generating electricity (Jiang et al. 2024). This seamless integration not only preserves the aesthetic appeal of the building but also enhances its functionality by producing clean, renewable energy on-site. In addition to their visual appeal, BIPV systems are often designed to blend with the architectural style of the building, offering flexibility in design and material choice.

One of the primary benefits of BIPV is its ability to reduce reliance on traditional energy sources. By generating electricity directly from the building structure, BIPV systems decrease the dependence on external power grids and fossil fuels. This helps to lower the carbon footprint of the building and contributes to energy self-sufficiency. The electricity generated by BIPV can be used to power various functions within the building, including lighting, heating, cooling, and other electrical systems. In some cases, excess energy produced during peak sunlight hours can be stored in batteries or sold back to the grid, providing additional financial benefits. BIPV systems also help offset operational costs by providing renewable energy directly to the building. While the initial investment in BIPV technology may be higher than traditional building materials, the long-term savings on energy bills and the potential revenue from selling excess energy can make BIPV a cost-effective solution over time (Esfahani et al. 2021). Additionally, many countries offer incentives such as tax credits or subsidies, for buildings that incorporate renewable energy systems like BIPV, further enhancing the economic viability of these systems. The durability and low maintenance requirements of solar panels also contribute to reducing long-term operational expenses, making BIPV an attractive option for both residential and commercial buildings. As interest in sustainable building practices continues to grow, BIPV is expected to play an increasingly important role in the design and construction of energy-efficient, eco-friendly buildings.

(a) **Thin Film Solar Cell Applications in Buildings**

Thin-film solar cells have garnered considerable attention in recent years due to their affordability, lightweight design, and flexibility. These characteristics make them highly suitable for integration into various parts of buildings, including rooftops, facades, and even windows. Their adaptability to different surfaces allows for easy installation on curved or irregular structures, making them an ideal choice for modern architecture (Li et al. 2020). Unlike traditional rigid silicon-based panels, thin-film solar cells can conform to unique building shapes providing greater design flexibility for architects and builders. This versatility enables seamless incorporation into diverse architectural styles, from sleek urban skyscrapers to more organically shaped structures. Thin-film solar cells are also known for their lower material consumption and ease of production, contributing to reduced overall costs compared to traditional silicon-based solar cells. Since thin-film cells require significantly less semiconductor material to produce energy, the manufacturing process is faster and more economical. This results in lower production costs, making thin-film solar cells a more economically viable option for large-scale deployment. These cost advantages, combined with the light weight, make thin-film technology particularly appealing for retrofitting older buildings or adding solar capacity without needing extensive structural reinforcements.

Recent innovations in transparent solar cell technology have opened new possibilities for integrating photovoltaic into buildings. Transparent thin-film solar cells allow sunlight to pass through while still generating electricity, making them perfect for use in windows and glass facades (Kim et al. 2020). This capability not only contributes to energy generation but also maintains the building aesthetic appeal and access to natural light. Transparent solar cells are particularly advantageous for creating multilayer and tandem configurations that combine both energy generation and architectural aesthetics. These tandem systems stack multiple thin-film layers with different absorption properties, enhancing overall efficiency by capturing more of the solar spectrum. In building applications, transparent solar modules can serve as part of the exterior shell providing clean energy while maintaining an unobstructed view. This integration not only enhances the aesthetic appeal of modern buildings but also promotes energy efficiency by utilizing otherwise unused surfaces. Such innovative uses of thin-film technology help transform previously untapped areas into power-generating surfaces, contributing to the overall energy efficiency of the building.

The increased visibility and integration of photovoltaic technology in public buildings are expected to encourage broader adoption and strengthen public confidence in solar energy systems. As more high-profile projects demonstrate the benefits of incorporating solar technology into building design, public awareness of its potential is likely to increase, driving greater acceptance and widespread use. In addition to urban and commercial building applications, thin-film flexible PV sheets are highly effective for powering solar water pumps in remote regions. These flexible panels can be rolled up and transported easily, making them particularly useful in areas without access to a stable electricity grid. Thin-film technology is especially beneficial in arid areas where farmers require water for drinking and irrigation (Zaky et al. 2020). In such regions, solar-powered water pumps using thin-film PV technology can provide a reliable, sustainable solution. The high efficiency of flexible PV sheets allows them to generate sufficient energy during short operational periods to meet daily water demands. Moreover, the portability of roll-up PV sheets enables farmers to transport the system easily to different locations, ensuring continuous use. This mobility represents a practical solution for addressing water access issues in remote and off-grid areas, offering significant potential for improving agricultural productivity and sustainability. Overall, the versatility, affordability, and lightweight nature of thin-film solar cells make them an increasingly popular choice for both building-integrated applications and off-grid solutions. Their ability to function on unconventional surfaces and in challenging environments highlights their importance in promoting sustainable energy use across different sectors, from urban construction to agriculture.

(b) **Photovoltaic Thermal (PV/T) collectors**

The efficiency of solar cells is significantly impacted by their operating temperature. As temperatures increase, the electrical output of the solar cells decreases, leading to a reduction in overall efficiency. This phenomenon occurs because the Maximum Power Point (MPP), which represents the optimal operating point for maximum energy generation, shifts to the left as the temperature rises. As a result, the solar cells are unable to maintain peak efficiency, causing a drop in their energy conversion capability. The reason for this decrease in performance is that elevated temperatures increase the recombination of charge carriers (electrons and holes) within the solar cell, leading to energy losses through heat dissipation (Herrando et al. 2023). This recombination reduces the number of electrons available to generate electricity, thus lowering the output of the system. In some BIPV systems, the primary objective of incorporating cooling methods is to lower the solar cell temperature and improve electrical efficiency. However, more advanced systems have been developed that not only improve efficiency but also harness the waste heat generated by the solar cells. These systems, known as photovoltaic thermal (PV/T) collectors, are designed to generate both electricity and usable heat (Prasetyo et al. 2023). By capturing the excess heat produced during the operation of solar cells, PV/T collectors provide an additional energy source for purposes such as space heating, water heating, or even industrial processes. This dual functionality allows buildings to maximize the energy yield from their solar installations, improving overall energy efficiency.

In certain configurations, PV/T collectors are enhanced using Compound Parabolic Concentrators (CPC) which are designed to concentrate solar radiation onto the PV panels. These concentrators capture more sunlight and focus it onto the solar cells, increasing the intensity of the light and improving both electrical output and heat production (Masood et al. 2022). By concentrating sunlight, CPC enable PV/T systems to generate more energy from the same amount of sunlight, making them highly efficient in climates where both solar power and heating are needed, as shown in Fig. 3.3. This makes PV/T collectors particularly valuable in colder climates, where space heating or water heating is a significant energy requirement, or in regions where combined solar power and heating systems are advantageous. The ability of PV/T collectors to utilize waste heat improves the overall efficiency of the system. Instead of letting the heat generated by the solar cells go to waste, PV/T systems capture and repurpose it, ensuring that the system extracts maximum value from the available solar energy. This combination of electrical and thermal energy generation makes PV/T collectors an attractive option for sustainable energy management in both residential and commercial applications. In residential buildings, PV/T systems can provide hot water for household use. Meanwhile in commercial or industrial settings, they can contribute to both electricity generation and large-scale heating needs such as space heating or process heating in factories. PV/T collectors represent a highly efficient solution for integrating renewable energy into buildings. Their ability to combine electricity and heat production in one system helps reduce reliance on traditional energy sources, lowers operational costs, and supports the shift towards greener energy solutions. As interest in sustainable building design grows, PV/T technology is expected to play an increasingly important role in maximizing energy efficiency and reducing carbon emissions across various sectors.

(c) **Solar Tracker**

Solar irradiation plays a critical role in determining the Maximum Power Point (MPP) of a solar cell. The amount of sunlight that reaches the surface of a photovoltaic (PV) panel directly affects the ability of the system to generate electricity, and optimizing

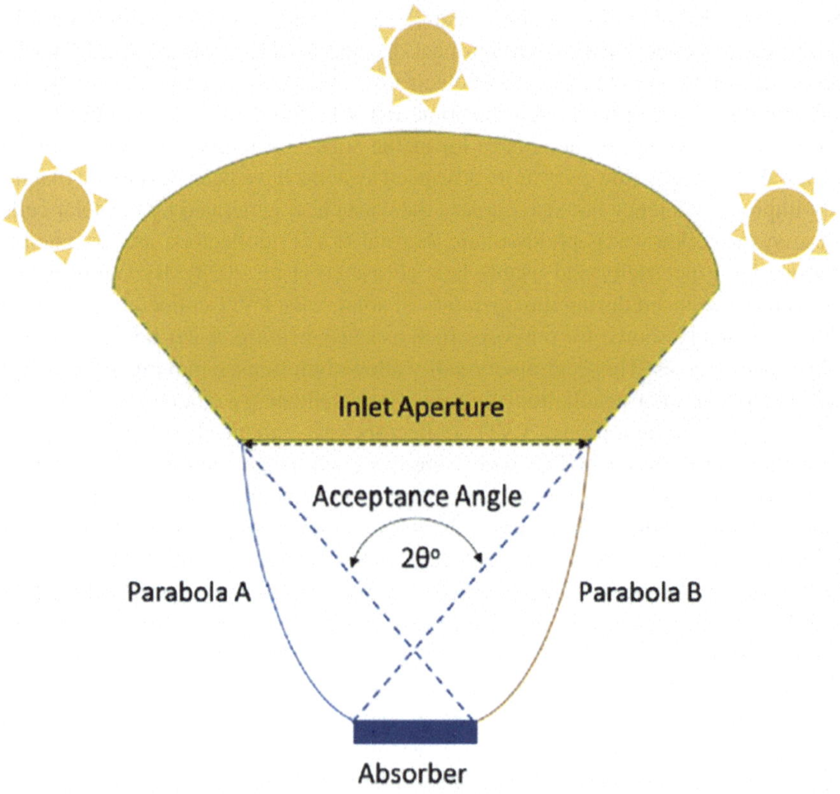

Fig. 3.3 Schematic diagram of a compound parabolic concentrators (Baig et al. 2018)

the panel orientation can significantly increase energy production. To maximize the amount of solar energy captured by photovoltaic cells, Compound Parabolic Concentrators (CPC) are often used. Alongside CPC, solar tracking systems are a well-established method for enhancing solar exposure. Solar trackers adjust the orientation of the solar panels to follow the movement of the Sun throughout the day. By continuously shifting the angle of the panels, solar trackers ensure that the PV system is always aligned to capture the maximum amount of sunlight. This dynamic adjustment improves the overall efficiency of the system by allowing the panels to receive optimal solar radiation at all times, even as the Sun moves across the sky (Kumar et al. 2021). Solar trackers are particularly useful in reducing the impact of shadowing and other environmental factors that can affect the performance of photovoltaic systems. For example, trackers can minimize the shading caused by nearby objects, ensuring that the panels maintain consistent exposure to sunlight. This helps to mitigate performance drops caused by partial shading, which can significantly reduce the output of a solar array.

By continuously optimizing the angle of the panels, trackers ensure optimal exposure to solar radiation, thereby increasing energy generation. Studies have shown that solar trackers can improve energy production by 10 to 25% compared to fixed solar installations, depending on the location and type of tracker used (Amelia 2020). While solar trackers offer considerable benefits in improving performance, they are not always implemented in standard installations because they can increase the initial setup costs and consume part of the electricity generated by the system. The motors and sensors that drive the tracking system require a portion of the energy produced by the solar panels, slightly reducing the net gain in energy. Additionally, the mechanical components add to the overall cost of installation and maintenance, which can be a deterrent for smaller-scale projects or areas with lower solar irradiance. However, in scenarios where maximizing output is essential such as in large-scale solar farms or when using expensive monocrystalline silicon cells the added investment in a solar tracker can be justified by the enhanced energy production (Mourad et al. 2022). For solar farms spread over vast areas, even a small percentage increase in efficiency can result in significant additional energy production, making solar trackers a valuable asset. Similarly, for installations using high-cost, high-efficiency monocrystalline panels, the extra energy generated by a tracker can lead to faster returns on investment. Moreover, solar tracking technology has evolved with the introduction of advanced dual-axis systems, which can further increase efficiency by adjusting the panels not only along a horizontal plane but also vertically. This dual-axis tracking ensures that the solar panels remain optimally oriented, even as the angle of the Sun changes throughout the year.

This ability to adjust to seasonal changes is a significant advantage in regions with high variability in solar angles. As the position of the Sun shifts throughout the seasons, a fixed solar panel setup may only be optimized for certain times of the year. A dual-axis tracker compensates for this by continuously adjusting to the optimal angle for energy generation, improving performance during both summer and winter months (Vargas et al. 2022). As a result, solar trackers are increasingly seen as a practical solution for maximizing energy yield, especially in high-irradiation regions where the benefits of additional solar exposure outweigh the costs. In areas that receive intense sunlight for most of the year, the increased energy production from a tracker-equipped solar array can significantly offset the higher installation and operational costs. An economic analysis is essential before implementing solar trackers to determine whether the potential increase in performance justifies the higher initial investment and operational energy consumption. Factors such as local solar irradiance, the size of the installation, and the type of tracker (single-axis or dual-axis) all play a role in determining the cost-effectiveness of solar tracking systems. While trackers can provide substantial benefits in some situations, careful consideration of the specifics project is necessary to ensure that the advantages outweigh the additional costs.

(d) **Solar Ventilation Preheating System**

Solar energy plays a critical role in reducing the heating demands of a building and one way to optimize this energy use is through solar ventilation preheating systems. These systems capture solar energy to preheat incoming ventilation air before it enters the HVAC system of the building, reducing the need for conventional heating sources (Huang et al. 2023). The efficiency of the ventilation preheating system is largely dependent on the amount of sunlight captured by the solar air collector. To optimize energy capture, solar air collectors are installed on the exterior of buildings, typically on sun-facing walls, where they can absorb the maximum amount of solar radiation throughout the day. In a typical system, solar radiation heats the air inside a collector, which is then drawn into the building ventilation system. This preheated air reduces the amount of energy needed to maintain comfortable indoor temperatures, improving overall efficiency (Guan et al. 2023). The process helps mitigate the need for conventional heating systems during cooler months, lowering energy consumption and reducing greenhouse gas emissions. Solar ventilation preheating systems are particularly effective in cold climates, where the heating demand is high, and maximizing solar energy use can significantly reduce operational costs.

By continuously harnessing solar energy throughout the day, solar ventilation preheating systems ensure that the air entering the building is always warmed to a certain degree, even during cloudy or low-light conditions. These systems are designed to optimize the capture of solar radiation while minimizing heat loss, ensuring that the heating needs are met more efficiently. Additionally, solar preheating systems can be paired with heat recovery systems, further improving energy savings by capturing heat from exhaust air and using it to warm incoming air. Studies have shown that solar ventilation preheating systems can reduce heating energy requirements by 20–50% depending on the building location, design, and the type of system installed (Guan et al. 2023). However, while solar preheating systems offer considerable benefits in improving energy efficiency, they are not always implemented in standard building designs due to higher initial setup costs. The solar collectors and installation requirements can add to the upfront investment, and the effectiveness of the system is depends heavily on the local climate and the building orientation.

However, in scenarios where reducing heating energy consumption is essential, such as in industrial buildings, greenhouses, or warehouses with high ventilation needs, the investment in a solar ventilation preheating system can be justified by the enhanced energy savings and improved indoor comfort (Gainza-Barrencua et al. 2021). In large buildings where the heating demand is significant, even a small reduction in energy consumption can lead to substantial cost savings over time. Furthermore, solar ventilation preheating systems are particularly beneficial in remote or off-grid areas where access to traditional energy sources is limited, making renewable energy a vital alternative. As interest in sustainable building design grows, solar ventilation preheating technology has evolved to offer more advanced solutions, including automated systems that adjust airflow and heating based on real-time temperature and solar irradiance data. These innovations ensure that buildings remain energy-efficient throughout the year, adjusting for seasonal variations in sunlight and temperature. As a result, solar ventilation preheating systems are increasingly seen as a practical solution for improving energy efficiency, especially in regions where reducing heating energy consumption is critical. An economic analysis is essential before implementing a solar ventilation preheating system to determine whether the potential increase in energy savings justifies the higher initial investment. Factors such as local solar irradiance, building size, and heating demand all play a role in determining the cost-effectiveness of these systems. While solar ventilation preheating can provide substantial benefits in energy savings and emission reductions, careful consideration of the specifics projects is necessary to ensure that the advantages outweigh the additional costs.

(e) **Solar-Powered Smart Windows**

Solar-powered smart windows represent a cutting-edge integration of photovoltaic technology and advanced window systems, designed to optimize both energy generation and energy efficiency in buildings. These smart windows incorporate transparent or semi-transparent solar cells that generate electricity while allowing natural light to pass through. It offering a dual function of solar energy production and dynamic control over light and heat transmission. The ability to generate electricity from sunlight without compromising visibility makes solar-powered smart windows a valuable addition to modern building designs, particularly for high-rise buildings and commercial properties where glass facades dominate the architecture. Solar-powered smart windows do more than just generate electricity. They are equipped with electrochromic or thermochromic coatings that adjust the transparency of the glass based on external conditions, such as sunlight intensity or outdoor temperatures (Jia et al. 2021). This automatic adjustment helps regulate the amount of heat and light entering the building, reducing the need for artificial lighting and mechanical heating or cooling systems. The combination of photovoltaic energy generation and dynamic shading control ensures that buildings can maximize energy efficiency while maintaining indoor comfort.

By capturing sunlight to power their own tinting functions, solar-powered smart windows eliminate the need for external power sources, making them self-sustaining and further enhancing their appeal as an energy-efficient solution. Studies have shown that the use of smart windows can reduce a heating and cooling energy consumption of building by up to 30% (Zhou et al. 2020). This also can be done while also cutting down on glare and UV exposure, which helps protect interior furnishings and occupants. The ability of these windows to intelligently respond to changing environmental conditions offers significant energy savings, particularly in climates with large temperature variations. Solar-powered smart windows are particularly useful in buildings where large glass surfaces are prevalent, such as office buildings, commercial complexes, and residential skyscrapers. In such structures, the windows not only provide aesthetic value and natural lighting but also contribute to energy generation, offering a multifunctional solution for green building design. These windows can be installed as part of an initial design of building or retrofitted onto existing structures, providing flexibility for both new construction and renovation projects.

While solar-powered smart windows offer considerable advantages, they are not without challenges. The initial installation costs of these systems tend to be higher than traditional windows due to the integration of photovoltaic cells and advanced glazing technologies. However, the long-term energy savings and reduced reliance on heating, cooling, and artificial lighting can offset the higher upfront costs, especially in regions with high solar irradiance. Additionally, the ability to generate renewable energy on-site contributes to a building sustainability profile, making smart windows an attractive option for achieving green building certifications and complying with energy-efficiency regulations. Moreover, recent advances in solar cell technology have further improved the performance of solar-powered smart windows. Innovations in materials such as perovskite solar cells and organic photovoltaics have increased the efficiency of these windows (Kumar and Naidu 2021). These advancements allow them to generate electricity more effectively while maintaining their transparency, ensuring visual clarity is not sacrificed. These advancements, along with automated control systems that adjust window tint based on real-time data, are expected to drive the adoption of solar-powered smart windows in both residential and commercial settings.

As the demand for sustainable building technologies grows, solar-powered smart windows are increasingly seen as a key solution for reducing energy consumption in urban environments. Their ability to provide both passive and active energy savings through dynamic shading and solar energy generation makes them particularly valuable in high-density urban areas where buildings face intense sunlight exposure. In regions with fluctuating climates, these windows help maintain optimal indoor temperatures throughout the year, reducing the strain on HVAC systems and enhancing occupant comfort. Before implementing solar-powered smart windows, an economic analysis is crucial to assess whether the potential energy savings justify the initial investment. Factors such as local sunlight availability, building size, and energy demand should be carefully considered to determine the cost-effectiveness of the system. While the upfront costs can be higher compared to conventional windows, the long-term benefits in energy efficiency, comfort, and sustainability make solar-powered smart windows an increasingly practical and attractive option for green building projects.

(f) **Solar-Integrated Sound Barriers**

Solar-integrated sound barriers combine noise reduction with renewable energy generation, offering a dual-purpose solution in urban areas and along transportation corridors. Traditionally, sound barriers are installed to reduce noise pollution from highways, railways, and airports. However, by integrating solar panels into these barriers, they can also serve as sources of clean energy, contributing to the sustainability efforts of cities and infrastructure projects (Soares and Wang 2024). These barriers are designed to absorb or deflect sound while simultaneously capturing sunlight through photovoltaic (PV) cells embedded into the structure, providing a multifunctional approach to urban development. The primary function of sound barriers is to mitigate noise pollution, improving the quality of life for nearby residential and commercial areas. By incorporating solar panels into the design, these barriers can turn previously passive structures into active energy producers, making use of otherwise underutilized surfaces. Solar-integrated sound barriers can be installed along highways, railways, and airport runways, where exposure to sunlight is typically high due to the open nature of these locations. This continuous exposure makes these barriers ideal for solar energy generation, as they can capture sunlight throughout the day without the shadowing issues often found in urban environments.

Solar-integrated sound barriers are particularly valuable in densely populated areas, where both noise pollution and energy demand are significant challenges. By installing these systems, municipalities can address two critical issues at once: reducing harmful noise levels and generating renewable electricity to power streetlights, nearby buildings, or even feed electricity back into the grid. This integration maximizes the utility of sound barriers, turning what was once a single-purpose infrastructure into a tool for sustainable energy production. Studies have shown that solar-integrated sound barriers can generate a substantial amount of electricity depending on their length, orientation, and the solar irradiance in the region (Soares and Wang 2024). For example, a 1-km stretch of solar-integrated sound barriers installed along a highway could potentially generate hundreds of megawatt-hours (MWh) of electricity per year, enough to power homes or public infrastructure in the surrounding area (Wang et al. 2023). This makes them particularly attractive in urban settings, where both space and energy are at a premium, and where noise pollution from traffic or industrial activity needs to be addressed.

Despite the numerous benefits of solar-integrated sound barriers, there are some challenges to their widespread implementation. One of the primary concerns is the initial cost of installation as the addition of photovoltaic technology increases the price of construction compared to standard sound barriers. Additionally, the efficiency of solar panels installed vertically or at steep angles may be lower than those installed on rooftops or flat surfaces due to suboptimal solar exposure. However, advances in photovoltaic technology, such as bifacial solar panels that can capture sunlight from both sides have been shown to improve energy yield in vertical installations, making these barriers more efficient over time. Moreover, the use of solar-integrated sound barriers can provide long-term financial benefits. While the upfront investment may be higher, the energy generated by these systems can reduce operating costs for municipalities, providing a steady source of renewable energy that offsets the initial expenses. In regions with strong incentives for renewable energy, such as tax credits or subsidies, the financial case for solar-integrated sound barriers becomes even more compelling. Additionally, by reducing noise pollution, these barriers contribute to improved public health outcomes, which can further justify their adoption in areas with high traffic or industrial activity.

Solar-integrated sound barriers are an innovative solution for maximizing the utility of urban infrastructure. They are particularly suited for deployment along highways, train tracks, and airports where there is ample space for long stretches of barriers and significant exposure to sunlight. In addition to providing cleaner energy, they offer an opportunity for cities to reduce their environmental footprint while addressing the growing demand for renewable energy sources. As with any major infrastructure project, an economic analysis is essential before implementing solar-integrated sound barriers. Factors such as local solar irradiance, the cost of installation, and the energy needs of the surrounding area should be considered to determine the cost-effectiveness of the system. In high-sunlight regions with strong demand for noise reduction, these barriers can offer significant benefits, both in terms of renewable energy generation and noise mitigation. However, careful planning and design are necessary to ensure that the system operates efficiently and provides maximum value for the investment.

(g) **Solar Street and Outdoor Lighting**

Solar-powered street and outdoor lighting systems have become an increasingly popular solution for energy-efficient urban and rural lighting. These systems integrate photovoltaic panels with LED lighting to provide illumination in streets, parks, parking lots, and other outdoor areas. By using solar energy to power outdoor lighting, these systems eliminate the need for grid electricity and reduce energy consumption, contributing to sustainability goals. The photovoltaic panels capture sunlight during the day and store the generated electricity in batteries, which is then used to power the LED lights at night. One of the main advantages of solar-powered street and outdoor lighting is their ability to operate independently of the electrical grid (Susilo and Setiawan 2021). This feature makes them particularly useful in remote or off-grid locations where conventional power infrastructure may be unreliable or non-existent. These systems are also valuable in urban areas, where they can reduce the burden on the local power grid and lower electricity costs. In addition, solar street lighting systems offer flexibility in placement as they do not require extensive wiring or trenching, making installation quicker and less invasive.

Solar street lighting systems are equipped with energy-efficient LED lights, which provide bright illumination while consuming significantly less power compared to traditional lighting technologies like halogen or fluorescent bulbs. The combination of solar power and LED technology ensures that these systems are not only environmentally friendly but also cost-effective in the long term. Studies have shown that solar street lighting systems can reduce outdoor lighting energy consumption by up to 80% compared to grid-connected lights, making them an ideal solution for municipalities looking to reduce energy expenses and carbon emissions (Bachanek et al. 2021). In addition to their environmental benefits, solar street and outdoor lighting systems enhance public safety and security. By providing reliable lighting in streets, parks, and public spaces, they help prevent accidents and reduce crime rates, especially in areas where grid-powered lighting may be limited or unavailable (Kim et al. 2024). The continuous operation of solar-powered lights even during power outages, ensures that critical outdoor areas remain illuminated, further improving safety in communities. Many systems also come equipped with motion sensors or dimming capabilities, which adjust the brightness of the lights based on the presence of pedestrians or vehicles, optimizing energy use and extending battery life.

However, the initial costs of installing solar-powered street lighting systems can be higher than traditional grid-connected lighting due to the need for photovoltaic panels, batteries, and control systems. Despite the higher upfront investment, the long-term energy savings and reduced maintenance costs often outweigh the initial expense, making these systems a financially viable solution over time. The maintenance of solar street lighting is generally minimal as LED lights have a long lifespan and the systems are designed to be durable and weather-resistant, ensuring reliable performance even in harsh environmental conditions. Solar street lighting is particularly beneficial in regions with high levels of sunlight, where the solar panels can generate ample electricity to power the lights throughout the night. For areas with less consistent sunlight, advanced battery storage systems and energy-efficient LED lights help ensure that the system operates effectively, even during cloudy or overcast days. In recent years, innovations in battery technology such as the use of lithium-ion batteries have further improved the efficiency and reliability of solar lighting systems, enabling longer operation times and faster charging capabilities.

The adoption of solar street and outdoor lighting is growing worldwide as governments and private developers recognize the value of renewable energy solutions in urban planning and infrastructure development. These systems are being implemented in both developed and developing regions as part of broader efforts to enhance sustainability, reduce carbon footprints, and improve public safety. In addition to public spaces, solar outdoor lighting is also being adopted in commercial and residential developments providing reliable and energy-efficient lighting solutions for parking lots, gardens, pathways, and recreational areas. Before implementing solar street lighting, an economic analysis is necessary to evaluate the cost-effectiveness of the system. Factors such as local solar irradiance, expected energy savings, installation costs, and maintenance requirements should be considered to ensure that the system provides the desired return on investment. In areas with consistent sunlight and high energy costs, the long-term benefits of solar-powered street lighting can quickly offset the initial installation costs, making it an attractive solution for sustainable urban and rural development.

(h) **Solar-Powered Green Roofs**

Solar-powered green roofs combine two innovative technologies green roofing and solar energy generation into one integrated system that supports both environmental sustainability and renewable energy production. Green roofs are designed to support vegetation and provide numerous environmental benefits such as improved insulation, reduced stormwater runoff, and enhanced biodiversity. When combined with solar panels, these roofs also generate electricity creating a multifunctional feature that maximizes the energy efficiency and ecological benefits of the building. A solar-powered green roof consists of solar panels mounted on an elevated structure above a layer of vegetation (Prasad et al. 2021). The raised solar panels provide partial shade to the plants below, creating a microclimate that can help certain plant species thrive, especially in hotter climates where full sun exposure could be detrimental to plant health. In return, the plants beneath the solar panels help cool the surrounding air, reducing the temperature around the solar panels and improving their efficiency. This symbiotic relationship enhances both the performance of the green roof and the solar panels, resulting in greater overall sustainability for the building (Alonso-Marroquin and Qadir 2023).

One of the key advantages of solar-powered green roofs is their ability to improve building insulation. Green roofs act as a natural insulator, reducing the heat transferred into the building during the summer and retaining warmth during the winter. By combining this with solar panels, which reduce reliance on external energy sources, buildings can achieve significant reductions in both heating and cooling costs. Additionally, the vegetation on green roofs helps reduce the urban heat island effect, which occurs when densely populated areas experience higher temperatures due to the extensive use of concrete and other heat-retaining materials. In terms of energy generation, solar-powered green roofs offer a sustainable solution for harnessing renewable energy while also providing the environmental benefits of green infrastructure. The electricity generated by the solar panels can be used to power the building or sold back to the grid, further contributing to energy savings and potential revenue generation. The combination of renewable energy production and ecological benefits makes solar-powered green roofs an attractive option for both commercial and residential buildings that aim to improve sustainability.

Solar-powered green roofs also contribute to improved stormwater management. The vegetation absorbs rainwater, reducing runoff and alleviating the strain on urban drainage systems. This is particularly important in cities where heavy rainfall can lead to flooding and water pollution. By integrating solar panels with green roofing systems, buildings can not only generate clean energy but also mitigate environmental challenges such as stormwater management and heat reduction. While solar-powered green roofs offer numerous benefits, they can be more complex and costly to install compared to traditional roofs or standalone solar panel systems. The structural load of a green roof combined with the weight of solar panels are often requires additional reinforcement to the framework of the building which can increase construction costs. Additionally, the need for specialized design and installation services can make these systems more expensive than conventional roofing or solar solutions. However, the long-term environmental and economic benefits, including energy savings, reduced maintenance costs, and potential tax incentives for green buildings, often justify the initial investment.

Moreover, solar-powered green roofs are particularly beneficial in urban areas where space for traditional solar installations may be limited. By utilizing the roof space for both energy generation and green infrastructure, buildings can maximize the functionality of available surfaces while contributing to urban sustainability efforts. These roofs are also ideal for buildings seeking to achieve green certifications such as LEED or BREEAM, as they enhance energy efficiency and reduce carbon emissions. Additionally, they provide environmental benefits like increased biodiversity and improved air quality, making them valuable for sustainable building design. In recent years, innovations in solar-powered green roof design have made these systems more efficient and adaptable. Lightweight solar panels and modular green roofing systems have simplified installation, making it easier for buildings of various sizes and types to implement this technology. Additionally, advancements in plant selection and irrigation systems have improved the longevity and resilience of green roofs, ensuring that they can thrive in diverse climates and conditions. Before implementing a solar-powered green roof, an economic analysis is essential to evaluate the cost-effectiveness of the system. Factors such as the structural capacity, local climate, expected energy savings, and environmental benefits should be considered to determine whether the long-term advantages outweigh the higher upfront costs. In regions with high energy costs or where green building incentives are available, the integration of solar panels with green roofs can offer a substantial return on investment, both financially and environmentally.

UK scenarios PV systems remain relatively expensive in the UK, currently costing around £6000 per kilowatt-peak. As a result, there has been substantial government support for the technology since 2000, which has fostered some growth in the sector. However, the UK only accounts for 0.3% of global PV installations, trailing far behind leaders like Germany, Japan, and the United States (Sovacool et al. 2022). The reasons for this are not entirely clear, considering the robust materials science base of the UK and its reputation for innovative manufacturing research. Yet, these strengths have not been fully leveraged to develop a concentrated PV research community. A lack of strong market support, combined with challenges in translating basic science research into commercial technology likely contribute to this issue. Furthermore, the UK lacks the centralized laboratory infrastructure that has been critical for driving PV research in countries like Germany, the USA, and Japan.

Despite these obstacles, the UK remains home to world-leading PV research. Institutions such as Imperial College London have established an international reputation for their research on quantum well devices and organic solar cells. Cambridge University has been at the forefront of developing conducting polymers for displays and later adapted these to photovoltaic applications. BP Solar once operated a leading laboratory for PV research in the UK, but following its closure, the expertise has dispersed across various universities and research institutions. This fragmentation is evident, with over 20 universities engaging in some level of PV research, though often with only one full-time academic per institution. Few groups have reached the critical mass necessary for significant breakthroughs.

To address this, the Engineering and Physical Sciences Research Council (EPSRC) at the UK introduced the SUPERGEN programme, aimed at structuring the research community and achieving critical mass. Two prominent initiatives include the PV Materials for the 21st Century and the Excitonic Solar Cells Consortium, which highlight the expertise at the UK in polymer and dye-sensitized solar cells, as well as its competencies in silicon and thin-film heterojunctions. PV system research is also strong with notable contributions from universities such as Loughborough, Southampton, and Northumbria. Both Loughborough and Northumbria are key players in the EU Integrated project PERFORM, which focuses on testing and improving the lifetime performance of PV systems. The School of the Built Environment (SBE) at the University of Nottingham (UoN) has also conducted extensive research on Building Integrated Photovoltaics (BIPV), where PV systems are installed on the roofs and windows of buildings like the Eco-house and the sustainable energy research centre.

3.3 Combine Heat and Power (CHP) Cogeneration

Combined heat and power (CHP) also known as cogeneration refers to the simultaneous production of electricity and useful heat from a single heat engine or power station. In traditional power plants, a significant amount of heat generated during electricity production is wasted, released into the environment via cooling towers, flue gases, or other methods. However, CHP systems capture this excess heat and repurpose it for useful applications such as domestic or industrial heating, thereby increasing the overall efficiency of energy use. These systems can either distribute the captured heat directly to nearby facilities or through pipelines in district heating networks (Bagherian et al. 2021). CHP is recognized as a highly efficient method of energy utilization as it effectively recycles waste heat that would otherwise be lost in conventional energy production systems. In addition to providing heat, CHP systems can be integrated with absorption chillers to utilize the captured heat for cooling purposes, a process referred to as "cold" production. This makes CHP systems versatile, allowing them to provide cooling as well as heating and electricity. When a plant produces electricity, heat, and cooling, it is often referred to as a trigeneration or, more broadly, a polygeneration plant. Trigeneration plants are particularly useful in climates or industries where both heating and cooling are essential, maximizing the utility of the system.

Figure 3.4 illustrates how a typical CHP system is integrated into a residential building, capturing heat from both the engine cooling system and flue gases through heat exchangers. This recovered heat is then used for space heating, hot water production, or other domestic applications. By utilizing both electricity and heat, cogeneration represents one of the most thermodynamically efficient uses of fuel. In typical electricity-only production, a substantial amount of energy is lost as waste heat, significantly lowering the overall efficiency of the system. In contrast, CHP) systems are designed to capture and utilize this waste heat, making them much more efficient.

By simultaneously producing electricity and useful heat, CHP systems optimize fuel use as well reducing energy losses that would otherwise occur in traditional power generation systems. This ability to repurpose waste heat not only improves fuel efficiency but also reduces the environmental impact, making CHP a more sustainable option compared to conventional methods.

3.3.1 Sustainability Evaluation of CHP Systems

The sustainability of CHP systems can be assessed through multiple criteria that cover economic, environmental, and social dimensions. Few key criteria for evaluating the sustainability of CHP systems include:

(a) **Efficiency**

The efficiency of CHP systems is one of their most significant advantages, particularly when compared to conventional power generation methods. In traditional power plants, energy is lost in the form of waste heat, with overall efficiency levels often below 50%. In contrast, CHP systems capture and repurpose this heat, enabling them to achieve efficiency rates of up to 75% or higher (Agun et al. 2021). This higher efficiency leads to a more economical use of fuel, as the same fuel source is used to generate both electricity and heat, maximizing its energy output. From an economic perspective, this increased efficiency translates into reduced fuel consumption, lowering operating costs over time, which is particularly important for industrial and commercial applications where large amounts of energy are required. From an environmental standpoint, the improved efficiency of CHP systems means fewer resources are needed to produce the same amount of energy, leading to a reduction in greenhouse gas emissions and other pollutants (Zhang et al. 2023a). By capturing waste heat that would otherwise be lost, CHP systems not only reduce the total fuel needed but also mitigate the environmental impact of energy production. This is crucial in efforts to lower the carbon footprint of industries, cities, and residential areas. Socially, higher efficiency in energy systems can lead to a more sustainable energy infrastructure, ensuring reliable access to energy for communities while reducing the strain on natural resources. The efficiency of CHP systems also supports broader sustainability goals by promoting energy conservation and reducing the societal costs associated with environmental degradation and energy shortages.

(b) **Installation Cost**

The installation cost of CHP systems is a critical factor in determining their economic feasibility, especially for large-scale industrial or residential applications. Initial setup costs for CHP systems can be higher than those for conventional energy systems due to the complexity of the technology and the need for specialized equipment, such as heat recovery units, heat exchangers, and the integration of multiple energy outputs (electricity, heat, and cooling). The upfront investment also varies depending on the

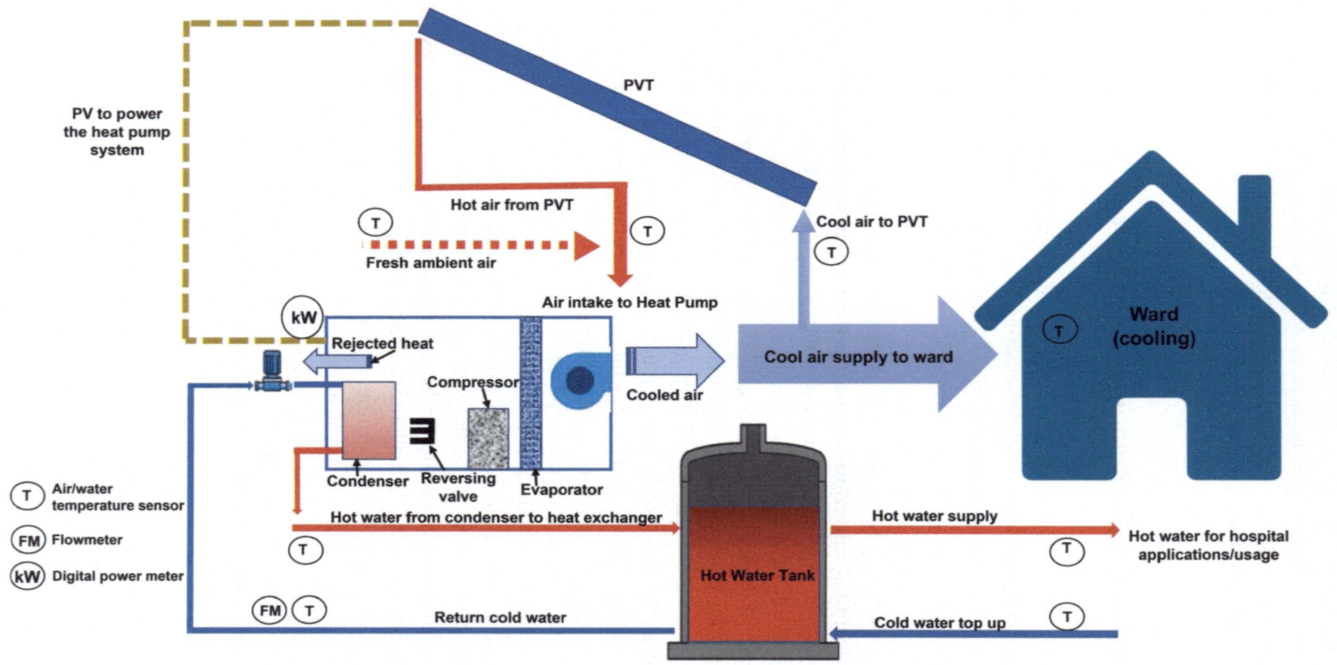

Fig. 3.4 CHP system Integrated with a residential building (Abd Rahman et al. 2022)

size and scope of the system, with large industrial plants requiring more capital for installation compared to smaller residential or commercial systems (Ji et al. 2021). Despite the higher initial costs, CHP systems offer long-term economic benefits through energy savings and increased efficiency. By producing both electricity and heat from a single fuel source, CHP systems reduce the need for separate heating and power systems, leading to lower operational and maintenance costs over time. Additionally, in some regions, governments offer financial incentives, tax credits, or subsidies for the installation of CHP systems, further enhancing their economic viability. These incentives can help offset the higher initial investment and making CHP more attractive for businesses and municipalities looking to reduce energy costs and improve sustainability. Proper economic analysis is essential to balance the upfront expenses against the long-term savings, ensuring that the system meets the financial goals of the project.

(c) **Fuel Cost**

Fuel cost plays a central role in determining the long-term economic sustainability of CHP systems. The type of fuel used, whether it is natural gas, biomass, coal, or renewable sources will directly impacts the operational expenses of the system. Natural gas is one of the most common fuels for CHP systems due to its relative abundance, lower carbon emissions compared to coal, and stable pricing in many regions. However, the volatility of fuel markets means that fluctuations in the price of natural gas or other fossil fuels can affect the overall cost-effectiveness of the system. In addition to conventional fuels, CHP systems can be powered by renewable sources such as biomass or biogas which have lower environmental impacts but may come with higher initial procurement and processing costs (Kavathia and Prajapati 2021). Biomass-fuelled CHP systems can help industries and municipalities reduce dependency on fossil fuels, but the availability and cost of biomass resources can vary depending on location and supply chains. The use of renewable or waste-derived fuels in CHP systems can reduce operating costs in the long term, especially when local or waste fuel sources are readily available. In cases where fuel costs are subsidized or offset by government policies such as carbon pricing or renewable energy incentives, CHP systems powered by alternative fuels become even more economically viable. Ultimately, selecting the appropriate fuel for a CHP system is critical to optimizing its economic performance. A thorough analysis of fuel availability, cost trends, and energy requirements should be conducted to ensure that the chosen fuel source not only meets the operational needs of the system but also aligns with long-term financial goals.

(d) **Electricity Cost**

The cost of electricity generated by a CHP system is a key factor in determining its economic viability, particularly when compared to traditional grid electricity. CHP systems offer the advantage of producing electricity on-site, which can often be cheaper than purchasing it from the grid, especially in regions where electricity prices are high. By generating electricity as part of a combined system, CHP users can avoid many of the transmission and distribution costs typically associated with grid electricity. This localized generation not only reduces reliance on external power sources but also shields the user from fluctuations in electricity prices due to market volatility or changes in energy policy. In many cases, the electricity produced by CHP systems can be used to power a facility or building directly, reducing overall energy expenses (Brown and Herrera 2021). In regions with favourable policies, excess electricity generated by CHP systems can be sold back to the grid, providing an additional revenue stream. This process known as net metering, allows the CHP system operator to offset their electricity costs by selling surplus power during times of peak production. This approach will further improving the economic efficiency of the system. The potential for revenue generation from excess electricity makes CHP systems especially attractive for industrial sites or commercial buildings with consistent energy demands.

However, it is important to note that the cost-effectiveness of CHP-generated electricity depends on several factors, including fuel prices, system efficiency, and local electricity rates. In regions where grid electricity is cheap or highly subsidized, the economic benefits of CHP may be less pronounced. Therefore, a thorough economic analysis that compares the cost of CHP electricity with local grid prices is essential for determining whether CHP is the right solution for a particular site. When electricity prices are high or expected to rise, CHP systems provide a stable, cost-effective alternative for both short-term savings and long-term energy security.

(e) **Heat Cost**

The cost of heat produced by a CHP system is a critical economic factor, particularly for industrial, commercial, and residential applications that require significant amounts of thermal energy. Unlike conventional power plants which discard heat as a by-product, CHP systems repurpose this waste heat for useful applications such as space heating, hot water production, and industrial processes. This dual output allows facilities to save on separate heating costs by utilizing the heat generated during electricity production, making CHP systems economically advantageous. The economic benefit of CHP-generated heat is especially significant in industries with high heat demand, such as manufacturing, food processing, and chemical production (Branchini et al. 2021). By using the heat produced on-site, these industries can drastically reduce their reliance on external fuel sources such as natural gas or oil for heating purposes. The cost of the heat output is directly linked to the efficiency of the system as well as the type and cost of the fuel used. The more efficiently a CHP system can capture and utilize waste heat, the lower the overall heating costs will be. In some cases, CHP systems can also be integrated with district heating networks, where the heat produced is distributed to multiple buildings or homes, further spreading out the cost savings.

Additionally, CHP systems that produce heat in combination with electricity can reduce operational costs in climates where heating is necessary for extended periods. In colder regions, the heat generated by CHP systems can significantly lower energy bills by reducing the need for stand-alone boilers or heating units. Even in temperate climates, CHP systems can provide hot water and heat for commercial buildings, hospitals, schools, or residential complexes, offsetting the costs of conventional heating systems. This reduction in heating expenses not only improves the economic viability of CHP but also contributes to energy security, as facilities are less dependent on fluctuating fuel prices for heating. The cost-effectiveness of CHP-generated heat depends on several factors, including the efficiency of the system, the local cost of fuel, and the specific heating needs of the facility or building. In facilities with high heat demands, the economic benefits of CHP are more pronounced as the heat produced can offset a significant portion of operational expenses. However, for locations with low heat requirements, the economic advantages may be less substantial, and careful analysis is needed to determine if the investment in CHP is justified.

(f) **Carbon dioxide emissions**

The reduction of carbon dioxide (CO_2) emissions is one of the most significant environmental benefits of CHP systems. Traditional power generation methods often produce large amounts of CO_2 due to inefficient fuel use, where much of the energy is lost as waste heat. In contrast, CHP systems maximize fuel efficiency by capturing and utilizing waste heat for productive purposes such as heating or cooling, thus reducing the overall fuel consumption needed to generate the same amount of energy. This leads to a substantial reduction in CO_2 emissions compared to conventional electricity-only systems, which require additional fuel to produce both electricity and heat separately. By improving fuel efficiency, CHP systems contribute to lower greenhouse gas emissions, making them a key technology in efforts to combat climate change. The reduced fuel consumption not only cuts CO_2 emissions but also decreases the release of other pollutants, such as sulfur dioxide (SO_2) and nitrogen oxides (NO_x), which are harmful to air quality (Broesicke et al. 2021). For industries and municipalities looking to meet strict environmental regulations or corporate sustainability goals, CHP systems offer an effective solution to reduce their carbon footprint. As previously mentioned, CHP systems can be integrated with renewable energy sources such as biomass or biogas, further lowering CO_2 emissions by utilizing carbon–neutral or low-carbon fuels.

The environmental benefits of CHP systems are even more pronounced when compared to separate power and heat generation. For instance, if a facility were to rely on grid electricity and a traditional boiler for heating, it would require more fuel and produce more emissions than a CHP system that generates both energy outputs from a single fuel source. By consolidating these processes, CHP systems can reduce CO_2 emissions by up to 30% or more, depending on the fuel and system efficiency (Pastore et al. 2022). This makes CHP systems an attractive option for regions or sectors facing carbon reduction targets or cap-and-trade regulations. Additionally, CHP systems that use renewable or low-carbon fuels can achieve near-zero CO_2 emissions, making them a crucial part of the global transition to cleaner energy systems. The ability of CHP systems to significantly lower CO_2 emissions not only benefits the environment but also provides economic advantages. Facilities that adopt CHP systems may qualify for carbon credits, tax incentives, or other financial benefits offered by governments seeking to encourage low-emission technologies (Salimi et al. 2022). In regions with carbon pricing mechanisms, reducing CO_2 emissions can translate into lower operating costs as companies pay less for their carbon output. This combination of environmental and economic benefits makes CHP a highly sustainable choice for reducing CO_2 emissions and promoting long-term environmental responsibility.

(g) **Energy Footprint**

The reduction of the overall energy footprint is a significant and often underappreciated benefit of CHP systems. While efficiency and fuel cost savings are important, the broader impact of reducing the energy footprint extends to national and societal levels. By optimizing fuel use and reducing waste, CHP systems lower the total energy demand on a macro scale, helping countries decrease their reliance on imported fuels and fossil energy sources. This shift not only strengthens energy security but also aligns with global efforts to reduce energy intensity and meet international climate goals. From a social perspective, a smaller energy footprint contributes to improved public health through the reduction of harmful emissions associated with traditional power generation (Renau et al. 2021). Communities benefit from cleaner air and reduced environmental degradation which can lower healthcare costs and improve overall well-being. Additionally, CHP systems reduce the strain on public infrastructure, as more efficient energy use results in lower pressure on national energy grids, reducing the frequency of blackouts and energy shortages. The energy footprint also plays a critical role in supporting regional resilience. By reducing overall energy consumption and increasing energy efficiency, CHP systems provide localized energy solutions that can operate independently from central grids. This autonomy is especially important in the face of climate change and natural disasters, where reliable access to energy is critical for recovery and survival. CHP systems build decentralized and resilient energy networks that are crucial for addressing future energy challenges.

3.3.2 Configurations of CHP Systems

CHP systems can be categorized into two main configurations: topping cycle and bottoming cycle, depending on the order in which electricity and heat are produced. Each configuration offers distinct advantages and is suited to different types of applications, ranging from industrial to commercial and residential.

(a) **Topping Cycle CHP Systems**

In topping cycle CHP systems, electricity is generated first and the waste heat from this process is captured and repurposed for heating. This configuration is the most common form of CHP and is widely used in commercial, industrial, and residential settings (Barbarelli et al. 2020). Topping cycle systems typically involve prime movers such as gas turbines, reciprocating engines, or steam turbines that produce electricity as the primary output. The waste heat generated during this process rather than being discarded is collected through heat exchangers and used for purposes such as space heating, water heating, or process heating in industrial facilities. The key benefit of the topping cycle is that it makes use of the waste heat that would otherwise be lost in conventional power generation systems, increasing overall energy efficiency. For commercial and residential buildings, the captured heat can be used for heating, ventilation, and air conditioning (HVAC) systems, reducing reliance on separate heating systems. In industrial applications, topping cycle CHP systems provide both electricity and heat for processes that require substantial energy input such as chemical production, food processing, and textiles. The flexibility of this configuration makes it suitable for a wide range of applications, particularly where electricity demand is the primary concern, and the available waste heat can be efficiently utilized.

(b) **Bottoming Cycle CHP Systems**

In contrast, bottoming cycle CHP systems produce heat first often as a by-product of high-temperature industrial processes. The waste heat generated in these processes is then recovered and used to produce electricity through a recovery boiler or steam turbine (Ahmed and Kumar 2023). Bottoming cycle systems are less common than topping cycle systems and are typically used in specialized industrial applications where large amounts of excess heat are produced such as in metal smelting, glass manufacturing, or cement production. The bottoming cycle is advantageous in situations where the heat generated during industrial processes is unavoidable and can be harnessed to produce electricity. This configuration allows industries to convert waste heat into useful power, thereby improving the overall efficiency of their operations. However, because the primary focus of the bottoming cycle is on heat generation, these systems are more specialized and less versatile compared to topping cycle systems. Bottoming cycle CHP systems are particularly beneficial in industries that operate at high temperatures and produce a significant amount of waste heat, allowing them to generate electricity as a secondary benefit without additional fuel consumption. Although bottoming cycle systems are less common, they provide a valuable solution for energy-intensive industries looking to improve their energy efficiency and reduce environmental impact by generating electricity from waste heat that would otherwise be lost.

3.3.3 Large-Scale CHP Systems

Large-scale CHP systems are designed to meet the substantial energy demands of industrial sites, urban districts, or even entire towns, providing both electricity and heating. These systems play a critical role in improving the overall energy efficiency of large facilities by utilizing waste heat that would otherwise be discarded in traditional power generation processes. Large-scale CHP plants are integral to district heating systems and industrial processes where both electrical power and thermal energy are required in significant quantities. Several types of large-scale CHP systems are commonly used, each with distinct technologies and advantages:

(a) **Gas Turbine CHP Plants**

Gas turbine CHP plants are widely used in industrial applications due to their ability to generate both electricity and heat with high efficiency. In these systems, a gas turbine burns natural gas or other fuels to produce mechanical power, which is then converted into electricity. The waste heat from the exhaust of the turbine, which can reach very high temperatures is captured and repurposed to generate additional power or provide heat for industrial processes or district heating networks (Matjanov 2020). Gas turbine CHP plants are especially beneficial for industries with continuous energy demands such as chemical processing, food production, and manufacturing. The heat recovered from the exhaust gases is used to produce steam or hot water, which can be fed into industrial processes or used for space heating in large buildings or urban districts. The versatility and high efficiency of gas turbine CHP systems make them a popular choice for large-scale energy applications, particularly in regions where natural gas is readily available and affordable.

(b) **Combined Cycle Power Plants**

Combined cycle power plants (CCPP) are another common type of large-scale CHP system. These plants are modified to capture and utilize waste heat from the electricity generation process to increase overall system efficiency. In a combined cycle power plant, electricity is generated in two stages: first, a gas turbine produces electricity and then the waste heat from the turbine is used to generate steam, which drives a steam turbine to produce additional electricity or heat (Zhao and Foong 2022). Combined cycle CHP plants are highly efficient because they maximize the energy output from a single fuel source. By recovering the waste heat from the gas turbine and using it to generate more electricity or heat, combined cycle plants can achieve efficiencies far greater than traditional power plants. These plants are widely used in large industrial facilities and urban district heating systems, where consistent electricity and heat demand make them an economically and environmentally sound choice.

(c) **Steam Turbine CHP Plants**

Steam turbine CHP plants are another type of large-scale cogeneration system that captures waste heat from steam after it has passed through the turbine. In these plants, steam is used to drive a turbine that generates electricity. After the steam has passed through the turbine, the remaining thermal energy in the steam is captured and used for industrial processes or heating purposes (Wang et al. 2022). These plants are particularly suitable for industries that require large amounts of steam for their operations, such as pulp and paper production, oil refining, and chemical processing. By capturing and repurposing the waste heat from the steam, these systems improve overall energy efficiency while reducing fuel consumption. Steam turbine CHP plants are a reliable option for large-scale energy production and are frequently integrated into district heating networks, where they provide both electricity and thermal energy to multiple buildings.

(d) **Molten-Carbonate Fuel Cells**

Molten-carbonate fuel cells (MCFC) represent a newer and innovative technology in the field of large-scale CHP systems. These fuel cells are highly efficient at producing both electricity and heat and are considered a clean energy solution due to their low emissions. MCFC operate at high temperatures around 600–700 °C allowing them to use various fuels, including natural gas, biogas, and hydrogen, while generating substantial amounts of heat (Contreras et al. 2021). The heat produced by molten-carbonate fuel cells can be captured and used for industrial heating or district heating applications, providing a twofold energy output. Additionally, these systems have the advantage of being scalable, making them suitable for both large industrial sites and urban energy projects. MCFC are gaining popularity in regions that prioritize reducing carbon emissions and integrating renewable or low-carbon energy sources into their energy mix. Their high efficiency and low environmental impact make them a promising solution for sustainable energy generation in large-scale applications.

3.3.4 Smaller Cogeneration Units

While large-scale CHP systems cater to industrial and urban needs, smaller cogeneration units are often employed for residential or small commercial applications. These smaller systems typically use reciprocating engines or Stirling engines, which burn natural gas, diesel, or other fuels to generate electricity (Schneider et al. 2020). The waste heat from the flue gases and the cooling water from these engines is captured and used to replace traditional gas or oil-fired boilers for heating water or providing space heating. The ability to efficiently utilize both electricity and heat in these smaller units makes them highly suitable for applications where energy demands are moderate yet still significant. Smaller cogeneration units are ideal for residential complexes, small businesses, and schools where heating and electricity demand are more modest but still significant. These systems offer energy savings and increased efficiency by making use of the heat produced during electricity generation, which would otherwise be wasted in conventional setups. In addition to fossil fuel-powered cogeneration units, some smaller CHP systems are powered by biomass or solar energy, making them a more environmentally sustainable option. Biomass-fuelled cogeneration systems convert organic materials such as wood chips or agricultural waste into energy, reducing reliance on fossil fuels and supporting local biomass supply chains. Meanwhile, solar thermal CHP systems capture sunlight to generate both heat and electricity, offering a clean energy solution for smaller applications. These renewable CHP systems contribute to reducing the carbon footprint of buildings and small industrial facilities, aligning with sustainability goals and energy conservation efforts. By utilizing renewable energy sources, smaller CHP units not only cut greenhouse gas emissions but also help future-proof buildings against rising energy costs and stricter environmental regulations.

3.3.5 Micro Combined Heat and Power

Combined Heat and Power (CHP) systems are highly efficient technologies designed to simultaneously produce electricity and heat from a single fuel source. The efficiency of these systems is maximized when the heat generated can be used on-site or in close proximity to the generation source. However, transporting heat over long distances significantly reduces efficiency due to the energy losses that occur during transmission. Long-distance heat transport requires heavily insulated pipes, which are both costly to install and prone to energy loss as heat dissipates over time (Jimenez-Navarro et al. 2020). These challenges have prompted a shift away from large-scale CHP systems, which may struggle with efficiency when the demand for heat is spread over large areas. Micro-CHP systems have emerged as a solution to this problem. As the demand for localized energy generation grows, these smaller distributed energy systems are gaining popularity. Micro-CHP systems are designed to generate both electricity and heat on-site, eliminating the need for long-distance transmission of heat. By focusing on localized energy needs, Micro-CHP systems are particularly well-suited for residential buildings, small businesses, remote areas, and urban centres that frequently experience power outages. Figure 3.5 illustrates how these systems operate at a household level, efficiently delivering both power and heat from a single source.

Micro-CHP systems offer numerous benefits particularly in scenarios where centralized electricity grids are unreliable or where the demand for both heat and electricity must be met at a localized level. These systems allow buildings to generate their own power and heat, making them an attractive option for homeowners, small businesses, and even off-grid communities. One of the major advantages of Micro-CHP systems is their ability to improve energy efficiency by capturing and utilizing the heat

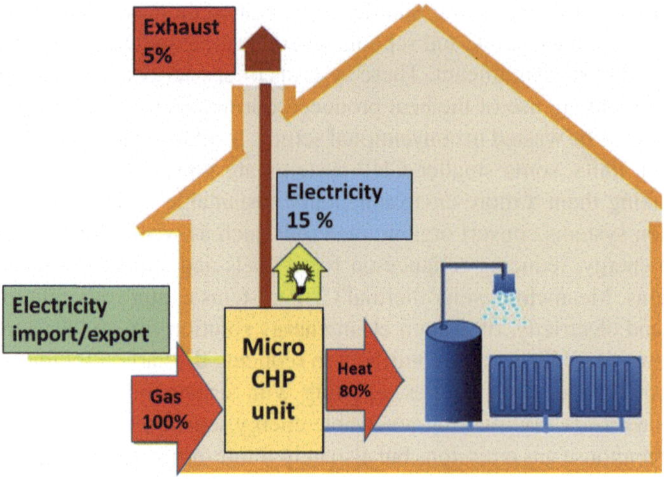

Fig. 3.5 Schematic energy flows in micro CHP system for house (Harrison 2004)

that would otherwise be wasted in conventional power generation. By producing both electricity and heat on-site, Micro-CHP systems reduce energy transmission losses. This is typically a significant problem for centralized power plants that must transport energy over long distances (Hammond and Titley 2022). The localized generation of energy enhances energy resilience by reducing dependence on external power grids. Buildings equipped with Micro-CHP units are better protected from outages or supply disruptions, ensuring a more reliable and consistent energy supply. In remote areas where access to reliable electricity is limited, Micro-CHP units provide an important alternative, ensuring consistent access to both heat and power.

In urban settings, Micro-CHP systems can help mitigate the risks associated with blackouts, offering backup power and heat generation during emergencies or grid failures (İncili et al. 2023). There is a growing trend towards the adoption of Micro-CHP systems as the preferred solution for decentralized energy generation. As energy policies and consumer preferences increasingly favour energy resilience and sustainability, the market for these systems is expected to expand significantly. By allowing users to generate their own power and heat, Micro-CHP systems offer greater control over energy consumption, which can also lead to cost savings. In regions where excess electricity can be sold back to the grid, Micro-CHP users may benefit from financial incentives or rebates, further enhancing the appeal of these systems.

Micro-CHP systems are often referred to as a type of distributed energy resource (DER), meaning that they generate energy at or near the point of use rather than relying on centralized power plants. This distributed approach is becoming increasingly important in the context of the global transition to more resilient, sustainable energy systems. Unlike conventional heating systems, which burn fuel solely to produce heat, micro cogeneration systems generate both heat and electricity, making them far more efficient than traditional boilers or furnaces. In a typical Micro-CHP system, electricity is produced through a prime mover such as an internal combustion engine or a fuel cell, which generates mechanical power. This mechanical power is then converted into electricity, which can be used on-site to meet the energy needs of the home or business. The excess heat generated during the electricity production process is captured and used for space heating or to produce hot water, which eliminates the need for a separate heating system. This dual generation of electricity and heat maximizes the fuel energy potential, reducing both energy costs and greenhouse gas emissions.

In some cases, Micro-CHP systems can be connected to the grid allowing users to sell any excess electricity back to the utility company. This process, known as net metering, provides financial incentives for users and helps reduce their overall energy bills. The ability to sell surplus electricity also makes Micro-CHP systems an attractive option for homeowners and businesses that want to reduce their reliance on the grid while also benefiting from potential revenue streams. Additionally, Micro-CHP systems can be designed to run on a variety of fuels, including natural gas, biogas, propane, and hydrogen, making them adaptable to different energy sources based on local availability.

The future of Micro-CHP systems looks promising as advances in technology make these units more efficient, affordable, and accessible. One of the key trends in Micro-CHP development is the integration of renewable energy sources such as solar or biomass into the systems. Solar-assisted Micro-CHP units use solar thermal energy to help power the system, reducing the amount of fuel required and lowering carbon emissions. Biomass-fuelled Micro-CHP systems are another sustainable option, converting organic materials into both heat and electricity, further reducing the carbon footprint of homes and businesses that adopt these systems. The environmental benefits of Micro-CHP systems are significant, especially when compared to traditional heating systems or centralized power generation. Micro-CHP units can reduce CO_2 emissions by up to one ton per household annually, contributing to national and international climate goals. As Micro-CHP technology continues to evolve, the market potential for these systems is expected to grow especially in areas where decentralized energy generation is becoming a priority. With the push for increased energy resilience, reduced emissions, and sustainable energy solutions, Micro-CHP systems are poised to become a key player in the future of home and business energy generation. These systems offer the flexibility, efficiency, and sustainability needed to meet the challenges of the modern energy landscape, making them an ideal solution for residential and small-scale commercial applications.

Micro-CHP System Components A Micro-CHP system consists of several key components that work together to generate both heat and electricity:

(a) **Prime Mover**

A typical Micro-CHP system consists of several essential components, with the prime mover being one of the most critical. The prime mover is the engine that drives the generator responsible for converting fuel into mechanical energy which is then used to produce electricity. Micro-CHP installations utilize four main types of prime mover technologies, each with distinct characteristics, advantages, and applications: internal combustion engines, Stirling engines, closed-cycle steam engines, and fuel cells. The choice of prime mover technology depends on the specific energy requirements, fuel availability, and efficiency goals of the installation, making it a flexible solution for a wide range of applications. Internal combustion engines (ICE) are among the most common prime movers used in Micro-CHP systems due to their robustness and well-established technology. ICE burn fuels such as natural gas, diesel, or biogas to drive a generator, producing both electricity and heat as by-products (Castaño Mesa et al. 2023). These engines are widely used in residential and small commercial applications, where reliability and fuel flexibility are key considerations. On the other hand, Stirling engines are external combustion engines that operate on a closed-cycle process. They offer high efficiency and the ability to run on various heat sources, including renewable fuels (González-Pino et al. 2020). Stirling engines are particularly valued for their quiet operation and low maintenance needs, making them an attractive option for residential use where noise and upkeep might be concerns.

In addition to these, closed-cycle steam engines are another option used in some Micro-CHP installations. These engines generate electricity by using steam produced from heated fluids within a closed system (Bataineh 2024). Closed-cycle steam engines are known for their high thermal efficiency and are often employed in applications where a significant amount of heat is needed, such as in combined space and water heating. Fuel cells represent a more advanced prime mover technology, generating electricity through an electrochemical process rather than combustion. Fuel cells are capable of achieving very high efficiencies, particularly when fuelled by hydrogen and have the added benefit of producing very low emissions (Yanda et al. 2023). In Micro-CHP systems, fuel cells are increasingly gaining popularity due to their clean energy potential and ability to integrate with renewable energy sources like hydrogen and biogas. The versatility of the prime mover component allows Micro-CHP systems to be tailored to different energy needs, ensuring that both electricity and heat can be generated in the most efficient and sustainable way possible. Each type of prime mover offers its own set of advantages, making it important for system designers to choose the technology that best suits the application, whether it is a residential home, a small business, or a community energy system.

(b) **Generator**

The generator in a Micro-CHP system is responsible for converting the mechanical energy produced by the prime mover into electricity. Once the prime mover creates mechanical energy, the generator transforms this energy into electrical power. The electricity generated is then supplied directly to the electrical distribution system of the building, where it can be used to power appliances. In most Micro-CHP systems, the generated electricity can either be used immediately on-site or, in some cases, fed back into the grid through a process known as net metering, which allows excess electricity to be sold to the utility provider. This not only reduces energy costs but also provides potential revenue for the system owner. The generator plays a crucial role in ensuring that the electrical needs of the building are met efficiently, while also making use of any excess heat that is produced during the electricity generation process.

(c) **Heat Recovery System**

The heat recovery system in a Micro-CHP unit is essential for capturing and repurposing the waste heat produced during electricity generation. While the prime mover generates mechanical energy to produce electricity, it also produces significant amounts of heat as a by-product. Instead of letting this heat go to waste, the heat recovery system captures it and makes it available for domestic heating or hot water production. This dual use of energy is what makes Micro-CHP systems highly efficient compared to conventional systems that only generate electricity. The captured heat can be used to meet various heating needs in residential or small commercial buildings, such as providing hot water for showers, heating water for central heating systems, or supporting space heating. In colder climates, the heat recovery system is especially valuable as it reduces the need for additional heating sources and as well further lowering energy costs and improving the overall energy efficiency of the building. By making use of the heat that would otherwise be wasted, the heat recovery system significantly boosts the overall efficiency of a Micro-CHP unit. This integration of heat and power generation ensures that nearly all the energy from the fuel is utilized, reducing fuel consumption and lowering greenhouse gas emissions. This makes the heat recovery system a key component in maximizing the environmental and economic benefits of Micro-CHP technology.

(d) **Control System**

The control system in a Micro-CHP unit plays a crucial role in ensuring the safe, efficient, and reliable operation of the entire system. This component monitors and regulates the performance of the prime mover, generator, and heat recovery system, ensuring that the system responds effectively to the energy demands of the building (Dashtdar et al. 2022). The control system automatically adjusts the output of both heat and electricity to match the real-time needs of the household or business, optimizing fuel use and maximizing efficiency. In addition to managing energy output, the control system also handles important safety functions. It monitors the operating conditions of the system such as temperature, pressure, and fuel supply, and takes action to prevent overheating, overloading, or other operational issues. If any part of the Micro-CHP system operates outside safe parameters, the control system can shut down the unit or adjust its performance to maintain safe operation. The control system also enables seamless integration with the electrical grid, allowing the system to export excess electricity or adjust power generation based on grid demand or pricing, if necessary. This smart functionality ensures that the Micro-CHP unit operates efficiently, safely, and economically, making it an essential component in the overall system.

(e) **Enclosure**

The enclosure of a Micro-CHP unit serves multiple important functions, ensuring the overall protection and performance of the system. First and foremost, it provides physical protection for the engine and other components, shielding them from potential damage caused by environmental factors such as dust, moisture, or extreme temperatures (Santon et al. 2005). By keeping the internal components of the system safeguarded, the enclosure helps to maintain the long-term reliability and efficiency of the unit. In addition to protecting the components, the enclosure also plays a crucial role in ensuring operator safety. It serves as a barrier between the high-temperature or high-pressure parts of the Micro-CHP system and the individuals who maintain or operate the unit. This safety measure reduces the risk of accidental contact with moving or heated parts, ensuring that the system can be operated without endangering personnel. Another critical function of the enclosure is noise reduction. Micro-CHP systems particularly those powered by internal combustion engines can generate significant noise during operation. The enclosure is designed with noise-dampening materials that help to minimize operational noise, making the system more suitable for use in residential or commercial environments where quiet operation is important. By reducing noise pollution, the enclosure enhances the overall user experience, allowing the system to operate unobtrusively.

UK scenarios Combined Heat and Power (CHP) is a well-established concept with a long history of use, both globally and in the UK. In the UK, CHP systems are increasingly seen as an important part of the national strategy to achieve energy efficiency and reduce carbon emissions. Engineers have long recognized the substantial efficiency gains achieved by integrating electricity generation with thermal loads in buildings and industrial settings. Interest in CHP has fluctuated over time due to shifting market conditions and government policies. As decarbonizing the energy sector and reducing CO_2 emissions become more pressing, initiatives like the Net Zero by 2050 target are driving this shift (Pye et al. 2017). As a result, CHP is gaining renewed attention as a practical solution for improving energy efficiency and promoting sustainability. In the UK, CHP systems offer several characteristics and benefits that make them attractive for various applications, from industrial plants to large-scale residential buildings. One of the primary advantages is on-site electricity generation, which reduces reliance on the national grid and improves energy security and resilience (Dong et al. 2021). By generating electricity locally and recovering waste heat, CHP systems maximize overall energy efficiency. This approach significantly reduces energy losses that are typically associated with transmission from distant power plants, which is especially important as the UK shifts toward more decentralized energy solutions.

Another critical feature of CHP systems is their high efficiency. This makes them far more efficient than traditional power generation methods, where significant amounts of energy are lost as waste heat. In settings with a year-round heat demand, such as hospitals, hotels, and industrial facilities, CHP systems provide consistent and reliable energy. The UK climate and heating needs further support CHP use, as these systems ensure the continuous generation of both heat and electricity, optimizing energy use in facilities with substantial heat requirements. The economic operation of CHP systems is also a key factor in their growing adoption. To be financially viable, these systems must operate for more than 4500 h per year, making them suitable for facilities with continuous energy demands (Koch et al. 2020). Before installation, a comprehensive feasibility study is critical to assess whether CHP is suitable for a particular application. The feasibility study should include an accurate demand profile to determine whether the system can meet the specific energy and heat needs of the facility while remaining economically viable.

3.3.6 Fuel Cell Combine Heat and Power (CHP)

A fuel cell is an advanced electrochemical device designed to convert the Gibbs free energy change from a chemical reaction directly into electrical power. Structurally, a fuel cell comprises two electrodes which are an anode and a cathode, separated by an electrolyte. Electricity is produced through the introduction of fuel, typically hydrogen at the anode, and an oxidant, such as oxygen at the cathode (Yanda et al. 2023). This electrochemical reaction, facilitated by a catalyst, allows hydrogen and oxygen to combine without combustion, producing electricity, with only water and heat as by-products. In the operation of fuel cell, oxygen is supplied to the cathode, while hydrogen is fed into the anode. A catalyst induces the hydrogen molecules to split into protons and electrons. The protons pass through the electrolyte, while the electrons are directed through an external circuit, generating an electrical current. This current can be utilized before the electrons return to the cathode, where they recombine with the protons and oxygen to form water. The chemical reactions occurring at both electrodes.

Fuel cells differ fundamentally from batteries in that they do not store energy but rather consume reactants that must be continuously replenished. The reactants flow into the cell, and reaction by-products flow out, while the electrolyte remains intact within the cell. As long as the fuel and oxidant flows are maintained, fuel cells can operate continuously. Another key distinction between fuel cells and batteries is that, while battery electrodes chemically change during charge and discharge cycles, fuel cell electrodes function as stable catalysts, sustaining their activity over extended periods without degradation (Hou et al. 2020).

Classification of Fuel Cell Classification of fuel cell can be categorised as follows:

(a) **Classification by Electrolyte**

Fuel cells can be classified based on the type of electrolyte they contain:

i. **Alkaline Fuel Cell (AFC)**: Utilizes an alkaline solution, typically potassium hydroxide (KOH) in water, as the electrolyte.
ii. **Phosphoric Acid Fuel Cell (PAFC)**: Uses phosphoric acid as the electrolyte.
iii. **Proton Exchange Membrane Fuel Cell (PEMFC)**: Contains a solid proton-conducting membrane as the electrolyte. This type of fuel cell is also commonly referred to as a Solid Polymer Fuel Cell (SPFC), Solid Polymer Electrolyte Fuel Cell (SPEFC), Polymer Electrolyte Fuel Cell (PEFC), or Polymer Electrolyte Membrane Fuel Cell (PEMFC).
iv. **Molten Carbonate Fuel Cell (MCFC)**: Utilizes a molten carbonate salt as the electrolyte.
v. **Solid Oxide Fuel Cell (SOFC)**: Employs a solid oxide ion-conducting ceramic as the electrolyte.

(b) **Classification by Operating Temperature**

Fuel cells can also be classified according to their operating temperature:

i. **Low Temperature Fuel Cells**: Operate below 100 °C under low pressures (up to approximately 5 atm), an example being the Proton Exchange Membrane Fuel Cell (PEMFC).
ii. **Medium Temperature Fuel Cells**: Operate within the temperature range of 100–500 °C, such as Phosphoric Acid Fuel Cells (PAFC) and Alkaline Fuel Cells (AFC).
iii. **High Temperature Fuel Cells**: Operate above 500 °C, including Molten Carbonate Fuel Cells (MCFC) and Solid Oxide Fuel Cells (SOFC).

In summary, six major types of fuel cells have been developed or are currently being developed for practical applications. Five of these types are classified based on the electrolyte used, as presented in Table 3.1 Additionally, Direct Methanol Fuel Cells (DMFC) are categorized based on the fuel utilized for electricity generation.

3.3.7 Hydrogen Production as Clean Energy

Hydrogen is an incredibly versatile and clean energy carrier offering vast potential for use in a variety of applications, including fuel cells, energy storage, and sustainable energy systems. As a zero-emission fuel when used in fuel cells, hydrogen is central to the global transition toward cleaner, more sustainable energy sources. The primary appeal of hydrogen lies in its ability to produce only water vapor as a by-product when used, making it an essential component of efforts to reduce carbon emissions and combat climate change (Malik et al. 2023). However, the environmental benefits of hydrogen depend heavily on how it is produced. While hydrogen is the most abundant element in the universe, it rarely exists in its pure form on Earth and must be extracted from other compounds, such as water or hydrocarbons. The production of hydrogen can have a significant environmental footprint, depending on the method of extraction (Hren et al. 2023). This makes the process by which hydrogen is produced a critical factor in determining whether it can truly serve as a clean energy solution. Currently, there are several established and emerging methods for hydrogen production, each with its own advantages, challenges, and applications. The main approaches include thermochemical methods, electrolysis, direct sunlight conversion, and biological hydrogen production. Each of these methods offers distinct pathways for producing hydrogen, with varying degrees of efficiency, environmental impact, and technological readiness.

(a) **Thermochemical Methods**

Thermochemical methods are among the most widely used processes for hydrogen production, primarily because they leverage existing hydrocarbon resources such as natural gas, coal, biomass, or waste materials. These methods rely on high-temperature chemical reactions to break down hydrocarbons and produce synthetic gas (syngas), which consists of a mixture of hydrogen (H_2), carbon monoxide (CO), carbon dioxide (CO_2), water (H_2O), and methane (CH_4) (Lee et al. 2022). Syngas is then further processed to increase its hydrogen content, after which the hydrogen is separated and purified for use. One of the most common thermochemical methods is steam methane reforming (SMR), which involves reacting methane (typically from natural gas) with high-temperature steam. This process produces hydrogen and carbon monoxide, which is then converted to additional hydrogen and carbon dioxide through a water–gas shift reaction (Batgi and Dincer 2024). SMR is widely used because it is relatively cost-effective and utilizes readily available natural gas resources. However, it also produces significant amounts of CO_2 emissions, making it less environmentally friendly unless coupled with carbon capture and storage (CCS) technologies.

Table 3.1 Operational characteristics and technological status of various fuel cells

Type of fuel cells	Mobile ion	Operating temperature (°C)	Power density (Mw/cm^2) (present) projected	Projected rated power level (kW)	Fuel efficiency (chemical and electric)	Lifetime projected (hours)	Capital cost projected (US$/kW)	Areas of application
AFC	OH^-	60–90	(100–200) > 300	10–100	40–60	> 10,000 (1.14 yrs)	> 200	Space, mobile
PAFC	H^+	160–220	(200) 250	100–5000	55	> 40,000 (4.57 yrs)	3000	Dispersed and distributed power
PEMFC	H^+	50–80	(350) > 600	0.01–1000	45–60	> 40,000 (4.57 yrs)	> 200	Portable, mobile, space, stationary
MCFC	$CO_3{}^{2-}$	600–700	(100) > 200	1000- 100,000	60–65	> 40,000 (4.57 yrs)	1000	Distributed power generation
SOFC	O^{2-}	800–1000	(240) 300	100–100,000	55–65	> 40,000 (4.57 yrs)	1500	Baseload power generation
DMFC	H^+	90	(230)	0.001–100	34	> 10,000 (1.14 yrs)	> 200	Portable, mobile

Other thermochemical methods such as coal gasification and biomass gasification are also produced hydrogen by reacting carbon-based materials with steam or oxygen. Coal gasification is a well-established technology, but it is often criticized for its high carbon emissions. Biomass gasification offers a more sustainable alternative by using renewable organic materials like wood chips, agricultural residues, or waste. Biomass-based hydrogen production can be considered carbon–neutral since the CO_2 released during the process is roughly equal to the amount of CO_2 absorbed by plants during their growth. Thermochemical methods provide a large-scale and reliable means of producing hydrogen, especially in industrial settings. But their overall environmental impact depends on the feedstock used and whether emission-reduction technologies like CCS are employed. As hydrogen production scales up globally, research and innovation in thermochemical processes continue to focus on making them more sustainable and efficient, particularly by incorporating biomass and improving carbon capture technologies.

(b) **Electrolysis of Water**

Electrolysis of water is a highly promising method for producing hydrogen, particularly because it can be powered by renewable energy sources, making it a zero-emission process. In electrolysis, electricity is used to split water (H_2O) into its constituent elements such as hydrogen (H_2) and oxygen (O_2). This process occurs within an electrolyze, a device that contains an aqueous electrolyte solution and two electrodes (Simoes et al. 2021). When an electric current is applied, water molecules break apart, releasing hydrogen gas at the cathode and oxygen gas at the anode. What makes electrolysis especially attractive is its potential to produce green hydrogen, which is generated without any carbon emissions. If the electricity used for electrolysis comes from renewable energy sources such as solar, wind, or hydropower, the entire process can be completely sustainable and environmentally friendly. In this case, hydrogen becomes a clean energy carrier that can be stored, transported, and used in various sectors including transportation, industry, and power generation, without contributing to greenhouse gas emissions.

There are three primary types of electrolysis technologies in use today:

i. **Alkaline Electrolysis (AEL)**: This is the most established method, using a liquid alkaline electrolyte such as potassium hydroxide (KOH). AEL is highly reliable and relatively low-cost, making it the most commonly used electrolysis technology.
ii. **Proton Exchange Membrane (PEM) Electrolysis**: This technology uses a solid polymer electrolyte and offers higher efficiency and faster response times compared to alkaline electrolysis. PEM electrolyzes are more compact and better suited for integration with intermittent renewable energy sources such as solar and wind. However, they are currently more expensive than AEL systems.
iii. **Solid Oxide Electrolysis (SOE)**: This emerging technology operates at high temperatures (around 700–1000 °C) and uses a solid oxide electrolyte. SOE is highly efficient and can directly utilize waste heat from industrial processes or power plants, making it a promising option for future hydrogen production.

The major advantage of electrolysis lies in its environmental benefits when paired with renewable energy. By converting surplus renewable energy into hydrogen during periods of excess generation such as sunny or windy days, electrolysis offers a way to store energy for later use, providing a solution for grid balancing and energy storage. This stored hydrogen can be used in fuel cells to generate electricity during periods of high demand or when renewable energy is less available, helping to stabilize the power grid. However, the widespread adoption of electrolysis for hydrogen production faces certain challenges. The primary barrier is cost. Producing hydrogen via electrolysis is currently more expensive than traditional methods like steam methane reforming (Younas et al. 2022). The cost of electricity which is the most significant input for electrolysis, plays a major role in the overall economics of the process. As the price of renewable electricity continues to fall and electrolysis technologies improve, the cost of producing hydrogen via electrolysis is expected to decrease, making it more competitive with fossil fuel-based methods. Electrolysis of water presents a clean, flexible, and sustainable pathway for hydrogen production, especially when powered by renewables. As technological advancements continue and the cost of renewable energy decreases, electrolysis is poised to become a central part of the global hydrogen economy.

(c) **Direct Sunlight Conversion**

Direct sunlight conversion represents an exciting and emerging technology for producing hydrogen. This method involves the direct use of solar energy to split water into hydrogen and oxygen, bypassing the need for electricity from external sources. This process is often achieved through photoelectrochemical (PEC) cells or photocatalysis (Singla et al. 2021). In these methods, sunlight acts as the driving force for water splitting offering a highly renewable and clean pathway for hydrogen production. In photoelectrochemical cells (PEC), sunlight is absorbed by semiconductor materials that generate electron–hole pairs, which are used to drive the chemical reaction that splits water (Siavash Moakhar et al. 2021). The photons from the sunlight excite the electrons in the semiconductor, causing them to move from a lower energy state to a higher one. These energized electrons can then reduce water molecules at the cathode, while the holes (positively charged gaps left behind by the electrons) oxidize water molecules at the anode, producing hydrogen and oxygen gases.

Similarly, photocatalysis uses specialized catalysts that can absorb solar energy and promote the water-splitting reaction (Mohsin et al. 2023). In this process, photocatalysts are typically immersed in water and directly exposed to sunlight. These catalysts, often made from materials like titanium dioxide (TiO_2), are capable of harvesting solar energy to generate hydrogen gas. While photocatalysis is still in the early stages of development, ongoing research is focused on improving the efficiency and durability of photocatalysts to make the process more viable for large-scale hydrogen production. The potential advantages of direct sunlight conversion are substantial. By leveraging solar energy, this method could offer a sustainable and environmentally friendly approach to hydrogen production with zero carbon emissions. The fact that sunlight is abundant and freely available makes this process highly attractive in regions with high solar irradiance, such as desert areas and sunny climates.

However, despite its promise, direct sunlight conversion faces several challenges that must be overcome before it can be widely adopted. One of the key obstacles is efficiency. Current PEC cells and photocatalysts have relatively low efficiencies in converting sunlight into hydrogen, making them less competitive with other hydrogen production methods. Additionally, the durability of the materials used in these systems is a concern, as prolonged exposure to sunlight and water can cause degradation, reducing the overall lifespan and effectiveness of the devices. Researchers are actively exploring ways to improve the material stability, efficiency, and scalability of direct sunlight conversion technologies. Advances in nanomaterials, semiconductor engineering, and photocatalyst design are expected to play a critical role in enhancing the performance of this hydrogen production method. As these technologies evolve, direct sunlight conversion could become an important component of a solar-driven hydrogen economy, particularly in regions where solar energy is plentiful and underutilized. In conclusion, direct sunlight conversion holds great potential as a clean, renewable, and sustainable method for producing hydrogen, although significant technological advancements are needed to improve its efficiency and cost-effectiveness. As research progresses, this approach could play a pivotal role in the transition to a zero-carbon hydrogen economy, especially in regions rich in solar resources.

(d) **Biological Hydrogen Production**

Biological hydrogen production is an innovative and environmentally friendly method that leverages the ability of certain microorganisms such as algae and bacteria to produce hydrogen under specific environmental conditions. This process is still in the developmental stage (Xu et al. 2022). It offers a sustainable and natural approach to hydrogen production, utilizing biological systems instead of energy-intensive industrial processes. While the technology holds great promise for the future, ongoing research is necessary to improve its efficiency, scalability, and economic viability. There are two primary biological pathways for hydrogen production: photo-fermentation and dark fermentation.

i. **Photo-Fermentation**

Photo-fermentation involves photosynthetic microorganisms such as green algae and certain bacteria that can produce hydrogen as a by-product of their metabolic processes when exposed to sunlight. These organisms contain enzymes known as hydrogenases that facilitate the production of hydrogen gas under anaerobic (oxygen-free) conditions (Xuan et al. 2023). When the microorganisms perform photosynthesis, they convert sunlight into chemical energy, which drives the production of hydrogen. This method has attracted interest due to its ability to harness solar energy in a natural biological system, creating a renewable pathway for hydrogen production.

ii. **Dark Fermentation**

Dark fermentation does not require sunlight and takes place in dark and anaerobic conditions. In this process, bacteria break down organic materials such as biomass, agricultural waste, or food waste to produce hydrogen along with other by-products like organic acids and gases (Jain et al. 2024). Dark fermentation can operate continuously and can use a wide variety of feedstock making it a versatile option for hydrogen production from renewable organic sources. While the hydrogen yield from dark fermentation is currently lower than other production methods, it offers a way to convert waste materials into a valuable clean energy source.

Biological hydrogen production holds significant potential due to its low environmental impact. Unlike thermochemical methods, biological production processes occur at low temperatures and do not require the combustion of fossil fuels, reducing carbon emissions. Additionally, using waste materials as feedstock for dark fermentation contributes to a circular economy by repurposing organic waste into clean energy, further enhancing its environmental benefits. Despite the promise of biological hydrogen production, there are several challenges that must be addressed to make it a commercially viable option. One of the primary issues is the low efficiency of hydrogen production in both photo-fermentation and dark fermentation processes. The microorganisms involved are limited by the rates of metabolism and hydrogenase activity, resulting in lower hydrogen yields compared to industrial methods like steam methane reforming or electrolysis. Furthermore, the bioreactors required for scaling up these processes are complex and can be costly to construct and maintain.

Research is currently focused on improving the efficiency of biological hydrogen production through genetic engineering and metabolic optimization. By modifying microorganisms to enhance their hydrogen-producing capabilities, scientists aim to increase hydrogen yields and make the process more efficient (Ramprakash et al. 2022; Sivaramakrishnan et al. 2021). Additionally, efforts are underway to develop more cost-effective and scalable bioreactor designs that could enable larger-scale implementation of biological hydrogen production. In summary, biological hydrogen production offers a sustainable and eco-friendly alternative to traditional hydrogen production methods, with the potential to integrate renewable resources and waste materials into the hydrogen supply chain. While challenges remain, continued advancements in biotechnology and process optimization could unlock the full potential of biological hydrogen production, making it a key contributor to the global hydrogen economy in the future.

3.3.8 Fuel Cell Combined Heat and Power in Buildings

Many fuel cell developers are focusing on the residential and small commercial power sectors, targeting systems within the 1–20 kWe range. These residential systems have a range of development goals, including grid-independent power generation, micro-scale CHP, and hybrid systems that combine electricity generation with absorption chillers for cooling purposes (Acha et al. 2020). Four distinct Total Energy System (TES) configurations have been developed for residential applications. The first TES configuration consists of a fuel cell sub-system (FCS) integrated with a vapor compression heat pump (Yadav et al. 2023), forming a tri-generation system (Yadav et al. 2023). In this setup, the fuel cell produces electricity to power household lights, appliances, and the compressor of the heat pump, while consuming fuel. Additionally, the fuel cell generates waste heat, which is used to fully meet domestic hot water needs and a portion of the space heating load. The remaining heating demand is supplied by the vapor compression heat pump, which also provides space cooling through its vapor compression cycle.

The second TES configuration builds on the first by adding an absorption cooling sub-system alongside the vapor compression heat pump (Sun et al. 2023). Absorption cooling based on the lithium bromide/water cycle is suitable for residential applications and can operate with low-temperature thermal energy. In this system, during cooling hours, the thermal energy from the fuel cell is primarily used for domestic water heating, with any excess energy directed to the absorption cooling system. Space cooling is provided by the combined operation of the vapor compression heat pump and the absorption cooling unit. During heating hours, the system functions similarly to the first configuration, with the fuel cell supplying thermal energy for heating, while electricity generated by the fuel cell powers the home during both heating and cooling periods.

The third TES configuration involves a combined system where both an absorption and vapor compression heat pump operate in parallel to provide space heating and cooling (Chen et al. 2021). This dual operation allows the vapor compression heat pump and the thermally driven absorption system to work together, offering flexibility in coordinating energy demands with the thermal and electrical outputs of the fuel cell. However, this design presents challenges, including optimizing the size of the space conditioning equipment, selecting the appropriate absorption/generation cycle, and ensuring efficient control of the parallel operation of heating and cooling systems.

The fourth Total Energy System (TES) option introduces a variation from the simpler initial design by incorporating a thermal storage tank to optimize the use of available thermal energy. In this setup, thermal energy generated by the fuel cell is transferred to the thermal storage tank, which then supplies both space heating and domestic hot water needs (Xing et al. 2021). During periods of low thermal demand, excess thermal energy is stored in the tank and utilized during times of higher demand. However, the stored thermal energy may not always be sufficient to meet the entire residential heating and hot water demands. As a result, space heating is provided through a combination of the thermal energy stored in the tank and the electric heat pump. Any shortfall in domestic hot water heating is compensated by the electrical output of the system. In this configuration, the space cooling load is entirely met by the electric heat pump, while lights and appliances of the household are powered by the electricity generated by the fuel cell.

(a) **PEM Fuel Cell Combined Heat and Power in buildings**

Proton exchange membrane fuel cells (PEMFC) also known as polymer electrolyte membrane fuel cells that are increasingly being developed for a wide range of applications, including transportation, stationary power generation, and portable system (Tellez-Cruz et al. 2021). These cells are particularly advantageous due to their ability to operate at lower temperatures typically around 80 °C and lower pressures compared to other fuel cell technologies, making them more adaptable for various uses. The core component of these systems is the proton exchange membrane (PEM), constructed from ionomers, which facilitates the conduction of protons while preventing gases like oxygen and hydrogen from passing through. One of the most widely used PEM materials is Nafion, a high-performance polymer which is positioned between bipolar plates to ensure proper functionality (Okonkwo et al. 2021).

PEM fuel cells generate electricity by converting chemical energy from the electrochemical reaction between hydrogen and oxygen into electrical energy, producing only water and heat as by-products. This clean energy output makes them highly desirable for applications focused on reducing carbon emissions. However, managing the water and heat within the cell remains a critical operational challenge. The PEM must remain adequately hydrated for efficient proton conduction. If the membrane becomes too dry, resistance increases, potentially causing damage to the membrane. On the other hand, if excess water accumulates, the electrodes can become flooded, preventing hydrogen and oxygen from reaching the catalyst and halting the reaction. As a result, precise control over water and heat management is essential to ensure the efficiency, longevity, and safety of PEM fuel cells. PEM fuel cells are highly suitable for residential applications due to several advantages that make them ideal as fuel cell systems for domestic use:

i. **Compact size and weight**: PEMFC systems are appropriately sized and lightweight, making them suitable for residential environments, such as a typical 7 kW PEMFC system.
ii. **Flexible power supply range**: PEMFC systems can provide a wide range of power outputs, making them adaptable to varying domestic power requirements. This flexibility ensures that PEMFC systems can meet diverse energy needs in homes.
iii. **Efficient heat recovery**: PEMFC systems typically operate at moderate temperatures between 50 and 80 °C, making them highly compatible with heat recovery systems, enabling the provision of both space heating and hot water in residential settings.
iv. **Rapid start-up capabilities**: PEMFC systems can meet fast start-up demands, making them efficient for use in scenarios requiring quick responses to shifts in power needs, ensuring minimal delays in power generation.
v. **Environmental benefits**: PEMFC systems offer numerous ecological advantages over traditional utility power generation. The operation of a fuel cell combines hydrogen and oxygen, producing only water as a by-product. Fuel processors, or reformers, convert a range of fuels into hydrogen, with CO_2 as the primary by-product. The high electrical efficiency of PEMFC systems results

in lower greenhouse gas emissions compared to other non-renewable energy sources, and the integration of waste heat recovery for water and space heating further reduces CO_2 emissions.

vi. **Low noise levels**: One of the key advantages of PEMFC systems is their quiet operation, as they lack combustion or moving parts. When integrated into a complete system, the only potential noise sources come from air pumps or fans, which operate at relatively low volumes.

vii. **Decreasing costs**: Over time, the cost of PEMFC systems has significantly decreased. In the early 2000s, the catalyst content for PEMFC systems was much higher, but advancements in technology have reduced these costs, making PEMFC systems increasingly affordable for residential use.

(b) **SOFC Combined Heat and Power in Buildings**

Solid oxide fuel cells (SOFC) operate by transferring oxygen ions through a solid oxide electrolyte at high temperatures, where they react with hydrogen at the anode. Unlike proton-exchange fuel cells, SOFC do not require expensive catalysts such as platinum, as their high operating temperatures allow for more flexibility in fuel use (Peng et al. 2021a). This eliminates the risk of carbon monoxide poisoning, making SOFC highly versatile. They can operate on a wide range of fuels, including methane, propane, butane, fermentation gas, gasified biomass, and even paint fumes. However, sulfur compounds present in the fuel must be removed before entering the fuel cell which easily accomplished using activated carbon beds or zinc absorbents.

The limited availability of a hydrogen infrastructure and the complexity of on-board hydrogen production which is necessary for PEMFC systems have shifted attention towards SOFC applications. The high operating temperatures of SOFC typically ranging between 600 and 1000 °C has allow for a broad variety of fuels to be used. While currently, only small-scale stationary SOFC units ranging from 1 kW to under 300 kW have been developed, these systems are capable of achieving electrical efficiencies of up to 50% and overall system efficiencies of 70–90%. SOFC technology is increasingly being integrated into small combined heat and power (CHP) systems, making it ideal for residential applications, particularly in single-family homes. Market deployment is already underway, and large-scale production of SOFC systems in the coming years is expected to result in significant cost reductions. SOFC CHP systems are also gaining attention for their ability to provide high-quality power in areas with unstable grids, addressing issues such as daily phase anomalies, weekly voltage dips and surges, and periodic power outages.

3.4 Wind Power

Wind power is the process of harnessing the energy from wind and converting it into more usable forms, most commonly electricity, through the use of wind turbines. The global capacity for wind-powered electricity generation has seen significant growth. As of 2005, wind-powered generators contributed 58,982 megawatts globally. Although wind power currently accounts for less than 1% of the electricity supply of the world, its role is much more substantial in several countries. For instance, it supplies 23% of electricity needs in Denmark, 6% in Germany, and approximately 8% in Spain, highlighting its potential in regions with favourable wind conditions. The principle behind wind power generation is straightforward: wind turbines harness the kinetic energy from wind by converting the rotation of the turbine blades into electrical energy through a generator (Peng et al. 2021b). Wind turbines are generally categorized into two types based on their rotor orientation: horizontal axis wind turbines (HAWT) and vertical axis wind turbines (VAWT). HAWT are the most widely used in modern wind farms (Yang et al. 2021). They typically feature a three-bladed rotor, which is optimized for efficiency and balance. However, there are variations in design with some models utilizing two or four blades. HAWT are well-suited for areas with consistent wind patterns and are typically installed at heights that capture steady, high-velocity winds. These turbines are particularly common in large-scale wind farms that contribute to national energy grids. For residential applications, smaller HAWT can be installed in areas with adequate wind resources, allowing homeowners to generate electricity on-site.

VAWT in contrast, have a rotor that spins around a vertical axis. These turbines have a range of designs, such as the Darrieus wind turbine and the Vortex Style wind turbine (Rajpar et al. 2021). VAWT offer several advantages, especially for urban environments and small-scale applications. Unlike HAWT, VAWT can capture wind from any direction without needing to rotate into the wind, making them ideal for locations where wind direction changes frequently or for installations in confined spaces such as rooftops in urban areas (Zhao et al. 2022). Furthermore, VAWT tend to be quieter than their horizontal counterparts and can operate at lower wind speeds. One unique feature of VAWT is that, as their rotational speed increases, they appear as a solid object, which can help reduce bird strikes. Advances in technology have also led to improvements in the durability and efficiency of small wind turbines. Modern wind turbines are designed to operate with minimal maintenance over extended periods. Additionally, innovations in blade design, materials, and energy storage systems have further increased the viability of small-scale wind power for individual homes. Some models now incorporate hybrid systems that integrate solar power, further enhancing energy generation capabilities during periods of low wind. Despite the promising outlook, challenges remain. The initial investment for wind turbine installation can be high particularly for smaller-scale residential systems. Additionally, the efficiency of wind turbines is highly dependent on the local wind climate, and not all regions are suitable for wind energy production. For homeowners to maximize the benefits of wind power, a thorough assessment of the wind potential at the site is necessary.

3.4.1 Small-Scale Wind Power for Domestic Application

Small-scale wind power systems particularly for residential use have gained popularity as an alternative to grid electricity. Households in regions with adequate wind resources can install small wind turbines to generate power for personal use contributing to energy independence and reducing reliance on non-renewable energy sources (Zhang et al. 2023b). These systems can either be standalone units, where the electricity generated is stored in batteries for later use, or they can be connected to the grid, allowing excess electricity to be sold back to utility companies, providing economic benefits. The design of small wind turbine systems differs considerably from that of larger grid-connected turbines. Most small wind turbine systems are direct-driven, variable-speed systems equipped with permanent magnet generators. As a result, a power converter is often required to maintain a constant frequency when necessary. Unlike large wind turbines, small wind systems do not require a gearbox, as they typically operate with much higher rotor speeds, making this approach more efficient for smaller-scale applications.

Small wind turbines have long been utilized for household electricity generation, particularly in conjunction with battery storage systems in remote locations. These turbines range in size from small 400-W units designed for residential use to multi-megawatt machines employed in wind farms and offshore applications. The smaller turbines often feature direct drive generators, direct current (DC) output, aeroelastic blades, lifetime bearings, and a wind vane to orient the turbine into the wind. In urban environments where high wind speeds is limited, smaller wind turbine systems can still be useful for powering low-demand equipment. Rooftop-mounted wind turbines can also help alleviate power distribution issues and offer resilience during power outages. Due to their lightweight design, small-scale wind turbines are particularly suited for urban settings, responding quickly to wind gusts and easily mounted, similar to a television antenna. These systems are designed to operate quietly, and it is claimed that they are inaudible even when standing a few feet beneath the turbine.

One key factor in the success of small-scale wind turbines for household applications is siting. Effective wind turbines require a consistent and strong wind flow to operate efficiently. Therefore, it is essential to assess local wind patterns, average wind speeds, and potential obstacles such as buildings or trees, which could disrupt airflow. For homeowners in regions with high wind availability, wind turbines can significantly reduce electricity bills while contributing to a more sustainable energy mix. To address the challenges associated with the variable power output of wind energy, hybrid generation systems are being explored and developed to enhance the efficiency and reliability of wind turbines. These systems integrate wind power with other renewable technologies, such as wind power combined with photovoltaic (PV) systems, wind power with fuel cell systems, and even configurations that combine wind power, PV, and fuel cell systems. These hybrid systems aim to provide a more stable and consistent energy supply by compensating for fluctuations in wind energy production.

(a) **Wind Power/Photovoltaic (PV) System**

The combination of wind turbines and photovoltaic (PV) panels is one of the most widely explored hybrid systems. Wind and solar energy complement each other well because their availability often peaks at different times. While solar panels generate electricity during the day, wind turbines can operate at any time particularly during periods when sunlight is unavailable such as at night or in cloudy conditions (Costoya et al. 2022). This hybrid approach ensures more consistent energy production throughout the day and night, reducing reliance on one single energy source. Additionally, wind and solar systems can be integrated with battery storage to further stabilize energy availability and reduce dependence on grid electricity.

(b) **Wind Power/Fuel Cell System**

Fuel cells can be integrated with wind turbines to supplement energy generation during periods of low wind speed or fluctuating wind conditions. Fuel cells generate electricity through an electrochemical reaction, typically using hydrogen as a fuel source and they can provide a continuous and reliable energy supply when wind power is insufficient (Kadri et al. 2020). In this hybrid configuration, excess wind energy can also be used to produce hydrogen via electrolysis, which is then stored and later used by the fuel cell when wind energy is not available. This combination improves energy security and reduces intermittency, making the system more resilient and reliable for both grid-connected and off-grid applications.

(c) **Wind Power/Photovoltaic (PV)/Fuel Cell System**

A fully integrated system that combines wind power, photovoltaic panels, and fuel cells provides a highly flexible and resilient energy solution. In this hybrid setup, wind and solar energy work together to supply electricity during periods of high availability, while fuel cells step in when both renewable sources are insufficient (Khan and Javaid 2020). By leveraging the strengths of each technology, this configuration maximizes energy production while minimizing the impact of fluctuating renewable sources. The addition of a fuel cell further enhances the ability of the system to store energy and provide a backup power supply, ensuring that electricity is available during prolonged periods of low wind or solar generation. This comprehensive hybrid system is ideal for locations with variable weather patterns or off-grid areas where energy storage and reliability are critical.

UK scenarios As the UK boasts the highest wind potential in Europe, wind power emerges as a key renewable energy technology for the nation. Despite its potential, the development of wind energy in the UK has progressed at a slower pace than in other European countries, as indicated by global comparative data. Presently, wind power generates electricity for nearly a quarter of a million homes, with approximately 1000 wind turbines operational across the country. However, several challenges, particularly related to installation costs and infrastructure, continue to impact the growth of wind power. The costs associated with installing wind energy systems are highly site-specific, often influenced by geographical terrain, distance to grid connections, and access to transportation routes. Currently, the capital cost of wind energy installation in the UK ranges between £500 and £750 per kilowatt (kW) of installed capacity. The bulk of these costs roughly two-thirds are attributable to the wind turbines themselves, while around a quarter is allocated for the development of access roads, foundations, and electrical cabling. The remaining costs are associated with installation, planning, legal fees, and other miscellaneous expenses. On average, the operation and maintenance costs are relatively modest, amounting to approximately 0.5 pence per kilowatt-hour (p/kWh).

Nevertheless, when factoring in the environmental and societal costs associated with carbon emissions, wind power becomes even more competitive, as it is a clean, renewable energy source that generates no harmful emissions. Looking ahead, it is projected that wind energy could supply approximately 6% of the total electricity demand in UK by 2010. This projection underscores the significant potential of wind power to contribute to the renewable energy goals, reduce greenhouse gas emissions, and enhance energy security. As the country continues to invest in wind energy infrastructure and technology, further reductions in both capital and operational costs are expected, positioning wind power as an increasingly viable and sustainable energy source.

3.5 Stirling Engine

The Stirling engine is a type of external combustion piston engine that operates based on the principles of thermodynamics. Unlike internal combustion engines, where fuel is combusted inside the cylinders, the Stirling engine uses an external heat source to heat a working fluid, typically air, helium, or hydrogen which is sealed within the system (Perozziello et al. 2021). This design allows the engine to be powered by a variety of heat sources, such as solar energy, geothermal energy, or waste heat from industrial processes, offering significant flexibility in its applications. The Stirling engine relies on a unique thermodynamic cycle that involves cyclic compression and expansion of the working fluid at different temperature levels (Durcansky et al. 2020). This cycle consists of four phases: two isothermal processes, where the gas exchanges heat at a constant temperature, and two isometric (or isochoric) processes, where heat is exchanged at a constant volume. During the isothermal expansion phase, heat is applied to the gas, increasing its pressure and driving the pistons of the engine (Honorio et al. 2021). Conversely, during the cooling phase, the gas is allowed to cool, reducing the pressure and completing the cycle. These cyclical changes in pressure and temperature are what enable the engine to perform mechanical work.

One of the notable advantages of the Stirling engine is its high thermal efficiency, which can surpass that of conventional internal combustion engines. This is largely due to its external heat source, which can be carefully managed for optimal efficiency, and the closed system design, which minimizes energy loss. Additionally, since there is no internal combustion, the engine operates with fewer moving parts and less noise, resulting in longer lifespans for components and quieter operation (Daoud and Friedrich 2020). The absence of combustion also makes the Stirling engine a cleaner technology, generating fewer emissions compared to traditional engines. The flexibility of the Stirling engine extends to its power output, making it suitable for a wide range of applications. It can be scaled down for small tasks like powering household appliances or upscaled for larger applications such as combined heat and power (CHP) systems for industrial or residential use. In recent years, interest in Stirling engines has grown due to their potential role in renewable energy systems especially in solar-thermal power generation where they can efficiently convert solar heat into electricity without relying on fossil fuels. However, despite these benefits, the Stirling engine does have limitations. Its reliance on external heat sources can result in slower start-up times compared to internal combustion engines, which can ignite fuel directly to produce power almost instantaneously. Additionally, the complexity of the design and higher upfront costs have been barriers to widespread commercial adoption, though ongoing advancements in materials and manufacturing processes continue to improve its feasibility. Table 3.2 provides a comparison of the process in the Stirling engine with other types of heat engines.

A well-engineered Stirling engine can achieve between 50 and 80% of its theoretical efficiency in converting thermal energy into mechanical work, with the primary constraints arising from frictional losses and material limitations. The operation of an idealized displacer-type Stirling engine, demonstrates the fundamental principles of the cycle (Zheng et al. 2021). In this type of engine, the compression and expansion of the working fluid are driven by the motion of the power piston, while the displacement of the gas between the hot and cold chambers is facilitated by the movement of the displacer. The isochoric heat transfer processes occur within a regenerator, positioned between the hot and cold spaces. Typically, this regenerator comprises a dense wire mesh capable of storing a substantial portion of the thermal energy retained in the working fluid following the expansion process. This stored thermal energy is then utilized during the next cycle to heat the working fluid as it moves from the compression chamber to the expansion chamber.

A key distinction between the Stirling engine and internal combustion engines, such as gasoline or diesel engines, lies in the closed-cycle operation of the working fluid. In a Stirling engine, the working fluid remains sealed within the system for the

Table 3.2 Heat engine processes

Cycle/process	Compression	Heat addition	Expansion	Heat rejection
Carnot	Adiabatic	Isothermal	Adiabatic	Isothermal
Otto (Petrol/gasoline)	Adiabatic	Isometric	Adiabatic	Isometric
Diesel	Adiabatic	Isobaric	Adiabatic	Isometric
Brayton (Jet)	Adiabatic	Isobaric	Adiabatic	Isobaric
Stirling	Isothermal	Isometric	Isothermal	Isometric
Ericsson	Isothermal	Isobaric	Isothermal	Isobaric

duration of its operation, continuously undergoing heating and cooling cycles. This contrasts with open-cycle internal combustion engines, where the working fluid is replaced after each cycle. This feature contributes to the efficiency and longevity of the Stirling engine, as it eliminates the need for continuous fuel and air intake and exhaust.

Apart from the displacer engine, where the power piston and displacer share the same cylinder commonly referred to as the "beta" type, two other fundamental configurations of Stirling engines can be distinguished based on the arrangement of their working spaces. In gamma type engines, the piston and displacer are housed in separate cylinders, which simplifies sealing and allows for the use of less complex drive systems for the displacer rod, as it no longer passes through the piston. However, this configuration results in increased dead space, leading to reduced efficiency and power output compared to the beta type. Stirling engines that utilize a second piston instead of a displacer are classified as alpha type machines (Khatke et al. 2020). These can be constructed in both single-acting and double-acting configurations. In the double-acting variant, an additional Stirling process occurs in the space beneath the pistons. Single-acting engines share numerous components with existing mass-produced reciprocating engines or pumps, making them cost-effective to manufacture. Additionally, these engines offer greater flexibility in the arrangement of heat exchangers. However, they have a lower compression ratio compared to beta engines, and maintaining effective sealing can be challenging due to the presence of two pistons per cycle. Double-acting machines, although more complex as they are more compact and lightweight and have a favourable balance between friction losses and shaft power. Conversely, beta type engines allow for higher compression ratios and reduced dead space, resulting in improved power density. While this leads to a more compact design, beta engines require more intricate drive systems, such as rhombic drives, to coordinate the strictly linear and coupled movements of the piston and displacer.

3.5.1 Applications of Stirling Engine

Stirling engines exhibit considerable versatility with applications spanning sectors such as water pumping, agricultural machinery, industrial operations, and electricity generation. Fundamentally, Stirling engines convert thermal energy from an external heat source into mechanical energy by utilizing the circulation of hot air within the engine. When applied for electricity generation, this mechanical energy is transferred to an alternator, where it is converted into electrical power. The generated electricity can then be stored in batteries for later use. The reliance on an external heat source from Stirling engine provides significant operational flexibility, as it can utilize a broad range of energy inputs. These include conventional fossil fuels like oil or natural gas, as well as renewable energy sources such as solar energy, biomass, biogas, and geothermal energy. This adaptability makes Stirling engines particularly suitable for a variety of energy systems, especially in scenarios where diverse fuel sources are available. Furthermore, Stirling engines are known for their environmentally friendly operation as they emit no harmful gases, and the excess heat produced can be repurposed for heating applications, including residential heating, water heating, or greenhouse climate control. Although Stirling engines have a wide array of industrial and commercial uses, this section specifically focuses on their role in electricity generation for residential applications. In such contexts, their ability to harness renewable energy sources presents a promising solution for sustainable energy generation, particularly in off-grid or decentralized energy systems. The flexibility stirling engines in utilizing various energy inputs allows them to operate efficiently with renewable energy sources, such as biomass and solar thermal energy, making them highly suitable for sustainable electricity generation.

(a) **Biomass-Fired Stirling Engine**

Biomass is a renewable energy source that can be used to power Stirling engines by providing the necessary heat. In biomass-fired systems, organic material such as wood, agricultural waste, or energy crops is burned to produce heat. This heat is then transferred to the Stirling engine, driving its cycle by expanding and compressing the working fluid inside. This process generates mechanical energy, which is subsequently converted into electricity (Schneider et al. 2020). A similar biomass-powered Stirling engine for electricity generation highlights the adaptability of such systems for various scales. These engines are suitable for rural areas or off-grid communities where biomass is readily available. Their ability to provide a sustainable energy alternative makes them a valuable tool in regions lacking access to conventional power sources. Larger-scale biomass Stirling engines have also been developed to meet higher energy demands in industrial or commercial settings. This engine can be adapted to run on low-grade fuels, such as landfill gas or biogas from anaerobic digesters. This feature makes biomass-fired Stirling engines particularly useful in waste-to-energy systems, where they contribute to converting waste into renewable energy, reducing dependence on fossil fuels, and lowering greenhouse gas emissions.

(b) **Solar-Powered Stirling Engine**

Solar thermal energy is another effective source for powering Stirling engines. Solar-powered Stirling engines convert solar heat directly into mechanical energy. This is achieved by focusing solar radiation onto the hot end of the displacer within the engine, using concentrators such as parabolic mirrors. The heat drives the Stirling cycle, generating mechanical work that can be transformed into electricity through an alternator. The advantage of solar-powered Stirling engines is their use of an external, renewable heat source (Zhu et al. 2021). Stirling engines are known for their efficiency and when combined with solar thermal energy, they provide a clean and effective solution for power generation. Unlike photovoltaic systems, which convert sunlight directly into electricity, Stirling engines use solar heat, making them ideal for applications where thermal energy is abundant. A challenge with solar energy is its intermittent availability, as solar power can only be harnessed during daylight hours and is affected by weather conditions. To address this, hybrid systems that combine solar energy with backup fuel sources such as natural gas or biomass have been developed. These systems ensure a continuous supply of heat, allowing the Stirling engine to operate even when solar energy is unavailable. Additionally, integrating thermal energy storage systems with solar-powered Stirling engines allows excess heat captured during peak sunlight hours to be stored and used when solar radiation is low or absent (Boretti and A. al Maaitah, 2023). This feature improves the overall reliability and efficiency of solar Stirling systems, enabling them to provide power during both day and night. Due to their ability to operate at high temperatures, Stirling engines are well-suited for concentrated solar power (CSP) applications, which use mirrors or lenses to focus sunlight and achieve the required temperatures for efficient engine performance.

References

Abd Rahman NM et al (2022) Field study of hybrid photovoltaic thermal and heat pump system for public hospital in the tropics. Case Stud Therm Eng 30:101722

Acha S et al (2020) Fuel cells as combined heat and power systems in commercial buildings: a case study in the food-retail sector. Energy 206:118046

Agun L et al (2021) Sterilization of oyster mushroom crop residue substrate by using cold plasma technology. Mater Today: Proc 39:903–906

Ahmed MI, Kumar R (2023) A systematic review on optimal placement of CHP. Smart Sci 11(1):171–191

Alavy M, Li T, Siegel JA (2020) Energy use in residential buildings: Analyses of high-efficiency filters and HVAC fans. Energy Build 209:109697

Al-Ezzi AS, Ansari MNM (2022) Photovoltaic solar cells: a review. Appl Syst Innov 5(4):67

Alonso-Marroquin F, Qadir G (2023) Synergy between photovoltaic panels and green roofs. Energies 16(13):5184

Amelia A et al (2020) Technologies of solar tracking systems: a review. In: IOP conference series: materials science and engineering. IOP Publishing

Bachanek KH et al (2021) Intelligent street lighting in a smart city concepts—a direction to energy saving in cities: an overview and case study. Energies 14(11):3018

Bagherian MA et al (2021) Classification and analysis of optimization techniques for integrated energy systems utilizing renewable energy sources: a review for CHP and CCHP systems. Processes 9(2):339

Baig H et al (2018) Conceptual design and performance evaluation of a hybrid concentrating photovoltaic system. Energy 147:547–560

Barbarelli S et al (2020) An externally fired micro combined-cycle, with largely adjustable steam turbine, in a CHP system. Procedia Manuf 42:532–537

Bataineh K (2024) Hybrid fuel-assisted solar-powered Stirling engine for combined cooling, heating, and power systems: a review. Energy 131506

Batgi SU, Dincer I (2024) A study on comparative environmental impact assessment of thermochemical cycles and steam methane reforming processes for hydrogen production processes. Comput Chem Eng 180:108514

Becker C et al (2013) Polycrystalline silicon thin-film solar cells: status and perspectives. Sol Energy Mater Sol Cells 119:112–123

Benda V, Černá L (2020) PV cells and modules—state of the art, limits and trends. Heliyon 6(12)

Benghanem M et al (2023) Evaluation of the performance of polycrystalline and monocrystalline PV technologies in a hot and arid region: an experimental analysis. Sustainability 15(20):14831

Boretti A, al Maaitah A (2023) Dispatchable power supply from beam down solar point concentrator coupled to thermal energy storage and a Stirling engine. Energy Harvest Syst 10(1):25–33
Branchini L et al (2021) Cogeneration supporting the energy transition in the Italian ceramic tile industry. Sustainability 13(7):4006
Broesicke OA et al (2021) Combined heat and power may conflict with decarbonization goals—air emissions of natural gas combined cycle power versus combined heat and power systems for commercial buildings. Environ Sci Technol 55(15):10645–10653
Brown MA, Herrera VS (2021) Combined heat and power as a platform for clean energy systems. Appl Energy 304:117686
Castaño Mesa ES, Quintana SH, Bedoya ID (2023) Development of a dual fuel ICE-based micro-CHP system and experimental evaluation of its performance at light loads using natural gas as primary fuel. Energies 16(17):6281
Chen E et al (2021) A solar-assisted hybrid air-cooled adiabatic absorption and vapor compression air conditioning system. Energy Convers Manage 250:114926
Contreras RR, Almarza J, Rincón L (2021) Molten carbonate fuel cells: a technological perspective and review. Energy Sour Part A: Recov Util Environ Effects 1–15
Costoya X et al (2022) Combining offshore wind and solar photovoltaic energy to stabilize energy supply under climate change scenarios: a case study on the western Iberian Peninsula. Renew Sustain Energy Rev 157:112037
Dambhare MV, Butey B, Moharil S (2021) Solar photovoltaic technology: a review of different types of solar cells and its future trends. J Phys Conf Ser 1913:1–012053
Daoud JM, Friedrich D (2020) A new duplex Stirling engine concept for solar-powered cooling. Int J Energy Res 44(7):6002–6014
Dashtdar M, Bajaj M, Hosseinimoghadam SMS (2022) Design of optimal energy management system in a residential microgrid based on smart control. Smart Sci 10(1):25–39
Di Sabatino M, Hendawi R, Garcia AS (2024) Silicon solar cells: trends, manufacturing challenges, and AI perspectives. Crystals 14(2):167
Dong S et al (2021) Establishing the value of community energy storage: a comparative analysis of the UK and Germany. J Energy Storage 40:102709
Durcansky P, Nosek R, Jandacka J (2020) Use of stirling engine for waste heat recovery. Energies 13(16):4133
Efaz ET et al (2021) A review of primary technologies of thin-film solar cells. Eng Res Express 3(3):032001
Enaganti PK et al (2020) Study of solar irradiance and performance analysis of submerged monocrystalline and polycrystalline solar cells. Prog Photovoltaics Res Appl 28(7):725–735
Esfahani SK et al (2021) Analysing the role of roof mounted BIPV system optimization on decreasing the effect of duck curve in Perth, Western Australia: an experimental case study. Sustain Energy Technol Assess 47:101328
Fan Q et al (2020) A non-conjugated polymer acceptor for efficient and thermally stable all-polymer solar cells. Angew Chem Int Ed 59(45):19835–19840
Gainza-Barrencua J et al (2021) Energy savings using sunspaces to preheat ventilation intake air: experimental and simulation study. J Build Eng 40:102343
González-Pino I et al (2020) Modelling and experimental characterization of a stirling engine-based domestic micro-CHP device. Energy Convers Manage 225:113429
Guan J et al (2023) Performance of a collector-storage solar air heating system for building mechanical ventilation preheating in the cold area. Energy Built Environ 4(6):639–652
Hammond GP, Titley AA (2022) Small-scale combined heat and power systems: the prospects for a distributed micro-generator in the 'net-zero' transition within the UK. Energies 15(16):6049
Harrison J (2004) Micro combined heat & power (CHP) for housing, pp 28–30
Hendawi R (2021) Reusable crucible materials and coatings for photovoltaic silicon applications
Herrando M et al (2023) A review of solar hybrid photovoltaic-thermal (PV-T) collectors and systems. Prog Energy Combust Sci 97:101072

Honorio T, Masara F, Benboudjema F (2021) Heat capacity, isothermal compressibility, isosteric heat of adsorption and thermal expansion of water confined in CSH. Cement 6:100015

Hou J et al (2020) Platinum-group-metal catalysts for proton exchange membrane fuel cells: from catalyst design to electrode structure optimization. EnergyChem 2(1):100023

Hren R et al (2023) Hydrogen production, storage and transport for renewable energy and chemicals: an environmental footprint assessment. Renew Sustain Energy Rev 173:113113

Hua W et al (2022) Applications of blockchain and artificial intelligence technologies for enabling prosumers in smart grids: a review. Renew Sustain Energy Rev 161:112308

Huang K et al (2023) Optimization of a collector-storage solar air heating system for building heat recovery ventilation preheating in the cold area. Energy Build 284:112875

İncili V et al (2023) Energy and exergy analyses of a coal-fired micro-CHP system coupled engine as a domestic solution. Energy 274:127400

Jain R et al (2024) Bio-hydrogen production through dark fermentation: an overview. Biomass Convers Biorefinery 14(12):12699–12724

Ji L et al (2021) Optimal design and sensitivity analysis of the stand-alone hybrid energy system with PV and biomass-CHP for remote villages. Energy 225:120323

Jia X et al (2021) Selectively absorbing small-molecule solar cells for self-powered electrochromic windows. Nano Energy 89:106404

Jiang H et al (2024) Roofing highways with solar panels substantially reduces carbon emissions and traffic losses. Earth's Fut 12(7):e2023EF003975

Jimenez-Navarro J-P et al (2020) Coupling the heating and power sectors: the role of centralised combined heat and power plants and district heat in a European decarbonised power system. Appl Energy 270:115134

Jin J et al (2021) Highly efficient and stable carbon-based perovskite solar cells with the polymer hole transport layer. Sol Energy 220:491–497

Kadri A et al (2020) Energy management and control strategy for a DFIG wind turbine/fuel cell hybrid system with super capacitor storage system. Energy 192:116518

Kavathia K, Prajapati P (2021) A review on biomass-fired CHP system using fruit and vegetable waste with regenerative organic Rankine cycle (RORC). Mater Today: Proc 43:572–578

Khan A, Javaid N (2020) Optimal sizing of a stand-alone photovoltaic, wind turbine and fuel cell systems. Comput Electr Eng 85:106682

Khatke K et al (2020) Thermodynamic analysis of stirling engine and its performance challenges: a review. J Automob Eng Apl 7(1):28–36

Kim G et al (2020) Transparent thin-film silicon solar cells for indoor light harvesting with conversion efficiencies of 36% without photodegradation. ACS Appl Mater Interfaces 12(24):27122–27130

Kim KH, Hwang T, Kim G (2024) The role and criteria of advanced street lighting to enhance urban safety in South Korea. Buildings 14(8):2305

Koch K, Höfner P, Gaderer M (2020) Techno-economic system comparison of a wood gas and a natural gas CHP plant in flexible district heating with a dynamic simulation model. Energy 202:117710

Kuhn TE et al (2021) Review of technological design options for building integrated photovoltaics (BIPV). Energy Build 231:110381

Kumar NS, Naidu KCB (2021) A review on perovskite solar cells (PSCs), materials and applications. J Materiomics 7(5):940–956

Kumar K et al (2021) Solar tracker transcript—a review. Int Trans Electr Energy Syst 31(12):e13250

Kumari N, Singh SK, Kumar S (2022) A comparative study of different materials used for solar photovoltaics technology. Mater Today Proc 66:3522–3528

Lee JE et al (2022) A review on integrated thermochemical hydrogen production from water. Int J Hydrogen Energy 47(7):4346–4356

Li M et al (2020) Indoor thin-film photovoltaics: progress and challenges. Adv Energy Mater 10(28):2000641

Liu W et al (2022) Light-induced activation of boron doping in hydrogenated amorphous silicon for over 25% efficiency silicon solar cells. Nat Energy 7(5):427–437
Machín A, Márquez F (2024) Advancements in photovoltaic cell materials: silicon, organic, and perovskite solar cells. Materials 17(5):1165
Malik FR et al (2023) Overview of hydrogen production technologies for fuel cell utilization. Eng Sci Technol Int J 43:101452
Masood F et al (2022) The compound parabolic concentrators for solar photovoltaic applications: opportunities and challenges. Energy Rep 8:13558–13584
Matjanov E (2020) Gas turbine efficiency enhancement using absorption chiller. Case Study for Tashkent CHP. Energy 192:116625
Mohsin M et al (2023) Semiconductor nanomaterial photocatalysts for water-splitting hydrogen production: the holy grail of converting solar energy to fuel. Nanomaterials 13(3):546
Mourad A-HI et al (2022) A state-of-the-art review: solar trackers. In: 2022 advances in science and engineering technology international conferences (ASET). IEEE
Okonkwo PC et al (2021) Nafion degradation mechanisms in proton exchange membrane fuel cell (PEMFC) system: a review. Int J Hydrogen Energy 46(55):27956–27973
Pastore LM, Basso GL, de Santoli L (2022) Can the renewable energy share increase in electricity and gas grids takes out the competitiveness of gas-driven CHP plants for distributed generation? Energy 256:124659
Peng J et al (2021a) Solid oxide fuel cell (SOFC) performance evaluation, fault diagnosis and health control: a review. J Power Sources 505:230058
Peng X, Liu Z, Jiang D (2021b) A review of multiphase energy conversion in wind power generation. Renew Sustain Energy Rev 147:111172
Perozziello C, Grosu L, Vaglieco BM (2021) Free-piston stirling engine technologies and models: a review. Energies 14(21):7009
Prasad AR et al (2021) Performance enhancement of solar photovoltaic system for roof top garden. Environ Sci Pollut Res 28(36):50017–50027
Prasetyo SD, Prabowo AR, Arifin Z (2023) The use of a hybrid photovoltaic/thermal (PV/T) collector system as a sustainable energy-harvest instrument in urban technology. Heliyon 9(2)
Pye S et al (2017) Achieving net-zero emissions through the reframing of UK national targets in the post-Paris agreement era. Nat Energy 2(3):1–7
Rahman KS (2022) Cadmium telluride (CdTe) thin film solar cells. Comprehensive guide on organic and inorganic solar cells. Elsevier, pp 65–83
Rajpar AH et al (2021) Recent development in the design of wind deflectors for vertical axis wind turbine: a review. Energies 14(16):5140
Ramprakash B et al (2022) Current strategies and future perspectives in biological hydrogen production: a review. Renew Sustain Energy Rev 168:112773
Renau J et al (2021) Novel use of green hydrogen fuel cell-based combined heat and power systems to reduce primary energy intake and greenhouse emissions in the building sector. Sustainability 13(4):1776
Salimi M et al (2022) Environmental aspects of the combined cooling, heating, and power (CCHP) systems: a review. Processes 10(4):711
Santon RC, Ivings MJ, Pritchard DK (2005) A new gas turbine enclosure ventilation design criterion. In: Turbo expo: power for land, sea, and air
Schneider T, Müller D, Karl J (2020) A review of thermochemical biomass conversion combined with Stirling engines for the small-scale cogeneration of heat and power. Renew Sustain Energy Rev 134:110288
Sendrayaperumal A et al (2021) Energy auditing for efficient planning and implementation in commercial and residential buildings. Adv Civil Eng 2021(1):1908568
Shah A (2020) Amorphous silicon solar cells. Solar Cells Modules 139–161
Siavash Moakhar R et al (2021) Photoelectrochemical water-splitting using CuO-based electrodes for hydrogen production: a review. Adv Mater 33(33):2007285

Simoes SG et al (2021) Water availability and water usage solutions for electrolysis in hydrogen production. J Clean Prod 315:128124
Singh D, Chaudhary R, Karthick A (2021) Review on the progress of building-applied/integrated photovoltaic system. Environ Sci Pollut Res 28:47689–47724
Singla S et al (2021) Photocatalytic water splitting hydrogen production via environmental benign carbon based nanomaterials. Int J Hydrogen Energy 46(68):33696–33717
Sivaramakrishnan R et al (2021) Insights on biological hydrogen production routes and potential microorganisms for high hydrogen yield. Fuel 291:120136
Soares L, Wang H (2024) Design study and potential implementation of photovoltaic noise barriers for sustainable highways. Transp Res Rec 2678(3):855–864
Sovacool BK et al (2022) Towards improved solar energy justice: exploring the complex inequities of household adoption of photovoltaic panels. Energy Policy 164:112868
Sreekanth M, Rahaman A, Sumangala T (2023) Advances in biodegradable polymer nanocomposites for solar cell applications: current status and future directions. In: Biodegradable and Biocompatible Polymer Nanocomposites, pp 397–433
Sun C et al (2022) Temperature effect of photovoltaic cells: a review. Adv Compos Hybrid Mater 5(4):2675–2699
Sun F et al (2023) Configuration optimization of solar-driven low temperature district heating and cooling system integrated with distributed water-lithium bromide absorption heat pumps. Sol Energy 253:401–413
Susilo SH, Setiawan A (2021) The assembly and installation of solar powered public street lights (PJU) at Panti Asuhan Putri Aisyiah Dau Malang. Int J Commun Engage Payungi 1(2):52–58
Fraunhofer Institute for Solar Energy Systems (2019) Photovoltaics report. Fraunhofer ISE, p 490
Tang J et al (2023) A review on energy conversion using hybrid photovoltaic and thermoelectric systems. J Power Sources 562:232785
Tellez-Cruz MM et al (2021) Proton exchange membrane fuel cells (PEMFCs): advances and challenges. Polymers 13(18):3064
Vargas AN et al (2022) Low-cost dual-axis solar tracker with photovoltaic energy processing for education. Sustain Energy Technol Assess 53:102542
Wang C et al (2022) Combined heat and power plants integrated with steam turbine renovations: optimal dispatch for maximizing the consumption of renewable energy. Energy Convers Manage 258:115561
Wang H, Guo L, Soares L (2023) Energy harvesting on New Jersey Roadways. 2023, New Jersey. Department of Transportation. Bureau of Research
Xing H et al (2021) Fuel cell power systems for maritime applications: progress and perspectives. Sustainability 13(3):1213
Xu X, Zhou Q, Yu D (2022) The future of hydrogen energy: Bio-hydrogen production technology. Int J Hydrogen Energy 47(79):33677–33698
Xuan J et al (2023) Hydrogenase and nitrogenase: key catalysts in biohydrogen production. Molecules 28(3):1392
Yadav AK, Sinha S, Kumar A (2023) Comprehensive review on performance assessment of solid oxide fuel cell-based hybrid power generation system. Therm Sci Eng Progr 102226
Yanda L, Yuwei Z, Razmjooy N (2023) Optimal arrangement of a micro-CHP system in the presence of fuel cell-heat pump based on metaheuristics. Int J Ambient Energy 44(1):2128–2139
Yang J et al (2021) Review of control strategy of large horizontal-axis wind turbines yaw system. Wind Energy 24(2):97–115
Younas M et al (2022) An overview of hydrogen production: current status, potential, and challenges. Fuel 316:123317
Zaky AA et al (2020) Energy efficiency improvement of water pumping system using synchronous reluctance motor fed by perovskite solar cells. Int J Energy Res 44(14):11629–11642
Zeina A, Almaz A (2023) The use of architectural treatments for optimal utilization of solar energy. Int des J 13:273–286

Zhang D et al (2023a) Carbon emission reduction analysis of CHP system driven by biogas based on emission factors. Energy Built Environ 4(5):576–588
Zhang Z et al (2023b) Overview of the development and application of wind energy in New Zealand. Energy Built Environ 4(6):725–742
Zhao Y, Foong LK (2022) Predicting electrical power output of combined cycle power plants using a novel artificial neural network optimized by electrostatic discharge algorithm. Measurement 198:111405
Zhao Z et al (2022) A review: Approaches for aerodynamic performance improvement of lift-type vertical axis wind turbine. Sustainable Energy Technol Assess 49:101789
Zheng P et al (2021) Effect of displacer on performance of Stirling-type pulse tube refrigerator with work recovery. Sci Technol Built Environ 27(5):567–577
Zhou Y et al (2020) Liquid thermo-responsive smart window derived from hydrogel. Joule 4(11):2458–2474
Zhu S et al (2021) A review of Stirling-engine-based combined heat and power technology. Appl Energy 294:116965

Chapter 4
Energy-Efficient Lighting Technologies for Building Applications

Contents

Abstract With the increasing demand for energy-efficient solutions, lighting technologies have evolved to play a pivotal role in reducing energy consumption in building applications. This chapter presents an in-depth analysis of energy-efficient lighting technologies tailored for building applications, with a particular focus on sustainable solutions that reduce energy consumption and enhance indoor environmental quality. The chapter explores natural lighting methods, such as solar daylighting, and advanced electric lighting technologies, including Compact Fluorescent Lamps (CFL), Light Emitting Diodes (LED), and Cold Cathode Fluorescent Lamps (CCFL). Each benefit of the technology, including energy conservation, extended lifespan, and improved light quality, are discussed in detail. Additionally, the integration of smart lighting systems and sensors for optimized lighting control, along with hybrid solar lighting systems, demonstrates how modern lighting solutions contribute to significant energy savings in residential, commercial, and institutional buildings. By combining daylighting with novel energy-efficient lamps and advanced lighting controls, the chapter outlines a path toward minimizing energy use while enhancing visual comfort and environmental sustainability in built environments.

S. Riffat et al., *Sustainable Energy Technologies and Low Carbon Buildings*, Lecture Notes in Energy 45, https://doi.org/10.1007/978-3-031-78853-6_4

4.1 Solar Daylighting

Electricity costs are continuously rising, prompting individuals and businesses alike to seek more efficient lighting solutions. One highly effective alternative is utilizing natural sunlight, which not only reduces energy consumption but also offers a superior quality of light. Daylight includes both sunlight and skylight, both of which are freely available, renewable resources that closely match the spectrum that human vision has evolved to perceive. This spectrum peaks at a wavelength of 555 nm, aligning with the highest sensitivity of the eyes to green-yellow light, which is optimal for human vision (Wang et al. 2024). Because of its superior spectral quality, daylight typically requires less intensity than artificial lighting to accomplish the same tasks. For example, reading with daylight from a side-lighting system requires only about 35% of the light intensity needed by a fluorescent light system (Kazemi et al. 2018). When the light source is positioned in front or above the reader, approximately 80% of daylight suffices compared to fluorescent lighting.

Colour rendering is another critical consideration when comparing daylight with artificial lighting. Many individuals have experienced the inconvenience of mismatched colours in clothing or objects under artificial lighting, only to find them appearing differently in natural daylight (Zhao et al. 2020). This occurs due to variations in colour rendering between different light sources. Artificial lighting often lacks the full spectrum of visible light, leading to inaccurate colour perception. The human eye adjusts to the spectral composition of different light sources, which affects how objects appear. Daylight recognized as the ideal light source for accurate colour rendering, creates a superior visual environment which enabling better differentiation between objects and their backgrounds. A good visual environment ensures that critical visual elements stand out from their background, while non-essential surfaces and objects remain visually subdued. The quality of illumination often plays a more crucial role than intensity in creating an effective visual environment. Poor visual environments especially those reliant on low-quality artificial lighting can cause insufficient contrast, making it difficult to distinguish important details.

By tapping into natural sunlight for daylighting, buildings can achieve dual benefits such as energy conservation and enhanced indoor light quality. Daylighting influences several aspects of building design, including layout, occupant comfort, structural considerations, and overall energy usage (Yu et al. 2020). When natural light is prioritized during the design process, it shapes the entire building from its inception. Daylight can reduce the need for electric lighting during daylight hours, decrease peak energy demands, and influence the selection of electric lighting systems. In commercial buildings, where interior lighting can account for 30–50% of total energy consumption, reducing reliance on electric lighting through daylighting can lead to substantial energy savings (Wang et al. 2022). For instance, traditional incandescent bulbs are known to be highly inefficient, converting only a small portion of energy into light while the majority is lost as heat. Fluorescent lights, although more efficient, still do not perform as optimally as natural light. While natural light does contribute to the thermal load of a building, it is highly effective for illuminating spaces, as nearly

Table 4.1 Luminous efficacy of lamps and daylight

Source	Efficacy (lm/w)
Candle	0.1
Oil lamp	0.3
Original Edison lamp	1.4
1910 Edison lamp	4.5
Incandescent lamp	14–20
Tungsten Halogen lamp	16–20
Fluorescent Lamp	50–80
Mercury lamp	30–60
Metal-Halide lamp	60–80
Low-pressure sodium	90–100
High-pressure sodium	120–140
Sunlight	92
Sky light, average	125
Sky light, very clear	175
Clear Sky + Sun, average	111
Overcast sky	110

half of the solar energy falls within the visible light spectrum (Souza et al. 2019). Table 4.1 illustrates the differences in luminous efficacy between various lamps and natural daylight.

Incorporating daylight into architectural design requires thoughtful planning, as it influences the placement of windows, skylights, light wells, and even the orientation of the building itself. This integration must balance the advantages of natural lighting with challenges such as potential glare, heat gain, and uneven light distribution. Excessive sunlight in workspaces can cause glare, resulting in discomfort for occupants. In such cases, the use of shading devices, louvers, or reflective surfaces may help diffuse light evenly while minimizing heat gain. In commercial and residential applications, daylighting strategies often complement electric lighting systems by using advanced control systems such as dimmers and daylight sensors to optimize the combination of natural and artificial light. This reduces energy consumption during periods when natural light is sufficient, lowering internal heat gains and decreasing the demand for mechanical cooling and ventilation systems. The initial investment in enhancing daylighting designs such as installing large windows, skylights, and solar tubes can be significant. However, these costs are often offset by long-term energy savings and lower operational expenses due to reduced reliance on electric lighting and climate control systems.

The strategic use of daylight in building design offers a multitude of benefits, including substantial energy savings and improved visual comfort for occupants. The quality of daylight surpasses that of artificial lighting in terms of colour rendering and visual clarity, and its integration into building designs can significantly reduce energy consumption. As electricity costs continue to rise, the role of daylighting in sustainable building practices becomes increasingly important. However, successful daylighting requires careful planning and integration with electric lighting systems to ensure optimal performance throughout different times of the day and seasons.

4.1.1 Daylighting Classification

Daylighting can be introduced into buildings using various methods, each suited to specific architectural needs and building types. One of the most common approaches is through spaces adjacent to roofs and exterior walls, where windows, clerestories, and skylights can easily capture and distribute natural light. Skylights are often utilized in areas such as atriums, hallways, or large open-plan spaces, providing a direct path for daylight to enter. These structures are particularly effective in ensuring that natural light penetrates deep into the interior of the building. In addition to traditional windows and skylights, fabric membranes have gained popularity especially in temporary structures like tents, exhibition halls, and large sports facilities. These membranes allow diffused natural light to permeate indoor environments providing adequate lighting while controlling glare and heat. They are also cost-effective and can be tailored to different shapes and sizes, offering flexibility in design. Another innovative solution for daylighting is the use of light pipes also referred to as light tubes or solar tubes. These devices are highly efficient at transporting daylight from the exterior of a building to its interior spaces. Light pipes are particularly useful for areas that are difficult to light naturally, such as corridors, underground parking garages, and rooms without windows. They consist of a dome or collector at the roof level, which captures sunlight and channels it through a reflective tube into the desired area. This technology allows buildings to benefit from natural light even in spaces that are traditionally dependent on artificial lighting.

Daylighting, while beneficial is technically complex and requires careful planning and execution to ensure high-quality illumination that complements the specific characteristics of the space. The placement, size, and orientation of windows, skylights, and light tubes must be carefully considered to optimize the balance between light intensity, distribution, and visual comfort. Additionally, factors such as heat gain, glare, and energy efficiency must be managed to avoid negative impacts on the thermal comfort and energy performance of the building. By selecting the appropriate daylighting strategies and integrating them with the building overall design, architects and engineers can create well-lit, energy-efficient spaces. This approach enhances the comfort and well-being of occupants while significantly reducing reliance on artificial lighting.

(a) **Skylights**

Skylights are versatile architectural features designed to allow natural light to penetrate building interiors from above. They are available in a wide range of configurations to meet diverse architectural needs and design preferences. Skylights are commonly constructed from materials such as glass or composite sections, which are typically flat and provide a clean, modern aesthetic. These materials offer durability and transparency, making them a popular choice in both residential and commercial buildings (Shirzadnia et al. 2023). In contrast, plastic skylights offer more flexibility in design. Reinforced plastic materials can be moulded into various shapes, allowing for curves and complex geometries that may not be possible with glass. This adaptability makes plastic skylights suitable for custom architectural projects where specific shapes or designs are required. Additionally, the lightweight nature of plastic compared to glass can make installation easier and reduce the overall load on the building structure.

Windows while traditionally placed on walls can also act as skylights when integrated into roofs or higher parts of a structure to channel daylight into interior spaces. These windows come in various designs and sizes, allowing architects to tailor the amount of daylight entering the building. Furthermore, the use of exterior shades, as illustrated in Fig. 4.1, provides an effective way to control daylight entry. These shades can be adjusted seasonally to reflect excessive sunlight during the summer months, minimizing heat gain, and to allow more sunlight in during the winter, improving energy efficiency and occupant comfort. Skylights not only improve the aesthetic appeal of buildings by creating bright, open spaces but also contribute significantly to energy savings by reducing the need for artificial lighting during daylight hours. Proper placement and thoughtful design of skylights can lead to an optimal balance between light and heat management, making them a crucial element in sustainable building practices.

(b) **Solar Light Pipe**

Solar light pipes also known as light tubes or daylight pipes are advanced systems designed to transport and distribute natural or artificial light within buildings. When utilized specifically for daylighting, they are often referred to as solar tubes, solar

Fig. 4.1 Shades mounted on the top of windows in Marmont centre at SBE of University of Nottingham (UoN)

pipes, or solar light pipes. These systems provide a highly efficient method for channelling daylight into interior spaces that lack direct access to windows or skylights such as basements, corridors, or rooms with opaque walls. A typical solar light pipe consists of three primary components: a dome, a tube body, and a diffuser, as illustrated in Fig. 4.2. The dome typically located on the roof or an external wall acts as a collector for sunlight. It is usually constructed from impact-modified acrylic, a material that is ten times stronger than standard acrylic, making it highly durable and capable of withstanding extreme temperature variations (Sharma et al. 2021). This material is also UV stabilized, absorbing up to 99% of harmful ultraviolet rays while maintaining optical clarity. The transparent cover of the dome improves light transmission and is less prone to dust accumulation compared to flat glass, which ensures long-term performance without frequent maintenance. Additionally, the dome can be enhanced with infrared filter films and reflection-reducing coatings to further optimize light collection, making it more effective in varying environmental conditions.

Once sunlight is captured by the dome, it is channelled through a round tube lined with a highly reflective material such as silver-coated aluminium, which maximizes light transmission through the building. The efficiency of this process is highest when the tube is short and straight, as these conditions allow the light to travel with minimal loss in intensity. In contrast, longer or angled tubes experience greater losses in light intensity, which can diminish the overall effectiveness of the system. To mitigate these losses, manufacturers often claim reflectivity levels between 98% and 99.5%, which helps to preserve light quality even over longer distances (Song et al. 2023). At the endpoint where light enters the room, a diffuser ensures that the

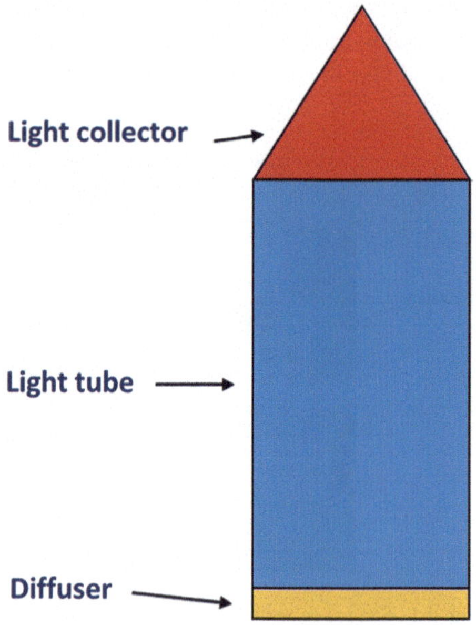

Fig. 4.2 Schematic diagram of light pipe daylighting system (Song et al. 2023)

light is distributed evenly across the space. This prevents the formation of glare or concentrated spots of light, which can be distracting or uncomfortable for occupants. Diffusers are available in a range of styles, including soft white or prismatic designs, and they fit securely into a trim ring that seals out dust and moisture, maintaining the integrity of the system over time.

For optimal performance, particularly in environments where sunlight availability fluctuates, a heliostat can be installed along with the dome. A heliostat is a device that tracks the movement of the sun throughout the day, continuously adjusting its position to direct the maximum amount of sunlight into the light tube (Thalange et al. 2017). This system ensures that even when the position of the Sun shift, the light pipe continues to collect and channel sunlight efficiently. The heliostat enhances the flexibility of light pipes, making them suitable for use in a variety of conditions and orientations. In certain designs, mirrors or other reflective elements can be integrated to further control and direct the light path, allowing for greater customization in light distribution. Interestingly, some advanced heliostats can even be adjusted to capture moonlight at night, providing a subtle and energy-efficient source of illumination for low-light environments.

(c) **Light Rod**

A light rod also referred to as a light fiber is an optically transparent solid rod or flexible fiber, typically made from glass or plastic that transmits light through the principle of multiple internal reflections. This optical mechanism enables light to travel considerable distances with minimal loss in intensity or quality. Light rods use total internal reflection, a phenomenon in which light is confined within the rod by bouncing off the internal surfaces without escaping (Mathiasen et al. 2022). This allows the light to be efficiently transferred from one end of the rod to the other, making it an effective medium for light transmission in various applications. Although light rods may resemble optical fiber communication cables in their structure, their primary function is different. While both systems rely on the principles of internal reflection, light rods are optimized for lighting applications rather than for data transmission. Specifically, light rods operate within the wavelength range of 0.3–0.8 μm, which includes much of the visible light spectrum (Austin, et al. 2021; Krishnamoorthi et al. 2023). This range is ideal for applications requiring effective illumination. Additionally, light rods often incorporate infrared filters to block unwanted wavelengths, preventing degradation in light quality and enhancing the clarity of the transmitted light.

In contrast, optical fiber communication systems are designed to transmit data and typically operate within a longer wavelength range, from 0.8 to 1.6 μm. This extended range is optimized for data transmission over long distances, where maintaining signal strength and minimizing data loss is crucial. While the technology behind light rods and fiber optic cables is similar, the differences in wavelength usage and intended purpose distinguish these two systems. Light rods prioritize the efficient transfer of visible light for illumination, while optical fiber cables focus on data transmission with minimal interference or loss. The use of light rods in lighting applications offers several distinct advantages. One of the key benefits is their ability to provide focused and directional light without requiring electrical components at the light-emitting end. This makes light rods especially useful in specialized lighting scenarios such as in medical environments, where precise and directed lighting is necessary without the interference of electrical equipment. Additionally, light rods are often employed in architectural lighting, where they can be used to create ambient or accent lighting effects in buildings, art galleries, or public spaces. Their ability to deliver light over long distances also makes them ideal for illuminating areas that are difficult to reach with conventional lighting systems. In art installations, light rods are frequently used to create striking visual effects by channelling light into specific locations (Kushner et al. 2021). They emphasize design elements or highlight focal points without the need for visible fixtures or wiring. The flexibility of light rods allows them to be easily adapted to the unique requirements of various artistic and architectural projects.

4.1.2 Daylighting Evaluations

Daylighting presents both notable benefits and potential challenges that must be carefully weighed when incorporating it into building designs. On the positive side, it offers several advantages such as enhancing the quality of indoor environments, improving occupant well-being, and significantly reducing energy consumption. However, the successful implementation of daylighting requires thoughtful planning and meticulous design to ensure its benefits are maximized and its drawbacks minimized. Without careful consideration, daylighting can introduce challenges that, if not properly addressed, may reduce its effectiveness or lead to unforeseen issues (Tabadkani et al. 2021). Once daylighting systems such as skylights or solar light pipes are installed, the natural light they provide remains constant during daylight hours, regardless of the immediate needs of the occupants. This can lead to persistent challenges such as glare, uncomfortable heat gain during the summer months, heat loss in winter, condensation, and the risk of water leakage. Thus, a well-considered design and installation process is crucial to optimize the performance of daylighting systems. Rushing through the design phase without adequately addressing factors such as light distribution, heat management, and condensation prevention can result in long-term problems that outweigh the initial energy savings. It is equally essential to synchronize daylighting with electric lighting systems to achieve a balance that adapts to varying light conditions throughout the day. Without effective control systems, the potential energy savings from daylighting can be lost due to inefficient overlap with artificial lighting. Below, a detailed evaluation of the merits and drawbacks of daylighting is provided to guide designers and engineers in making informed decisions.

(a) **Merit of Daylighting**

i. **Energy Conservation**

One of the primary merits of daylighting is its potential to significantly reduce energy consumption in both new and existing buildings. By harnessing natural light, buildings can decrease their reliance on artificial lighting, leading to substantial energy savings. Daylighting systems such as solar light tubes and optical fiber lighting, offer flexible solutions that are particularly well-suited for retrofitting older structures. These technologies allow buildings that were not originally designed with energy efficiency in mind to take advantage of modern daylighting methods without extensive structural modifications. The energy savings from daylighting extend beyond just lighting. By reducing the need for artificial light, buildings also experience lower internal heat gains which is a common by-product of electric lighting systems (Nundy et al. 2021). This reduction in heat load can subsequently decrease the demand for cooling systems during warmer months, resulting in further energy conservation. In addition, natural light offers a free and renewable light source, which decreases the building reliance on non-renewable energy for lighting, making it an environmentally sustainable option.

Furthermore, daylighting proves especially advantageous during peak energy demand periods of times when energy consumption is at its highest such as midday in summer. Since daylighting directly reduces the amount of artificial lighting required during these periods, it helps alleviate strain on the electrical grid and leads to cost savings for building owners. These savings are not only immediate through reduced electricity bills but also provide long-term benefits. Energy-efficient buildings often experience lower operational costs and can qualify for green building certifications, which may offer additional financial incentives. By integrating daylighting systems into building design, energy consumption can be reduced across lighting, cooling, and overall operational demands, making daylighting an integral component of sustainable building practices.

ii. **Bringing Greenery Indoors**

Daylighting plays a crucial role in supporting the health and growth of indoor plants, contributing to both aesthetic and environmental benefits. Natural light is essential for the process of photosynthesis, through which plants produce energy, primarily from the visible and violet light that fuels chlorophyll production. By incorporating daylighting systems such as skylights or solar light pipes, indoor environments can support vibrant indoor landscapes without the need for artificial grow lights. The integration of plants into indoor spaces enabled by adequate natural light not only enhances the visual appeal of these spaces but also improves air quality. Plants act as natural air purifiers, filtering out toxins such as carbon dioxide, formaldehyde, and volatile organic compounds (VOC) from the air. This process improves the indoor air quality, creating healthier environments for occupants. Additionally, greenery can reduce indoor noise levels by absorbing sound, contributing to a more tranquil and comfortable atmosphere, particularly in office or residential settings. Another benefit of daylight-driven plant growth is the ability to cultivate indoor gardens or green walls, which can add a fresh and dynamic element to interior design. With sufficient daylight, there is no need to move plants to windows or outdoor areas to ensure they receive adequate light. This allows for more flexibility in plant placement, enabling the design of indoor landscapes in areas that would otherwise remain unused. Beyond the practical advantages, the presence of plants in well-lit spaces has been shown to offer psychological benefits. Research indicates that indoor plants, especially when naturally lit, can boost occupant mood, reduce stress levels, and enhance overall productivity. This makes daylighting an essential element in biophilic design, which seeks to integrate nature into built environments to improve human well-being. Overall, the synergy between daylighting and indoor greenery enhances both the functionality and liveability of interior spaces, making it a valuable asset in sustainable architecture.

iii. **True Colours of the Indoor Space with Natural Light**

Natural light is widely regarded as the optimal source for accurate colour rendering, which significantly enhances the visual appeal and functionality of indoor spaces. Unlike many artificial lighting sources which often lack the full spectrum of visible light, daylight contains a broader and more balanced distribution of wavelengths. This allows it to reflect the true colours of objects more accurately, making interiors appear vibrant and visually engaging. For instance, under natural light, colours tend to appear more consistent and lifelike. This is particularly important in environments where accurate colour perception is essential such as in art galleries, retail stores, or design studios. In these spaces, the correct rendering of colours can impact decision-making, aesthetics, and even user experience. Daylight also improves contrast and clarity, enabling better differentiation between objects and their backgrounds, which enhances both comfort and task performance.

To further optimize daylight distribution in indoor spaces, engineered diffusers are often used. These devices evenly disperse natural light throughout a room, minimizing the risk of harsh shadows or overly bright spots. This ensures that light is uniformly spread across the space, enhancing visual comfort and creating a more pleasant environment for occupants. Beyond the aesthetic and functional benefits, humans are biologically attuned to daylight. Exposure to natural light has been shown to provide psychological and physiological benefits that cannot be replicated by artificial lighting (Brown et al. 2022). For example, daylight helps regulate our circadian rhythms, which in turn influences mood, sleep, and overall well-being. Natural lighting creates a more dynamic and natural indoor environment, improving occupants productivity and mood especially in workspaces or living areas where people spend extended periods of time. Overall, the use of natural light enhances the true colours of indoor spaces, fostering both aesthetic quality and human comfort that artificial lighting systems often fail to provide.

iv. **Tubular Light Pipes**

Tubular light pipes also known as solar tubes or sun tunnels are highly efficient daylighting systems designed to channel natural light into interior spaces, especially those that are difficult to illuminate using traditional windows or skylights. These systems consist of a highly reflective tube that captures sunlight at the roof or exterior wall and transmits it into interior rooms, such as bathrooms, hallways, or closets, where direct access to windows is often not feasible. One of the major advantages of tubular light pipes is their compact design combined with high light transmission efficiency. The reflective surfaces inside the tube maximize the amount of sunlight that is delivered into the space, providing a significant amount of natural light even through small openings (Sambiagio and Noel 2020). Tubular light pipes are more versatile than traditional skylights which require larger structural changes, making them ideal for installation in almost any area of a building. Their efficiency and small footprint make them suitable for residential, commercial, and industrial applications.

In terms of installation, tubular light pipes offer practical benefits. Unlike traditional skylights, which often require extensive modifications such as new framing, drywall adjustments, and painting, tubular skylights can typically be installed in less than two hours without major renovations. This ease of installation translates to lower upfront costs, making them a cost-effective solution for daylighting both new and retrofitted buildings. Due to their innovative design, tubular light pipes are extremely energy efficient. By providing natural light to interior spaces during the day, they reduce the need for electric lighting, especially in areas where lights would typically be left on throughout the day. Additionally, tubular light pipes are designed to minimize heat transfer. They allow light to enter while preventing excess heat from the sun, ensuring that interior temperatures remain stable and comfortable. This feature reduces the need for additional cooling during hot weather, further lowering energy consumption. In summary, tubular light pipes are an efficient and versatile solution for delivering natural light to interior spaces. Their ease of installation, energy efficiency, and ability to reduce heat gain make them an excellent choice for sustainable building designs.

v. **Enhanced Spaciousness with Tubular Light Pipes**

Tubular light pipes not only provide energy-efficient lighting but also contribute to a sense of enhanced spaciousness in interior environments. By introducing natural light into otherwise dim or windowless areas, these systems create visual connections between indoor and outdoor spaces. This natural light infusion gives the impression that the room is larger and more open, even in compact or enclosed areas such as hallways, stairwells, or small offices. Daylight has the unique ability to expand the perceived boundaries of a room by making the space feel brighter and more inviting. In architectural design, the use of natural light is a well-known strategy to enhance the visual and spatial qualities of an environment (Emre Ilgın 2021). Tubular light pipes allow even small, enclosed spaces to benefit from the psychological and aesthetic advantages of daylight, creating an airy and open atmosphere. Additionally, because tubular light pipes are compact and can be installed in tight spaces, they provide the opportunity to bring daylight into parts of the building that traditionally rely on artificial lighting. By illuminating these spaces with natural light, the overall ambiance of the interior is improved, contributing to the feeling of expansiveness. This is particularly important in urban settings, where access to outdoor views and natural light may be limited. The introduction of daylight into a room also affects how colours, textures, and surfaces are perceived. Sunlight enhances the depth and dimension of a space, making architectural features stand out more prominently. As a result, rooms illuminated by tubular light pipes often feel more dynamic and visually engaging, even if they are relatively small in size. Incorporating tubular light pipes as part of a building daylighting strategy not only reduces energy consumption but also transforms the feel of the space. The enhanced sense of openness and connection to the outdoors creates a more pleasant and functional environment for occupants.

vi. **Daylight Provides Indoor People with a Sense of Cheeriness and Brightness**

Daylight has a profound effect on the emotional and psychological well-being of individuals, particularly in indoor environments. Spaces illuminated by natural light tend to feel more cheerful and vibrant, providing occupants with a sense of brightness and connection to the outdoors. This is especially relevant in workplaces and residential buildings, where occupants may spend extended periods of time indoors. Daylight introduces a natural variation in light levels and colour temperatures throughout the day, mimicking the cycles of the sun and enhancing the overall ambiance of a space. In office settings, employees often express dissatisfaction with artificial lighting, citing it as monotonous and lacking the natural changes that daylight provides. The constant and unvarying nature of artificial lighting can contribute to feelings of confinement, isolation, and even claustrophobia. In contrast, exposure to daylight allows individuals to observe weather changes and natural daylight shifts, fostering a deeper connection to the outside world. These subtle environmental cues, which are absent in artificially lit spaces, can significantly improve occupants mood and productivity.

Daylight has also been shown to have direct physiological effects. Studies suggest that exposure to natural light can help regulate circadian rhythms, the body internal clock, which influences sleep patterns, alertness, and overall energy levels. In spaces where individuals are exposed to daylight, there tends to be an improvement in both mental clarity and emotional well-being. This connection to the natural light cycle enhances not only the mood but also physical health, reducing stress levels and promoting a more relaxed atmosphere. In residential buildings, daylight can similarly create a sense of warmth and comfort, making homes feel more welcoming and enjoyable. The dynamic nature of natural light, with its subtle changes in brightness and colour throughout the day, contributes to a lively and uplifting environment. Whether it is the soft morning light or the warm glow of the afternoon sun, daylight helps to cultivate a positive and energizing atmosphere within indoor spaces. Incorporating natural daylight into interior spaces has far-reaching benefits for both mental and physical health. It enhances feelings of cheerfulness and brightness, providing a vital link to the outside world that artificial lighting systems simply cannot replicate.

(b) **Demerits of Daylighting**

Glare occurs when excessive or unwanted light enters the visual field, often causing discomfort and reducing visibility. This phenomenon is particularly common in spaces that receive direct sunlight especially during summer when the sun is higher in the sky and its intensity is at its peak. Discomfort glare can manifest in various ways including distraction, annoyance, eye strain, and even headaches. In extreme cases, glare can impair visual tasks, making it difficult for occupants to focus, which can negatively impact productivity in work environments or comfort in residential spaces. In buildings with large windows or skylights, the intensity of direct sunlight can create harsh contrasts between bright and shaded areas, further exacerbating the problem of glare (Eiz et al. 2021). This contrast makes it difficult for the human eye to adjust, leading to visual discomfort. Moreover, reflective surfaces inside the building such as glass tables, polished floors, or computer screens can amplify glare by reflecting sunlight into unintended areas, increasing the overall discomfort for occupants.

To mitigate glare, it is essential to incorporate shading devices into the design of daylighting systems. For example, blinds, curtains, or exterior louvers can be installed to control the amount of sunlight entering the space. These systems allow for adjustable light levels, enabling occupants to reduce glare without blocking natural light entirely. Diffusers are another effective solution, especially when used with tubular light pipes or skylights. By scattering the incoming light, diffusers reduce the intensity of direct sunlight and create a more balanced, even distribution of light throughout the room (Omishore et al. 2020). Additionally, light shelves horizontal surfaces installed inside or outside windows can be used to reflect sunlight upwards toward the ceiling, where it is diffused and spread evenly across the space. This method minimizes direct sunlight penetration into workspaces while maximizing the use of daylight. In combination with low-emissivity glass, which limits the amount of solar radiation passing through windows. These solutions can significantly reduce glare and improve visual comfort for building occupants. While intense sunlight can cause significant glare, implementing a combination of shading systems and diffusers can effectively manage light levels, allowing for the benefits of daylight without the associated discomfort.

ii. **Fading of Colours**

While daylight is highly regarded for its excellent colour rendering, it also carries the potential disadvantage of causing the fading of colours in interior furnishings over time. This issue primarily arises due to the high ultraviolet (UV) content present in natural sunlight. UV radiation which is part of the invisible spectrum of light can break down the chemical bonds in dyes and pigments causing them to deteriorate and lose their vibrancy (Alonso-Villar et al. 2022). As a result, prolonged exposure to direct sunlight can cause the colours in fabrics, artwork, wallpaper, and other materials to fade significantly. This phenomenon is particularly noticeable in curtains, carpets, furniture, and other indoor elements that are positioned near windows or under skylights, where they receive the most direct sunlight. Over time, the damage caused by UV radiation can become permanent, leading to significant discoloration and degradation of valuable or aesthetic items in residential, commercial, and cultural spaces. In environments where preserving the integrity of colours is important, such as retail stores, offices, or art galleries, this can be a major drawback of daylighting. One of the most effective ways to mitigate this issue is through the use of UV filters. These filters can be applied to windows, skylights, and solar light pipes to block a significant portion of UV rays while still allowing the visible spectrum of daylight to enter.

UV-blocking films or coated glass are commonly used in these applications to prevent UV radiation from reaching indoor spaces, reducing the risk of fading while maintaining the benefits of natural light (Alonso-Villar et al. 2022). This is especially useful in environments where valuable objects or textiles are at risk of deterioration. Additionally, UV-absorbing materials can be incorporated into the design of domes and diffusers for solar light pipes. These materials help prevent harmful UV radiation from passing through the daylighting system and entering the interior space, providing further protection for furnishings and artwork. By using these technologies, building occupants can enjoy the benefits of natural light without worrying about the long-term fading of colours. Despite these preventive measures, it is important to note that some level of fading may still occur over time, particularly in areas that receive constant or intense sunlight. However, by strategically placing shading devices, window treatments, and UV-protective elements, the impact of colour fading can be significantly reduced. This makes daylighting a more viable option for a wide range of building types and uses. While fading of colours remains a potential downside of daylighting, modern technologies like UV filters and UV-absorbing materials offer effective solutions to preserve the vibrancy and longevity of indoor decor.

iii. **Inactivity During Night or Overcast Conditions**

One of the key limitations of daylighting systems is their inherent reliance on natural sunlight, which becomes unavailable during the night or in heavily overcast conditions. Unlike electric lighting which provides a consistent and controllable light source, daylight fluctuates in both intensity and colour throughout the day and across seasons. As a result, spaces that rely heavily on daylighting may face periods where insufficient light is available, leading to the need for supplemental artificial lighting. To address the issue of inactivity at night, a practical solution is to integrate hybrid lighting systems that combine daylighting with artificial light sources (Lv et al. 2020). For instance, an incandescent or LED fixture can be mounted inside the light tube or skylight and connected to a wall switch or automated control system. This setup allows the lighting system to provide consistent illumination even after sunset or during cloudy weather. The hybrid system seamlessly transitions between natural light during the day and artificial light at night, ensuring continuous and reliable lighting for the occupants.

Another method to enhance the functionality of daylighting systems during overcast or night time conditions is to use daylight sensors and lighting controls. These sensors can detect the amount of available natural light and automatically adjust the electric lighting accordingly. When sufficient daylight is present, the sensors dim or turn off the artificial lights, reducing energy consumption (Plörer et al. 2021). Conversely, when natural light levels drop due to cloud cover, weather conditions, or the onset of evening, the sensors activate the electric lights to maintain adequate indoor illumination. This ensures that the building remains well-lit while still maximizing energy efficiency by prioritizing natural light when available. Although daylight fluctuates in both brightness and colour, some architects and designers embrace this variability as a dynamic and expressive element in their designs. The natural progression of sunlight throughout the day creates changing atmospheres, offering a unique aesthetic quality that artificial lighting cannot replicate. The interplay between light and shadow as well as the shifting hues of daylight can add visual interest to interior spaces, making them feel more organic and alive. For many, this unpredictability becomes a creative tool rather than a drawback, particularly in spaces designed to highlight the connection between the indoors and outdoors.

However, in environments where consistent lighting is crucial such as offices, educational institutions, or hospitals, the fluctuations in daylight may be viewed as an inconvenience. In such cases, hybrid systems that blend natural and artificial lighting provide the best of both worlds, ensuring a balance between energy efficiency and practicality. While the inactivity of daylighting systems during night or overcast conditions is a notable limitation, the integration of hybrid lighting solutions provides a practical solution. Additionally, sensor-based controls ensure consistent illumination, offering both energy savings and continuous lighting regardless of external conditions.

UK scenarios Daylighting research and industry in the UK are highly active. The Chartered Institution of Building Services Engineers (CIBSE) updated its lighting code in 2006 and published a comprehensive guide titled "CIBSE Lighting Guide LG10: Daylight and Window Design," (CIBSE and Design 1999). Several UK companies are also engaged in the development of solar light pipes. The School of Built Environment at The University of Nottingham is particularly active in researching daylighting and solar light pipes/rods. This institution has accumulated significant experience and research materials in this field, supported by the work of three Ph.D. students who conducted extensive research on these topics. Demonstrations of light pipes are installed in school buildings, as depicted in Fig. 4.3, and an experimental test platform has been constructed, as shown in Fig. 4.4. The funded projects for daylighting research is tabulated in Table 4.2

4.2 Novel Energy-Saving Lamps

As energy efficiency and sustainability continue to shape the future of lighting technology, fluorescent lights have emerged as one of the most widely adopted alternatives to traditional incandescent bulbs. Fluorescent lamps are known for their ability to consume significantly less electricity while offering a longer operational lifespan. These characteristics make them a more sustainable and cost-effective solution for both residential and commercial applications (Belessiotis et al. 2022). With their energy savings and durability, fluorescent lights have gained widespread use, particularly in offices, schools, and large commercial spaces, where they help to reduce overall operational costs. The benefits of fluorescent lights do not stop at energy efficiency. These lamps also emit less heat, contributing to a cooler indoor environment and reducing the need for air conditioning, which in turn further conserves energy (Hudallah 2021). This is especially important in commercial buildings where

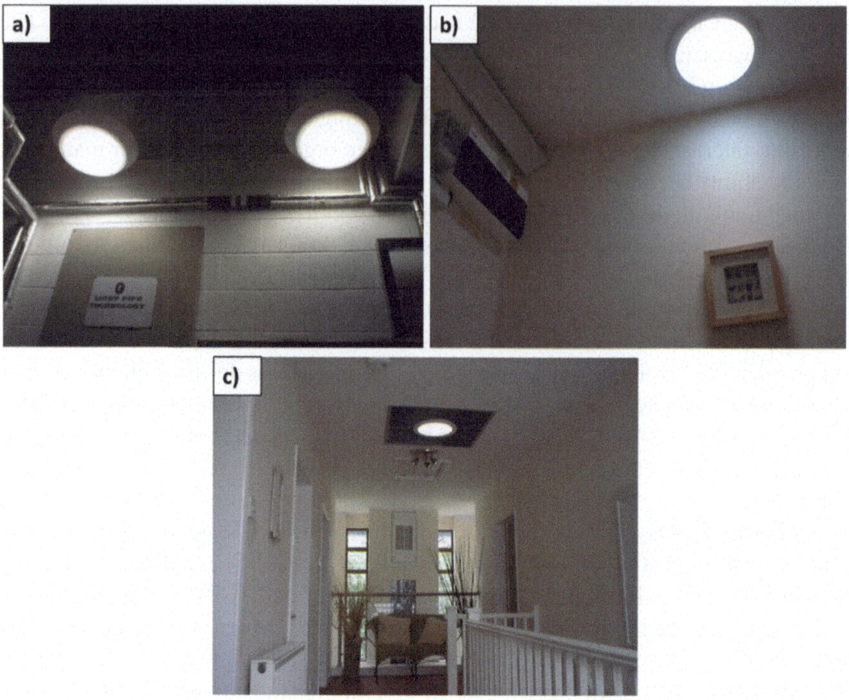

Fig. 4.3 Solar light pipes mounted in buildings at SBE of UoN. **a** Lightpipe in Marmont centre Lab, **b** Ecohouse toilet in ground floor and **c** a lightpipe in Ecohouse hallway

large numbers of lights are in operation for extended periods. The longer lifespan of fluorescent lamps typically lasting up to 10,000 h or more reduces the frequency of replacement, which translates to lower maintenance costs and fewer waste materials, enhancing their sustainability profile.

In recent years, the lighting industry has seen even greater advancements in energy-saving technologies, giving rise to more efficient options that outperform traditional fluorescent lamps. These innovations are designed to further reduce energy consumption while delivering higher-quality illumination. Compact fluorescent lamps (CFL) which are smaller versions of standard fluorescent lights, offer similar benefits but are compatible with most standard light fixtures, making them an easy upgrade for homeowners and businesses alike. CFL use less energy than incandescent bulbs and last up to 10 times longer, making them a popular choice for those looking to reduce both their energy bills and environmental footprint (Yadlapalli et al. 2020). Moreover, the introduction of light-emitting diodes (LED) has revolutionized the lighting industry, pushing the boundaries of efficiency and performance. LED are now considered the most energy-efficient lighting option available, using up less energy than traditional incandescent bulbs while offering lifespans of 25,000 to 50,000 h or more (Katzin et al. 2021). Unlike fluorescent lights, which contain small amounts of mercury and require careful disposal, LED are free of

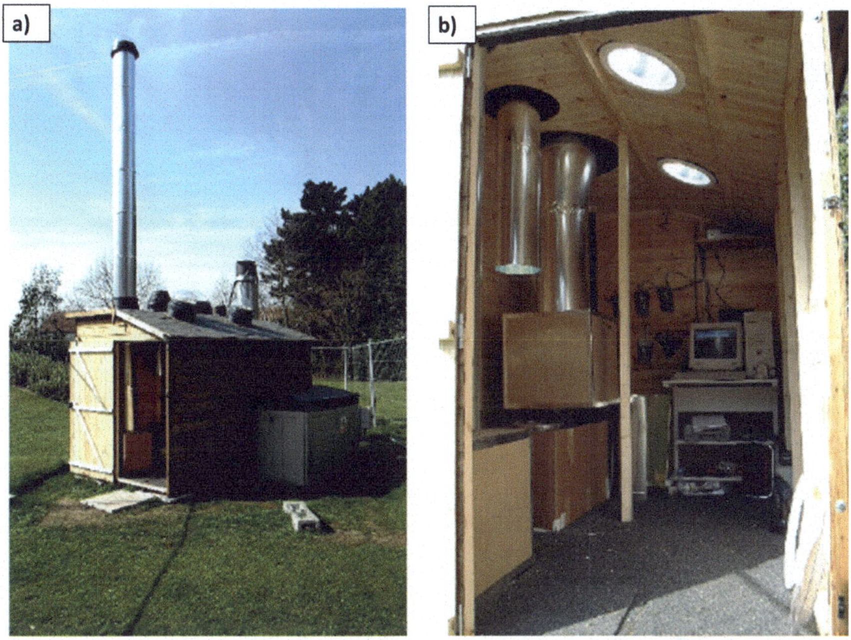

Fig. 4.4 Testing chamber for light pipes: **a** exterior and **b** interior (right) in SBE of UoN (Callow 2003)

hazardous materials, making them a safer and more environmentally friendly alternative. Additionally, the versatility of LEDs has expanded their application beyond standard lighting solutions to include everything from smart lighting systems to outdoor streetlights and architectural lighting.

A key factor that sets modern energy-saving lamps apart is their ability to provide superior illumination quality. Many traditional incandescent and fluorescent lamps suffer from poor colour rendering, which can distort how colours appear under artificial lighting. LED offer excellent colour accuracy, often measured using the Colour Rendering Index (CRI). With CRI values close to natural daylight, LED lighting provides clear, vibrant illumination that enhances visibility and comfort in indoor environments. In addition to improving efficiency and light quality, recent developments in lighting technology have focused on creating solutions that are more adaptive and intelligent. Smart lighting systems allow users to control lighting via smartphones or integrate lighting with home automation systems. These smart systems not only improve convenience but also optimize energy use by adjusting light intensity based on occupancy or natural daylight levels. These advancements help create lighting solutions that are not only energy-efficient but also responsive to the changing needs of modern buildings and users.

The shift toward novel energy-saving lamps reflects a growing global emphasis on sustainability, energy conservation, and environmental responsibility. As the demand

Table 4.2 Funded projects for daylighting

Topic classification	Project reference	Principal	Title	Department organization	Time	Value £
Building operations and management design engineering human–computer interactions	EPSRC EP/E027148/1	B Krausse	Quantifying Visual Discomfort in the Daylit Workplace using a High Dynamic Range (HDR) Imaging Approach	Institute of Energy and Sustainable Dev, De Montfort University	01/01/07–31/12/08	183,063
Building operations and management	EPSRC GR/S25074/02	S.A. Fotios	White Light: an investigation of the perception of exterior lighting	Architectural Studies, University of Sheffield	01/09/05–31/08/06	46,946
Sci. research, construction tech, energy saving, renewable sources of energy	EU project JOR3970172 (4th framework)	Kai Siren S.B.Riffat (partner)	Integrated system for daylighting, natural ventilation and solar heating	Helsinki University of Technology Univ of Nottingham	01/02/98–31/07/00	N/A
Economic aspects, energy saving	EU project ERK6-CT-1999–00007 (5th framework)	Michael Bowtell	European integrated daylighting design tool ('DIAL-EUROPE')	The chancellor, masters and scholars of the university of Cambridge	01/04/00–31/03/03	261386 EURO
Energy saving, Fossil Fuels, renewable sources of energy	EU project JOU20404 (3rd framework)	M Fontoynont	Properties of glazing materials with respect to daylighting applications	Centre National de la Recherche Scientifique (CNRS)	01/02/94–31/07/95	N/A
Construction technology, renewable sources of energy	EU project JOR3960010 (4th framework)	Berndt Schalin	Integrated daylighting system based on smart controls for user satisfaction	Helvar Oy	01/04/96–31/03/99	N/A

(continued)

Table 4.2 (continued)

Topic classification	Project reference	Principal	Title	Department organization	Time	Value £
Other energy topics, energy saving	EU project XVII/4.1031/ 94-153	Peter voit	Optimum integration of daylighting components at the rehabilitation process of commercial buildings	Transsolar Energietechnik GmbH	01/11/94–01/08/96	127880 EURO
Other energy topics, energy saving	EU project XVII/4.1031/ 94–138	Friedrich sick	Integration of daylighting technologies in demand side management programs: estimation of the energetical and peak load potential	Fraunhofer Institute for Solar Energy Systems	01/11/95–28/02/98	29804 EURO
Energy saving, Fossil Fuels, renewable sources of energy	EU project JOU20144 (3rd framework)	Dimosthenis Asimakopoulos	Availability of daylighting in Europe and design of an European Daylighting Atlas	National Observatory of Athens	01/01/93–30/06/95	N/A
Construction technology, renewable sources of energy	EU project JOR3950016 (4th framework)	Friedrich Sick	Evaluation and user assessment of lighting systems performance	Fraunhofer gesellschaft zur forderung der angewandten forschung e.v	01/01/96–31/12/98	N/A

(continued)

Table 4.2 (continued)

Topic classification	Project reference	Principal	Title	Department organization	Time	Value £
Coordination, cooperation, scientific research	EU project CIPD925033	D.N. Asimakopoulos	Modelling of sky luminance for the combined daylighting and discomfort glareprediction: evaluation methods applicate to collected data with draft standart models as output	National Observatory of Athens	01/06/93–30/11/95	N/A
Construction technology, renewable sources of energy	EU project JOR3960097 (4th framework)	Volker Wittwer	Innovative glazings for high solar gains and daylighting	fraunhofer gesellschaft zur forderung der angewandten forschung e.v	01/09/96–31/08/98	N/A
Renewable sources of energy	EU project XVII/4.1030/95–094	Athanassios Argiriou	Creation of an educational structure on the use of daylighting techniques in buildings	NATIONAL OBSERVATORY OF ATHENS (NOA)	22/11/95–21/05/97	93,000 EURO
Energy saving, Fossil Fuels, Renewable sources of energy	EU project JOU20282 (3rd framework)	Poul-Erik Kristensen	Daylighting design of European buildings	Esbensen - Rådgivende Ingeniører ApS	01/09/94–31/08/97	N/A

(continued)

Table 4.2 (continued)

Topic classification	Project reference	Principal	Title	Department organization	Time	Value £
Economic aspects, construction technology, energy saving, renewable sources of energy	EU project NNE5/550/ 2001 (5th framework)	Thomas Schott	Light Weight PV-louvres For Multifunctional Solar Control And Daylighting Systems With Improved Building Integration - Target Action E	zentrum fuer sonnenenergie- und wasserstoff - forschung baden-wuerttemberg	01/02/03–31/12/05	1294526 EURO
Scientific research, construction technology, energy saving, renewable sources of energy	EU project JOE3971004 (4th framework)	Anna Gazzini	Development of a natural light illuminator with flexible light duct for wide scale application in dark spaces of buildings	Solarconvert Srl Processi e Sistemi Energetici	31/07/97–31/03/98	N/A
Economic aspects, construction technology, energy saving	EU project JOE3980096 (4th framework)	H.A.L Van dijk	Daylighting products with Redirecting Visual properties	Netherlands organisation for applied scientific research - tno	01/10/98–31/12/00	N/A

for energy-efficient solutions continues to rise, the lighting industry is likely to see further innovations, with new technologies emerging that deliver even greater efficiency, versatility, and performance. Energy-saving lamps like fluorescent lights, CFL, and LED have transformed the lighting landscape, offering significant energy savings, longer lifespans, and superior illumination quality. As new technologies continue to evolve, the lighting industry will play a key role in reducing global energy consumption and contributing to a more sustainable future.

4.2.1 Compact Fluorescent Lamp (CFL)

A compact fluorescent lamp (CFL) as shown in Fig. 4.5, commonly referred to as an energy-saving light bulb, is a type of fluorescent lamp specifically engineered to fit into standard light bulb sockets or smaller lighting fixtures (Ishak et al. 2023). CFLs offer distinct advantages over traditional incandescent bulbs, such as a longer operational lifespan and reduced electricity consumption, as shown in Table 4.3. On average, a CFL can last up to 10,000 h, which is significantly longer than the typical 1,000 h of incandescent bulbs. Although CFLs may have a higher upfront cost, they typically offset this expense within approximately 500 h of use due to the energy savings they provide over time. One of the key features of CFLs is their availability in both AC and DC versions, with DC CFLs being particularly advantageous in off-grid scenarios. In settings such as recreational vehicles, remote cabins, or other off-the-grid dwellings, DC CFLs are powered by renewable energy sources like small solar panels or car batteries (Hofmann and Rasch 2020). This makes CFLs a practical alternative to conventional lighting methods, particularly in developing countries where access to stable electricity is often limited. In these regions, DC CFLs are increasingly replacing kerosene lanterns by offering a safer, cleaner, and more energy-efficient lighting solution that reduces the risk of fire hazards and indoor air pollution.

Beyond their energy-saving benefits, CFLs contribute significantly to environmental sustainability by lowering the greenhouse gas emissions associated with electricity production. Traditional lighting sources, such as incandescent bulbs, contribute to higher carbon emissions due to their energy inefficiency, while CFLs use less electricity to achieve the same level of illumination. However, CFLs do contain a small amount of mercury, which requires careful disposal to avoid environmental contamination. In response to this concern, advancements in recycling initiatives and mercury recovery technologies have made it easier to mitigate the environmental impact of CFL disposal (Viana et al. 2022).

Additionally, the development of newer, more efficient CFL designs has further enhanced their performance while minimizing their environmental impact. These improvements include reducing the mercury content in newer models and increasing the efficiency of light output. CFL are now available in a wide range of colour temperatures, from warm to cool light, allowing users to customize their lighting preferences to suit different environments. This versatility makes CFL adaptable

Fig. 4.5 A spiral type compact fluorescent

Table 4.3 Electric consumption comparison between compact fluorescent lamp and incandescent lamp

Incandescent (W)	Compact fluorescent (W)
25	5–6
40	10
60	11–13
75	18–20
100	25–27
125	28–35
150	40–42

for a wide range of applications, from residential lighting in homes to commercial and industrial settings where energy efficiency is a priority. Dimmable CFL have also become available, providing users with greater flexibility in creating the desired lighting ambiance. This feature allows for more control over light intensity, making CFL suitable for both task lighting and ambient lighting in various settings. With the ability to create different lighting atmospheres, CFL offer a level of customization that enhances both comfort and aesthetic appeal. As lighting technology continues to evolve, CFL remain a popular choice for consumers seeking to reduce energy consumption without compromising on illumination quality. Their long lifespan, energy efficiency, and adaptability make them a valuable component of sustainable lighting solutions, especially in regions where reducing energy costs and environmental impact is a key concern.

4.2.2 *Light Emitting Diodes (LED)*

Light Emitting Diodes (LED) represent a significant advancement in lighting technology, characterized by their efficiency, longevity, and versatility. Unlike traditional incandescent bulbs, which generate light by heating a filament, LED produce light through the movement of electrons across a semiconductor material, a process known as electroluminescence (Chang et al. 2012). This process is highly efficient, resulting in much less wasted energy in the form of heat. LED can convert up to 90% of the energy they consume into light, making them far superior to incandescent bulbs, which only achieve around 10% efficiency. The applications of LED are extensive and continue to grow rapidly. In residential settings, LED are commonly used in everything from general-purpose bulbs to specialized fixtures such as under-cabinet lighting, outdoor floodlights, and decorative accents. Their ability to offer bright, focused illumination with minimal energy consumption has made them a preferred choice for homeowners looking to reduce their energy usage. Furthermore, LED have found widespread use in commercial and industrial sectors, where they are used in environments that require consistent, long-term lighting. In these settings, LED power office lighting systems, warehouse illumination, and even street lighting, proving to be durable and highly reliable for prolonged use.

In addition to general lighting, LED are integral to electronic displays and screens, providing the backlighting for smartphones, televisions, and computer monitors. Their compact size and ability to operate efficiently across a wide range of temperatures make them ideal for use in outdoor and automotive lighting applications as well. Their robustness against vibrations and environmental extremes such as extreme cold or heat has allowed LED technology to become the standard for many industries where durability is critical. Beyond their energy efficiency, LED boast an impressive lifespan, often lasting between 25,000 to 50,000 h or more, depending on the application and environmental conditions. This extraordinary longevity significantly reduces the need for frequent replacements, leading to lower maintenance costs over time. For large facilities or outdoor installations, where access to lighting fixtures can be challenging and costly, the durability and longevity of LED provide a substantial economic benefit. Additionally, the extended lifespan contributes to reduced environmental impact, as fewer bulbs are disposed of compared to traditional lighting options, aligning with sustainability goals in both residential and commercial contexts.

One of the key benefits of LED technology is its versatility. LED are available in a wide range of colour temperatures from warm to cool, allowing users to select lighting that suits their specific needs and preferences. This flexibility is particularly useful in environments where ambient lighting is important such as in restaurants, retail stores, or residential living spaces. In addition, many LED are now dimmable, offering greater control over lighting intensity and atmosphere. This makes them suitable for creating both bright task lighting and soft, ambient lighting, depending on the desired effect. Despite the higher initial cost of LED compared to incandescent or compact fluorescent lamps (CFL), the total cost of ownership is significantly lower due to the substantial energy savings and reduced frequency of replacement. Over time, the investment in LED technology is recouped through lower electricity bills and maintenance costs, making it a highly cost-effective solution. As a result, LED have become the lighting option of choice in sectors focused on sustainability and energy conservation.

As LED technology continues to evolve, further advancements in light output and colour rendering have made these systems even more attractive. Improvements in the Colour Rendering Index (CRI) allow LED to more accurately replicate the vibrancy of natural light, enhancing visibility and colour accuracy in spaces such as galleries, offices, and retail environments. With ongoing innovations, LED are expected to dominate the lighting market, offering solutions that meet the demands of both energy efficiency and high-quality illumination. LED have revolutionized the lighting industry with their unparalleled efficiency, versatility, and long lifespan. Their broad range of applications, combined with their low operating costs and minimal environmental impact, positions them as the future of lighting. As the technology continues to improve, LED will remain at the forefront of energy-saving lighting solutions, driving further reductions in global energy consumption and setting new standards for lighting quality and sustainability.

4.2.3 Cold Cathode Fluorescent Light (CCFL)

The Cold Cathode Fluorescent Light (CCFL) represents a significant advancement in compact fluorescent lighting technology. Unlike traditional Compact Fluorescent Lamps which rely on filaments to generate light, CCFL operate without the use of filaments. Instead, they rely on electrodes to produce light, a design innovation that allows them to function at a voltage that is approximately five times higher than that of standard CFL, while drawing only about one-tenth of the current (Alberts et al. 2010). This makes CCFL a unique and energy-efficient solution, offering distinct advantages over other forms of fluorescent lighting. One of the standout features of CCFL is their slender diameter, typically around 3 mm, which makes them ideal for use in applications that require compact, thin lighting solutions. For example, CCFL are commonly employed as backlighting in thin monitors, displays, and various electronic devices such as laptops and televisions. Their small size and ability to produce uniform light across a surface make them well-suited for these applications, where space constraints and consistent light output are critical.

Despite their compact size, CCFL boast an impressive lifespan of up to 50,000 h, making them highly durable and long-lasting. However, one of the trade-offs for this longevity is that CCFL tend to be less efficient in terms of lumens per watt compared to traditional CFL. On average, CCFL achieve about half the efficiency of CFL, meaning they produce less light per unit of energy consumed. Nonetheless, their extended lifespan and reliability make them an attractive option for specific lighting applications, particularly in electronics and signage, where long-lasting performance is prioritized over raw light output. Although CFL and CCFL technologies offer substantial energy savings and environmental benefits, their adoption has faced some resistance from both consumers and professionals. This reluctance persists despite the payback period for these energy-saving technologies, which typically ranges from three to twelve months due to the reduction in energy costs. One of the main reasons for this resistance is the higher initial cost of CCFL compared to incandescent or standard fluorescent bulbs. Consumers often hesitate to invest in CCFL because of the upfront expense, even though the long-term savings on electricity bills would outweigh the initial costs. In addition, some professionals who install lighting systems may overlook the potential savings, viewing the reduction in electricity costs as insignificant or not worth the investment.

To address these challenges, governments and environmental organizations around the world have introduced initiatives aimed at promoting the adoption of CFL and CCFL technologies. Many of these programs involve the distribution of energy-efficient bulbs to consumers at little or no cost, with the goal of raising awareness about the environmental and economic benefits of these lighting solutions. For instance, in the United Kingdom, environmental activists have successfully lobbied Parliament to consider measures such as taxing or outright banning incandescent bulbs (Koretsky 2021). These efforts have sparked public debate and encouraged campaigns like "Ban the Bulb", which advocate for the transition to more sustainable and energy-efficient lighting options. Moreover, CCFL offer unique advantages in specialized applications where their compact size and extended lifespan are highly valued. In fields such as electronics, signage, and architectural lighting, CCFL are preferred for their reliability and efficiency. In architectural settings, for example, CCFL can be used to create accent lighting or illuminate complex structures where traditional lighting fixtures may not fit. The ability to operate in various environments and under different conditions makes CCFL a versatile choice for niche lighting applications. As the lighting industry continues to evolve, it is likely that CCFL technology will see further advancements, particularly in improving efficiency and broadening its applicability in both residential and commercial settings. Ongoing research and development in this area, driven by the global emphasis on sustainability and energy conservation, will likely position CCFL as a crucial element in the future of energy-efficient lighting. With their unique properties and potential for improvement, CCFL are poised to play a significant role in addressing the growing demand for sustainable lighting solutions.

4.3 Smart Lighting Systems

Smart lighting systems represent a significant convergence of advanced lighting technology and the Internet of Things (IoT), allowing users to control and manage lighting through various devices such as smartphones, tablets, or voice-activated assistants like Amazon Alexa or Google Home (Füchtenhans et al. 2021). These systems integrate technologies such as sensors, wireless communication, and cloud computing, creating an adaptable and automated lighting environment. Smart lighting responds to real-time data on factors like occupancy, daylight levels, and user preferences, offering users unprecedented control and customization over their lighting systems. The primary advantage of smart lighting systems is their ability to optimize energy use by automatically adjusting light levels based on current conditions. For instance, these systems can dim or brighten lights depending on the time of day, whether the room is occupied, or how much natural light is entering the space. This automated response reduces unnecessary energy consumption and enhances user comfort, providing the right amount of light only when it is needed (Martirano 2011). This capability makes smart lighting an essential tool for both residential and commercial settings, where reducing energy use is a key concern.

Smart lighting systems offer a variety of benefits to users, including the ability to set schedules, control lighting remotely, and create customized lighting scenes for different activities. For example, homeowners can program lights to gradually dim in the evening to promote relaxation or mimic natural daylight patterns to maintain a healthy circadian rhythm (Lamnatou et al. 2022). Lights can also be set to turn on and off based on occupancy ensuring that spaces are only lit when needed, thus further enhancing energy efficiency. Moreover, users can control their smart lighting systems from virtually anywhere, allowing them to adjust lighting levels even when they are not at home, offering both convenience and security. In commercial settings, smart lighting systems are increasingly integrated with building management systems (BMS) to optimize energy use across large spaces like offices, warehouses, or industrial facilities. These systems use data from occupancy sensors and daylight harvesting systems to automatically adjust lighting levels based on the presence of people and the availability of natural light (Uzair et al. 2022). By minimizing energy use in unoccupied areas or reducing light levels when daylight is sufficient, smart lighting systems help reduce operational costs and significantly lower the carbon footprint. This makes smart lighting an integral part of sustainable building practices.

Smart lighting also plays a critical role in the development of smart cities, where it is used to manage public lighting systems, improve energy efficiency, and enhance urban security (Sikder et al. 2018). Streetlights equipped with motion sensors can dim when no movement is detected, reducing both light pollution and energy use. In high-traffic areas, the lights can brighten when pedestrians or vehicles are present, enhancing safety without wasting energy. Furthermore, smart streetlights can be integrated with other urban systems to gather valuable data for city planners. They can track foot traffic patterns, measure air quality, or monitor noise levels, providing insights that help cities become more efficient and sustainable. As technology continues to evolve, the future of smart lighting will likely involve even greater integration with other smart home and smart building technologies. Advances in artificial intelligence (AI) and machine learning will make lighting systems even more intelligent. This will allow them to learn from user behaviours and preferences, adjust in real time, and optimize energy use with minimal input from users. Future smart lighting systems may be able to anticipate user needs based on historical data, adjusting lighting in anticipation of activities such as waking up, relaxing in the evening, or preparing for meetings in office spaces. Moreover, as cities continue to adopt smart infrastructure, smart lighting will play an increasingly important role in creating intelligent, energy-efficient, and responsive urban environments. Smart lighting systems represent the future of energy-efficient lighting. They provide users with unparalleled control, improve energy efficiency, and enhance urban planning by integrating with other smart technologies. As the world moves toward greater sustainability, smart lighting will play a pivotal role in reducing energy consumption, lowering carbon emissions, and improving overall quality of life.

4.4 Hybrid Solar Lighting

Hybrid Solar Lighting (HSL) systems offer a sophisticated approach to modern lighting design, combining natural sunlight with artificial lighting to provide consistent illumination throughout the day. These systems utilize optical fibers or other light-transport technologies to capture and direct sunlight into interior spaces, while supplementing it with electric lighting as needed (Al-Amayreh and Alahmer 2022). The primary advantage of HSL lies in its ability to reduce reliance on electric lighting during daylight hours, leading to significant energy savings and a reduction in the overall carbon footprint of a building. In HSL systems, sunlight is typically collected through rooftop concentrators or heliostats, which track the movement of the sun throughout the day to capture maximum sunlight. This sunlight is then directed into the building using optical fibers or reflective ducts, which distribute the natural light to various rooms and spaces. As daylight levels fluctuate due to weather conditions or the time of day, the system automatically adjusts the electric lighting to maintain consistent illumination. This ensures that occupants always have adequate lighting, regardless of external factors (Lapsa et al. 2007). The seamless integration of natural and artificial lighting not only enhances energy efficiency but also improves indoor environmental quality. Numerous studies have demonstrated that natural light positively impacts mood, productivity, and well-being. By allowing natural light to enter spaces that would otherwise rely solely on artificial light, HSL systems create a more pleasant and healthy indoor environment for occupants.

The implementation of HSL systems is particularly beneficial in commercial and institutional buildings, such as offices, schools, and museums, where lighting is required for extended periods throughout the day. By reducing the need for electric lighting during peak daylight hours, HSL systems can significantly lower energy costs and decrease the demand on HVAC systems, as less heat is generated from electric lights. This not only leads to lower operating costs but also enhances the overall energy efficiency of the building. Moreover, HSL systems contribute to the aesthetic appeal of buildings by bringing natural light into spaces that might otherwise be dim or rely entirely on artificial illumination. The introduction of natural light can transform the look and feel of interior spaces, making them more vibrant, inviting, and connected to the outdoors. As the demand for sustainable building solutions continues to grow, HSL is gaining traction as an innovative and effective strategy for reducing energy consumption and enhancing indoor environments. HSL systems offer an advanced lighting solution that blends the benefits of natural sunlight with artificial lighting. By optimizing energy use and improving the indoor environment, HSL is becoming an increasingly popular choice in the design of energy-efficient, sustainable buildings.

4.5 Photovoltaic (PV) Integrated Lighting

Photovoltaic (PV) integrated lighting systems represent a cutting-edge combination of solar power technology and energy-efficient lighting, creating a self-sustaining solution that reduces dependence on the electrical grid. These systems utilize photovoltaic panels to convert sunlight into electricity, which is then used to power LED lights or other energy-efficient fixtures. The integration of PV panels with lighting systems is particularly advantageous in regions with abundant sunlight, where it can significantly lower energy costs and reduce the carbon footprint of buildings and outdoor spaces (Elsayed et al. 2019). By harnessing solar energy, PV integrated lighting contributes to both energy savings and environmental sustainability, making it an increasingly popular choice for renewable energy solutions. The applications of PV integrated lighting are diverse, ranging from small-scale residential uses such as solar-powered garden lights, to large-scale urban installations, such as street lighting systems in cities. In residential settings, PV integrated lighting systems are commonly used for outdoor applications such as pathway lights, deck lights, and security lights. These systems typically consist of a PV panel, a battery for energy storage, and LED light fixtures, allowing them to operate independently from the electrical grid. This independence from grid power makes PV lighting ideal for areas where access to electricity may be unreliable or non-existent.

In commercial and municipal applications, PV integrated lighting has become a reliable, cost-effective solution for outdoor lighting in areas such as streetlights, parking lot lights, and public parks. By utilizing renewable solar energy, these systems can operate efficiently in remote areas without access to the electrical grid, providing an essential service in rural or isolated regions. As demonstrated in Fig. 4.6, PV integrated streetlights are equipped with solar panels and batteries, which store energy for use during periods of low sunlight or at night. This ensures that the lights remain operational regardless of sunlight availability, making them a dependable source of illumination in both developed and developing regions. One of the key advantages of PV integrated lighting is its ability to function in off-grid locations, making it an ideal solution for developing regions, rural communities, or disaster-prone areas where access to reliable electricity is limited (Duman and Güler 2019). In these situations, PV integrated lighting offers a sustainable and autonomous lighting solution, reducing the need for costly infrastructure development while still providing essential lighting services. The integration of energy storage systems, such as batteries, further enhances the reliability of these systems, ensuring continuous operation even during periods of low sunlight or at night.

Fig. 4.6 Photovoltaic (PV) integrated street lighting system showcasing solar-powered LED streetlights installed along a roadway (Liu 2014)

As solar technology continues to advance, with improvements in panel efficiency and reductions in the cost of production, PV integrated lighting systems are becoming increasingly viable for a wider range of applications. These advancements include more efficient solar cells, longer-lasting batteries, and improved energy management systems that optimize power consumption. With these improvements, PV integrated lighting systems are becoming more cost-effective, reliable, and adaptable to various environments and climates. The adoption of PV integrated lighting supports global efforts to transition to renewable energy sources and reduce greenhouse gas emissions (Gorjian et al. 2021). As part of broader sustainability strategies, these systems contribute to energy independence, reduced reliance on fossil fuels, and lower operational costs. By utilizing solar energy to power lighting systems, PV integrated solutions help reduce urban energy consumption and promote environmental stewardship. In the context of growing concerns over climate change, PV integrated lighting is emerging as a key component of sustainable lighting strategies which helping cities and regions meet their renewable energy targets and sustainability goals. PV integrated lighting offers a versatile, sustainable, and efficient solution for both residential and commercial lighting needs. As solar technology continues to evolve, these systems will play an increasingly important role in promoting energy efficiency, reducing carbon emissions, and supporting the global shift toward renewable energy.

4.6 Advanced Lighting Controls and Sensors

Advanced lighting controls and sensors represent a significant leap forward in the management and optimization of lighting systems, offering not only enhanced energy efficiency but also improved user convenience. These technologies encompass a variety of systems including motion sensors, daylight sensors, and programmable lighting controls. All of these work together to adjust lighting levels in real time based on changing conditions, such as occupancy, ambient light, and time of day (Pandharipande and Newsham 2018). By automating the control of lighting, these technologies help to minimize energy waste and ensure that lights are only used when necessary, leading to substantial energy savings for both residential and commercial buildings. Motion sensors are widely used in spaces such as hallways, restrooms, offices, and stairwells, where lights are often left on even when the space is unoccupied (Wagiman et al. 2020). These sensors detect movement and automatically turn lights on when someone enters the space and turn them off when the space is vacant. This reduces unnecessary energy consumption, especially in areas where people frequently come and go, preventing lights from being left on for extended periods. The installation of motion sensors can result in significant energy savings, particularly in high-traffic areas, and can also contribute to increased security by ensuring that spaces are illuminated when occupied.

Daylight sensors measure the amount of natural light in a space and adjust the electric lighting accordingly. In rooms with ample daylight, the electric lights may automatically dim or turn off completely, allowing the natural light to provide sufficient illumination. This not only conserves energy but also enhances visual comfort by balancing artificial and natural light levels. Daylight harvesting as this process is often called, allows buildings to take full advantage of available sunlight, reducing the demand on electric lighting systems during the day and contributing to lower operating costs. In addition to motion and daylight sensors, programmable lighting controls allow users to set schedules for lighting systems ensuring that lights are turned off during non-peak hours or when the building is unoccupied. This is particularly useful in commercial settings, where businesses can program lights to follow work hours or set automatic shutdown schedules for nights and weekends. By creating specific lighting schedules, building managers can ensure that energy is not wasted when spaces are not in use, further enhancing the efficiency of energy consumption in the building. The benefits of advanced lighting controls go beyond just energy savings. These systems also offer significant improvements in the user experience by providing customized lighting environments that enhance comfort and productivity. In office settings, lighting can be automatically adjusted to reduce glare on computer screens or to create a more relaxed atmosphere during meetings or collaborative work sessions. Similarly, in residential settings, homeowners can program lights to gradually dim in the evening, mimicking the natural progression of daylight and promoting better sleep patterns by signaling the body to wind down (Gorgulu and Kocabey 2020).

By automating the control of lighting systems, these technologies can contribute to the longevity of light fixtures. Reducing the frequency of on–off cycles extends the lifespan of light bulbs and reduces the need for maintenance and replacement. This helps to lower both the operational costs and the environmental impact associated with the disposal of used lighting equipment. In environments where lighting systems are in constant use, such as in commercial buildings or public spaces, extending the life of lighting components can lead to significant savings in maintenance costs and downtime. As smart building technology continues to evolve, advanced lighting controls and sensors will play an increasingly important role in optimizing energy use and enhancing the overall functionality of lighting systems. These systems are likely to become more integrated with other smart technologies such as HVAC systems, security systems, and building management platforms allowing for a more comprehensive approach to energy efficiency and building automation. For example, integrating lighting controls with occupancy sensors in an HVAC system can ensure that both lighting and temperature are adjusted according to real-time usage patterns, further optimizing energy consumption. Advanced lighting controls and sensors provide a sophisticated and dynamic solution for managing lighting systems, offering not only substantial energy savings but also improved user experience, comfort, and lighting system longevity. As buildings and homes become smarter and more energy-conscious, these technologies will continue to play a vital role in reducing energy consumption and enhancing overall sustainability.

References

Al-Amayreh MI, Alahmer A (2022) On improving the efficiency of hybrid solar lighting and thermal system using dual-axis solar tracking system. Energy Rep 8:841–847

Alberts IL, Barratt DS, Ray AK (2010) Hollow cathode effect in cold cathode fluorescent lamps: a review. J Display Technol 6(2):52–59

Alonso-Villar EM, Rivas T, Pozo-Antonio JS (2022) Sol-silicate versus organic paints: durability after outdoor and ultraviolet radiation exposures. Prog Org Coat 168:106843

Austin E et al (2021) Visible light. Part I: properties and cutaneous effects of visible light. J Am Acad Dermatol 84(5): 1219–1231

Belessiotis GV et al (2022) Universal electrolyte for DSSC operation under both simulated solar and indoor fluorescent lighting. Mater Chem Phys 277:125543

Brown TM et al (2022) Recommendations for daytime, evening, and nighttime indoor light exposure to best support physiology, sleep, and wakefulness in healthy adults. PLoS Biol 20(3):e3001571

Callow JM (2003) Daylighting using tubular light guide systems. University of Nottingham

Chang M-H et al (2012) Light emitting diodes reliability review. Microelectron Reliab 52(5):762–782

CIBSE D, Design W (1999) Lighting Guide LG10. Chartered institution of building services engineers

de Souza DF et al (2019) Efficiency, quality, and environmental impacts: a comparative study of residential artificial lighting. Energy Rep 5:409–424

Duman AC, Güler Ö (2019) Techno-economic analysis of off-grid photovoltaic LED road lighting systems: a case study for northern, central and southern regions of Turkey. Build Environ 156:89–98

Eiz HM et al (2021) The visual and thermal impact of skylight design on the interior space of an educational building in a hot climate. Eng J 25(1):187–198

Elsayed NM et al (2019) Photovoltaic applications for lighting load energy saving: case studies, educational building. In: 2019 International Conference on Innovative Trends in Computer Engineering (ITCE). IEEE

Emre Ilgın H (2021) Space efficiency in contemporary supertall office buildings. J Archit Eng 27(3):04021024

Füchtenhans M, Grosse EH, Glock CH (2021) Smart lighting systems: state-of-the-art and potential applications in warehouse order picking. Int J Prod Res 59(12):3817–3839

Gorgulu S, Kocabey S (2020) An energy saving potential analysis of lighting retrofit scenarios in outdoor lighting systems: a case study for a university campus. J Clean Prod 260:121060

Gorjian S et al (2021) A review on opportunities for implementation of solar energy technologies in agricultural greenhouses. J Clean Prod 285:124807

Hofmann D, Rasch E (2020) Fluorescent lamps. In: Revolution in lamps. River Publishers, pp 161–200

Hudallah N et al (2021) Comparison of CFL lights and led lights reviewed from the side of the price, strong light and heat caused. In: Iop conference series: earth and environmental science. 2021. IOP Publishing

Ishak R et al (2023) Widespread compact fluorescent lamp evaluations in 50 Hz electrical network. Bullet Electr Eng Inf 12(3):1267–1275

Katzin D, Marcelis LF, van Mourik S (2021) Energy savings in greenhouses by transition from high-pressure sodium to LED lighting. Appl Energy 281:116019

Kazemi R et al (2018) Comparing task performance, visual comfort and alertness under different lighting sources: an experimental study. EXCLI J 17:1018

Koretsky Z (2021) Phasing out an embedded technology: insights from banning the incandescent light bulb in Europe. Energy Res Soc Sci 82:102310

Krishnamoorthi A et al (2023) Ultrafast transient absorption spectra and kinetics of rod and cone visual pigments. Molecules 28(15):5829

Kushner S et al (2021) Levitating rigid objects with hidden rods and wires. In: Computer Graphics Forum. 2021. Wiley Online Library

Lamnatou C, Chemisana D, Cristofari C (2022) Smart grids and smart technologies in relation to photovoltaics, storage systems, buildings and the environment. Renew Energy 185:1376–1391

Lapsa MV et al (2007) Hybrid solar lighting provides energy savings and reduces waste heat. Energy Eng 104(4):7–20

Liu G (2014) Sustainable feasibility of solar photovoltaic powered street lighting systems. Int J Electr Power Energy Syst 56:168–174

Lv Y et al (2020) Techno-economic evaluation of an optical fiber based hybrid solar lighting system. Energy Convers Manage 225:113399

Martirano L (2011) A smart lighting control to save energy. In: Proceedings of the 6th IEEE International Conference on Intelligent Data Acquisition and Advanced Computing Systems, IEEE

Mathiasen N, Frandsen AK, Grønlund L (2022) Daylight conditions in housing–Its role and priority in Danish building regulations. Arch, Struct Const 2(1):23–37

Nundy S et al (2021) Electrically actuated visible and near-infrared regulating switchable smart window for energy positive building: a review. J Clean Prod 301:126854

Omishore A, Mohelník P, Míček D (2020) Light pipe comparative study. Select Sci Papers-J Civ Eng 14(1):17–26

Pandharipande A, Newsham GR (2018) Lighting controls: evolution and revolution. Light Res Technol 50(1):115–128

Plörer D et al (2021) Control strategies for daylight and artificial lighting in office buildings—a bibliometrically assisted review. Energies 14(13):3852

Sambiagio C, Noel T (2020) Flow photochemistry: shine some light on those tubes! Trends in Chemistry 2(2):92–106

Sharma K et al (2021) Comparative study of acrylic flat plate and dome shaped collector for summer and winter solstice conditions. Materials Today: Proceedings 45:5489–5493

Shirzadnia Z, Goharian A, Mahdavinejad M (2023) Designerly approach to skylight configuration based on daylight performance; Toward a novel optimization process. Energy and Buildings 286:112970

Sikder AK et al (2018) IoT-enabled smart lighting systems for smart cities. in 2018 IEEE 8th Annual Computing and Communication Workshop and Conference (CCWC). IEEE

Song J, Dessie BB, Gao L (2023) Analysis and Comparison of Daylighting Technologies: Light Pipe, Optical Fiber, and Heliostat. Sustainability 15(14):11044

Tabadkani S et al (2021) Daylight in buildings and visual comfort evaluation: The advantages and limitations

Thalange VC et al (2017) Design, optimization and optical performance study of tripod heliostat for solar power tower plant. Energy 135:610–624

Uzair M et al (2022) A low-cost IoT based buildings management system (BMS) using arduino mega 2560 and raspberry pi 4 for smart monitoring and automation. Int J Electr Comput Eng Syst 13(3):219–236

Viana LN et al (2022) Fluorescent lamps: a review on environmental concerns and current recycling perspectives highlighting Hg and rare earth elements. J Environ Chem Eng 10(6):108915

Wagiman KR et al (2020) Lighting system control techniques in commercial buildings: current trends and future directions. J Build Eng 31:101342

Wang Y, Yang W, Wang Q (2022) Multi-objective parametric optimization of the composite external shading for the classroom based on lighting, energy consumption, and visual comfort. Energy Build 275:112441

Wang Y et al (2024) Research on visual sensitivity characteristics of amorphous silicon photocells. Optoelectron Lett 20(7):385–392

Yadlapalli RT, Narasipuram RP, Kotapati A (2020) An overview of energy efficient solid state LED driver topologies. Int J Energy Res 44(2):612–630

Yu F, Wennersten R, Leng J (2020) A state-of-art review on concepts, criteria, methods and factors for reaching 'thermal-daylighting balance.' Build Environ 186:107330

Zhao E-Z et al (2020) The impact of LED color rendering on the dark adaptation of human eyes at tunnel entrances. Int J Environ Res Public Health 17(5):1566

Chapter 5
Zero-Energy and Low Carbon Buildings

Contents

Abstract As the global focus shifts toward combating climate change, zero-energy and low-carbon buildings have emerged as essential components of sustainable urban development. This chapter explores the innovative design and technological advancements driving the development of zero-energy and low-carbon buildings. It examines strategies aimed at reducing energy consumption while enhancing environmental sustainability, focusing on energy-efficient insulation, passive solar heating, and the integration of renewable energy systems such as solar and wind power. Additionally, it highlights the role of smart grids and building energy management systems (BEMS) in optimizing energy use and enabling buildings to contribute to larger energy networks. The chapter provides a global perspective on the adoption of these buildings, particularly in North America, Europe, and Asia, where rigorous energy policies and incentives support their growth. Economic and health benefits, including lower operating costs, increased property value, and improved indoor air quality, are discussed, showcasing the comprehensive advantages of zero-energy and low-carbon buildings. As cities worldwide prioritize sustainable development, these buildings are becoming central to urban planning and the creation of resilient, energy-efficient infrastructures.

S. Riffat et al., *Sustainable Energy Technologies and Low Carbon Buildings*, Lecture Notes in Energy 45, https://doi.org/10.1007/978-3-031-78853-6_5

5.1 Overview of Zero Energy and Low Carbon Buildings

The growing imperative to combat climate change has significantly accelerated the development and adoption of zero energy and low carbon buildings. These buildings are meticulously designed to minimize energy consumption and reduce carbon emissions, primarily by incorporating advanced energy-efficient technologies and leveraging renewable energy sources such as solar and wind power (Christopher et al. 2023). This is accomplished through a combination of highly efficient insulation, strategic orientation to maximize natural light, and the integration of on-site renewable energy generation systems (Ahmed et al. 2022). As these technologies continue to evolve, the potential for buildings to not only reduce their environmental impact but also contribute to the energy grid through net metering becomes increasingly viable. In addition to net metering, advancements in smart grid technologies and building energy management systems (BEMS) are further enhancing the potential of zero energy buildings to integrate with larger energy networks (Mariano-Hernández et al. 2021). These technologies allow buildings to dynamically respond to energy demand and supply fluctuations. This makes zero energy buildings not just passive consumers of energy but active participants in energy distribution systems.

The adoption of zero energy and low carbon buildings has gained considerable momentum, particularly in regions like the United States and Canada where there has been a noticeable increase in the number of projects meeting these stringent standards (González-Torres et al. 2022). In Europe, countries such as Germany and Denmark are also at the forefront of adopting zero energy building standards supported by rigorous energy efficiency policies and incentives for renewable energy integration (Wang et al. 2021a). In Asia, Japan and South Korea have initiated several pilot projects aimed at achieving zero energy targets in both urban and suburban environments, demonstrating that the potential for these buildings is not confined to Western regions (Fragkos et al. 2021). This surge is driven by several factors, including growing awareness of the environmental benefits, advances in building technologies, and a stronger regulatory push towards sustainable construction practices.

Moreover, the economic advantages associated with zero energy buildings such as lower operating costs, increased property value, and improved occupant health and comfort have become key motivators for developers and homeowners alike (Elaouzy and Fadar 2022). Studies indicate that zero energy buildings not only reduce operational costs but also increase market competitiveness, as demand for energy-efficient properties grows among environmentally conscious consumers (Moran et al. 2020). Additionally, the long-term health benefits associated with enhanced indoor air quality and temperature regulation make these buildings particularly appealing for residential developments. Governments and municipalities are also offering incentives and tax breaks to encourage the construction of these buildings, further fuelling their adoption. Beyond the individual benefits, the rise of zero energy and low carbon buildings is part of a broader movement towards sustainable urban development. This movement is characterized by a shift in how communities are planned and built, with a focus on reducing carbon footprints and enhancing the resilience of urban areas to climate change. Urban planners are increasingly adopting principles of "smart city" design, where zero energy buildings form the backbone of sustainable infrastructure (Karvonen et al. 2020). These cities integrate green spaces, energy-efficient public transport, and renewable energy generation, thus creating a holistic approach to reducing urban energy demand and mitigating the impacts of climate change.

As cities and regions adopt more stringent building codes and sustainability standards, zero energy buildings are becoming an integral part of the urban landscape. The ongoing development of regulatory frameworks plays a crucial role in guiding and supporting the construction of zero energy and low carbon buildings (Ade-Ojo 2022). These frameworks provide clear guidelines for achieving energy efficiency and sustainability, setting the stage for the widespread adoption of these innovative building practices. For example, the European Union Energy Performance of Buildings Directive (EPBD) has established ambitious goals for reducing the energy consumption of buildings, mandating that all new buildings must be nearly zero energy by 2021. Similarly, in the United States, the Department of Energy has set targets to ensure that all new federal buildings achieve net-zero energy status by 2030, creating a roadmap for public sector adoption.

5.1.1 Design and Technological Innovations

The design of zero-energy and low-carbon buildings incorporates a blend of advanced technologies and architectural strategies aimed at reducing energy consumption while maintaining occupant comfort. A key focus is on maximizing the energy efficiency of buildings through highly efficient insulation and materials that minimize heat loss and maintain internal temperatures with minimal reliance on mechanical heating or cooling systems. One common approach is the use of passive design techniques, which emphasize natural light, ventilation, and solar heating (Al-Shamkhee et al. 2022). These techniques are integrated into the building architecture, often through strategic orientation to optimize sunlight exposure in colder climates or reduce heat gain in warmer regions. For example, south-facing windows in the Northern Hemisphere capture more sunlight during winter months, reducing the need for artificial heating. Furthermore, innovative building materials play a crucial role in reducing the overall energy footprint. Advanced thermal insulation materials such as aerogels and phase-change materials are increasingly used to prevent heat transfer through walls and roofs (Mushtaha et al. 2021). These materials ensure that indoor temperatures remain stable, minimizing the need for energy-intensive HVAC (heating, ventilation, and air conditioning) systems.

Another important innovation is the integration of on-site renewable energy systems such as solar photovoltaic (PV) panels and small-scale wind turbines. These systems generate clean energy that can be used to power the building, further reducing its reliance on external energy sources. In many cases, the electricity generated on-site can exceed the building demand. This allows for net metering, where surplus energy is fed back into the grid, effectively making the building an energy producer rather than just a consumer. The use of intelligent building management systems also enhances energy efficiency by automating heating, cooling, and lighting based on real-time data (Merabet et al. 2021). Smart thermostats, occupancy sensors, and automated shading systems adjust the indoor environment according to occupancy and external weather conditions, ensuring energy is used efficiently. These systems not only reduce energy consumption but also improve comfort by maintaining optimal indoor conditions without human intervention.

Finally, green roofing systems and vertical gardens are becoming increasingly popular in zero-energy designs. These features not only provide insulation but also contribute to urban biodiversity and improve air quality. Green roofs help regulate temperatures by absorbing heat in the summer and providing insulation in the winter, while vertical gardens can reduce the building energy consumption by acting as natural shading. The design and technological innovations employed in zero-energy buildings are focused on leveraging natural resources, reducing energy demand, and integrating renewable energy systems. As these technologies continue to evolve, the efficiency and sustainability of zero-energy buildings will only increase, making them a critical component of future sustainable urban development.

5.1.2 Integration with Energy Networks

A key feature of zero-energy buildings is their ability to interact with broader energy networks, turning them from passive energy consumers into active participants in energy distribution systems. One of the primary mechanisms enabling this is net metering, which allows buildings to supply surplus energy back to the grid. By generating more energy than they consume, especially through on-site renewable energy systems like solar panels or wind turbines, these buildings can contribute to the local energy grid. This excess energy can offset their consumption during periods of low production, ensuring a continuous balance between energy generation and use. In addition to net metering, the development of smart grid technologies is further enhancing the role of zero-energy buildings within energy networks (Bandeiras et al. 2020). Smart grids are designed to communicate with buildings, adjusting energy flow based on real-time demand and supply conditions. This integration allows zero-energy buildings to dynamically respond to energy shortages or surpluses, helping to stabilize the grid. During peak demand periods, buildings can reduce their energy consumption or draw power from stored energy reserves, while at times of surplus energy production, they can feed excess electricity into the grid.

Moreover, Building Energy Management Systems (BEMS) are integral to the efficient operation of zero-energy buildings. BEMS use advanced monitoring and control technologies to optimize energy use within the building. By analysing data on energy consumption patterns, occupancy, and external factors such as weather conditions, these systems make real-time adjustments to reduce energy wastage (Digitemie and Ekemezie 2024). For example, BEMS can automatically adjust heating, cooling, and lighting systems based on occupancy, ensuring that energy is only used when necessary. This integration with energy networks also provides an opportunity for buildings to support demand response programs, where they adjust their energy consumption in response to signals from the utility grid. In exchange, building owners may receive financial incentives for reducing energy use during peak times, contributing to overall grid stability and reducing the need for additional power generation from fossil fuels. By leveraging these technologies, zero-energy buildings not only minimize their environmental impact but also play a proactive role in creating more resilient and efficient energy systems. The future of zero-energy buildings lies in their ability to communicate seamlessly with energy grids, enhancing both sustainability and energy security.

5.1.3 Global Trends in Adoption of Low Carbon Buildings

The adoption of zero-energy and low-carbon buildings has gained momentum across different regions of the world. This growth is driven by a combination of regulatory frameworks, technological advancements, and increasing awareness of environmental sustainability. While the push for energy-efficient buildings is evident in many developed regions, the approach and scale of adoption vary globally. In North America, the United States and Canada have been at the forefront of implementing stringent energy efficiency standards. In the U.S., the federal government, through agencies such as the Department of Energy (DOE) has set ambitious targets to achieve net-zero energy in all new federal buildings by 2030 (Heizer and Shelide 2020). Similarly, state and local governments have been developing building codes that require energy-efficient construction practices, with California being a prime example. California Zero Net Energy (ZNE) goals mandate that all new residential buildings meet net-zero energy standards by 2020, and commercial buildings follow by 2030 (Willingham 2020). In Canada, efforts to reduce carbon emissions through improved energy performance in buildings are also well underway, with various provinces adopting policies aligned with federal goals for energy-efficient buildings.

In Europe, countries such as Germany and Denmark have been pioneers in the adoption of zero-energy building standards. The European Union Energy Performance of Buildings Directive (EPBD) has established clear guidelines requiring nearly all new buildings to meet nearly zero-energy standards by 2021 (Monstvilas et al. 2023). These regulations are supported by strong policy incentives and financial support to encourage the integration of renewable energy systems in building designs. For example, in Germany, the KfW Development Bank offers low-interest loans and grants for energy-efficient buildings, incentivizing homeowners and developers to invest in zero-energy construction (Mazzucato and Macfarlane 2023). Denmark has similarly implemented policies that promote energy efficiency, with a focus on reducing energy consumption through advanced insulation, renewable energy integration, and innovative building technologies. In Asia, countries like Japan and South Korea are taking notable steps toward zero-energy construction, especially in urban environments. Japan has introduced several pilot projects aimed at achieving zero-energy targets in both residential and commercial buildings (Kuwahara et al. 2022). These initiatives are often supported by cutting-edge technologies, such as advanced solar photovoltaic (PV) systems and energy storage solutions. South Korea has also initiated programs that encourage the development of zero-energy buildings, particularly in urban areas where energy demand is high (Park and Chung 2023). The government focus on integrating renewable energy into new construction projects is part of a broader strategy to reduce greenhouse gas emissions and achieve sustainability goals.

Beyond these regions, zero-energy buildings are also gaining traction in developing countries, albeit at a slower pace. China and India are seeing a rise in energy-efficient construction as urbanization continues to expand (Jain and Goswami 2021; Saini et al. 2022). In these countries, the challenge lies in balancing rapid development with sustainability goals, but growing regulatory pressure and technological advancements are driving the adoption of energy-efficient designs in new construction projects. Despite the varying pace of adoption, the global trend toward zero-energy and low-carbon buildings is undeniable. As technological advancements make these buildings more feasible and cost-effective, and as regulatory frameworks continue to tighten, the construction of zero-energy buildings is set to increase worldwide. These buildings are not only helping to reduce carbon emissions and energy consumption but are also contributing to global efforts to combat climate change and promote sustainable urban development.

5.1.4 Economic and Well-Being Benefits

The economic advantages of zero-energy and low-carbon buildings are becoming increasingly evident as the adoption of these buildings grows. One of the most significant economic benefits is the reduction in operating costs. Zero-energy buildings are designed to be highly energy-efficient, reducing the need for external energy sources. By relying on renewable energy systems such as solar panels, these buildings can significantly lower electricity bills, resulting in long-term savings for homeowners, businesses, and developers (Galvin 2022). In many cases, the initial investment in energy-efficient technologies and renewable energy systems is offset by the substantial reduction in energy costs over time. Another key economic benefit is the increased property value associated with zero-energy buildings. As consumer demand for environmentally friendly and energy-efficient homes rises, properties that meet zero-energy standards are often seen as more desirable. These buildings are typically equipped with advanced technologies and sustainable features, which not only reduce energy consumption but also enhance comfort and durability. As a result, properties with zero-energy features tend to command higher resale values in the real estate market, making them attractive investments for developers and homeowners alike.

In addition to the economic benefits, governments and municipalities are offering a range of incentives and tax breaks to encourage the construction of zero-energy and low-carbon buildings. These incentives, which can include grants, low-interest loans, and tax deductions, help reduce the financial burden of the initial investment in energy-efficient building materials and renewable energy systems (Bertoldi et al. 2021). Such programs are designed to make zero-energy buildings more financially accessible, further fuelling their adoption. Beyond the economic advantages, zero-energy and low-carbon buildings offer considerable health benefits to their occupants. The design and construction of these buildings prioritize indoor air quality, with systems that provide proper ventilation and reduce exposure to indoor pollutants (Qian et al. 2024). For example, zero-energy buildings often feature high-performance HVAC systems equipped with air filtration, ensuring a constant supply of fresh, clean air. This results in improved respiratory health and overall well-being for occupants, particularly in urban areas where outdoor air pollution is a concern.

Moreover, the enhanced temperature regulation provided by these buildings contributes to occupant comfort. High-quality insulation, energy-efficient windows, and smart climate control systems maintain consistent indoor temperatures, reducing reliance on artificial heating and cooling systems. This not only leads to lower energy consumption but also creates a more comfortable and healthy living environment. In addition, maintaining stable indoor temperatures helps prevent temperature-related health problems, particularly for vulnerable populations such as children and the elderly. In summary, zero-energy and low-carbon buildings provide a range of economic and health benefits. From reduced operating costs and increased property values to improved air quality and occupant comfort, these buildings represent a sustainable and health-conscious approach to modern construction. As more governments provide incentives and consumers recognize the long-term advantages of zero-energy buildings, their adoption will likely continue to rise.

5.1.5 Zero-Energy Buildings and Sustainable Urban Development

Zero-energy and low-carbon buildings are playing a pivotal role in the transformation of urban landscapes, particularly as cities around the world prioritize sustainable urban development. As the global population continues to urbanize, cities are facing increased pressure to reduce their carbon footprints and enhance the resilience of their infrastructures. In this context, zero-energy buildings are becoming a cornerstone of the smart city movement, which focuses on integrating technology and sustainability into urban planning (Asghari et al. 2021). One of the primary contributions of zero-energy buildings to sustainable urban development is their ability to reduce energy demand in densely populated areas. By minimizing reliance on non-renewable energy sources and integrating on-site renewable energy systems such as solar panels, these buildings help reduce the overall energy consumption of cities. This decreases the strain on urban energy grids, particularly during peak demand periods, and supports the transition to more sustainable energy sources.

Furthermore, zero-energy buildings align with the broader goals of climate resilience. As cities become more vulnerable to the impacts of climate change, including extreme weather events and rising temperatures, there is a growing need for urban infrastructures that can adapt to these challenges (Tomrukcu and Ashrafian 2024). The advanced insulation, energy-efficient design, and smart building management systems employed in zero-energy buildings not only reduce their energy consumption but also improve their ability to withstand fluctuating climate conditions. For instance, buildings designed with passive solar heating or natural ventilation systems are better equipped to maintain comfortable indoor temperatures, even during extreme heatwaves or cold spells. Incorporating zero-energy buildings into urban planning also supports the creation of green infrastructure. Many of these buildings feature sustainable landscaping elements such as green roofs and vertical gardens, which contribute to urban biodiversity, improve air quality, and reduce the urban heat island effect. Green roofs provide insulation for buildings, reduce stormwater runoff, and offer habitats for urban wildlife, making cities more liveable and environmentally friendly.

Additionally, the adoption of zero-energy buildings is influencing the way urban planners approach community design. In many cities, there is a growing trend toward creating sustainable neighbourhoods, where zero-energy buildings are clustered together to form eco-districts. These neighbourhoods prioritize pedestrian-friendly designs, energy-efficient public transport, and shared renewable energy resources, fostering a sense of community and reducing the overall environmental impact of urban areas. By encouraging the development of these eco-districts, cities can achieve more sustainable growth while enhancing the quality of life for their residents (Echlouchi et al. 2022). As more cities and regions adopt zero-energy and low-carbon building standards, these structures are becoming integral to the urban landscape of the future. They not only reduce the environmental impact of urban development but also contribute to creating resilient, liveable cities that can adapt to the challenges posed by climate change and rapid urbanization. The ongoing shift toward sustainable urban planning, driven by the adoption of zero-energy buildings, is setting the stage for a future where cities can thrive without compromising the health of the planet.

5.2 Concept and Definition of Zero-Energy Buildings/ Net-Zero Energy Building

The pursuit of energy efficiency in buildings has long been integral to energy policy, focusing on the reduction of power consumption to lower the carbon footprint of residential and commercial spaces. This goal is achieved through the enforcement of strict standards in design, material selection, and the integration of renewable energy technologies (Faraj et al. 2020). Energy-efficient designs incorporate a variety of strategies such as passive solar heating, improved insulation, and advanced building materials that significantly reduce energy loss. Material selection often focuses on locally sourced or recycled products, reducing the overall embodied energy in the construction process. Furthermore, building energy management systems (BEMS) have become integral to managing and optimizing energy consumption within these buildings, ensuring minimal wastage and maximizing the use of renewable energy sources (Khan et al. 2022). These practices have laid the groundwork for the development of zero-energy buildings (ZEB), a concept that has faced various challenges in its path to widespread commercialization and adoption.

The idea of zero-energy buildings was first conceptualized in 1976, when Esbensen and Korsgaard at the Technical University of Denmark created the first examples during an experiment aimed at using solar energy for residential heating in winter (Yan et al. 2021). This pioneering research led to the coining of the term "zero energy house," which marked the beginning of a new era in building design aimed at achieving energy neutrality. This development was revolutionary for its time, as it demonstrated the potential of renewable energy to significantly reduce dependency on fossil fuels for residential heating. The early zero energy houses utilized rudimentary technologies compared to common standards today, such as basic photovoltaic (PV) panels and water heating systems. However, these innovations laid the foundation for modern advancements in solar energy and heat recovery systems, which now play a crucial role in ZEB.

Since 1977, numerous studies have proposed various definitions of ZEB. However, the lack of a universally accepted definition has created challenges in establishing a clear framework for the broader economy and hindered the development of a unified global strategy for energy-efficient construction. This absence of a standardized definition complicates efforts to develop international guidelines, as different countries prioritize varying aspects of energy efficiency depending on their climatic and economic conditions. Definitions and criteria for zero-energy buildings vary significantly across different countries, regions, and communities. For example, terms like "nearly zero-energy building" in Europe (Attia et al. 2022), "net ZEB" in the United States (D'Agostino et al. 2021a), "zero-emission building" in Australia (Brown et al. 2024), and "zero carbon emission building" as promoted by the C40 Cities Climate Leadership Group (Nguyen et al. 2020), reflect the diversity of interpretations of the concept. Despite these variations, the overarching goals remain consistent to minimize the use of fossil fuels, maximize the utilization of renewable energy, and enhance the energy efficiency of buildings. As a result, policy and regulation frameworks differ widely. As previously mentioned, the EPBD requires nearly zero energy standards for all new buildings by 2021. Whereas in the United States, the emphasis lies on achieving net-zero energy (NZEB) status in commercial buildings by 2025. Australia focuses heavily on carbon emissions reductions, reflecting its prioritization of climate mitigation strategies. These variations highlight the need for flexibility in the approach to defining and implementing ZEB principles across different regions, as a "one-size-fits-all" model would fail to account for geographical and economic disparities.

The EPBD stipulates that near-zero energy buildings must achieve exceptionally high energy efficiency, with energy consumption on-site being either negligible or entirely offset by renewable energy sources (Barrutieta et al. 2023). Nevertheless, significant barriers to the implementation of ZEB principles persist, as identified by the Sustainable Building Institute Europe. While the EPBD requires near-zero energy buildings to maintain very high energy efficiency, the cost-optimality concept ensures that energy-saving measures are implemented in the most economically efficient way over the lifecycle of the building. This balance does not specify a fixed primary energy consumption level but encourages the most cost-effective reduction in energy use. However, near zero-energy buildings as defined by the EPBD are expected to have very low energy consumption with the remaining energy needs covered by renewable sources. This cost-optimality approach balances upfront investment with long-term savings, aiming to make ZEB adoption economically viable. However, the high initial costs of renewable energy technologies particularly in developing regions, continue to be a significant barrier. Solutions such as government subsidies, low-interest loans, and tax incentives are critical in bridging this economic gap, allowing for broader adoption of ZEB principles in both residential and commercial sectors. Despite these guidelines, a common definition of zero-energy buildings has yet to be established across EU nations.

The concept of ZEB is multifaceted, with various existing approaches emphasizing different aspects of the ZEB framework. Calculating the energy balance of a facility that incorporates on-site and/or off-site renewable energy generation and interacts with the utility grid to achieve a "zero" target is inherently complex, as shown in Fig. 5.1 (D'Agostino et al. 2021b). Furthermore, there is no universally accepted standard for calculating this "zero" energy balance, which adds to the challenge. One of the major complexities is the variability in energy demand throughout the year, with fluctuations due to seasonal changes and occupant behaviour. Effective energy balance calculations must account for these factors, using sophisticated energy modelling tools that can predict energy consumption patterns and align them with renewable energy production. While voluntary environmental assessment methodologies such as Leadership in Energy and Environmental Design (LEED) and Building Research Establishment Environmental Assessment Method (BREEAM) offer a broader scope than current ZEB frameworks, the lack of a standardized calculation method for ZEB remains a significant barrier (Marszal et al. 2011). If ZEB is to be considered a future benchmark for buildings, it is imperative to establish a physically robust, communicative, and scientifically sound calculation method. This would involve the collaboration of international bodies to develop a unified approach, ensuring that ZEB certification is consistent and reliable across borders. Moreover, architects, engineers, and policymakers would benefit from a clear set of guidelines that provide practical solutions for balancing energy efficiency with renewable energy integration.

The definition of Net-Zero Energy Buildings (NZEB) is crucial as it reflects a commitment of the country to achieving energy or carbon neutrality. However, there is no uniform definition for NZEB largely due to varying objectives among organizations, governments, and entities in addressing the environmental impacts of buildings

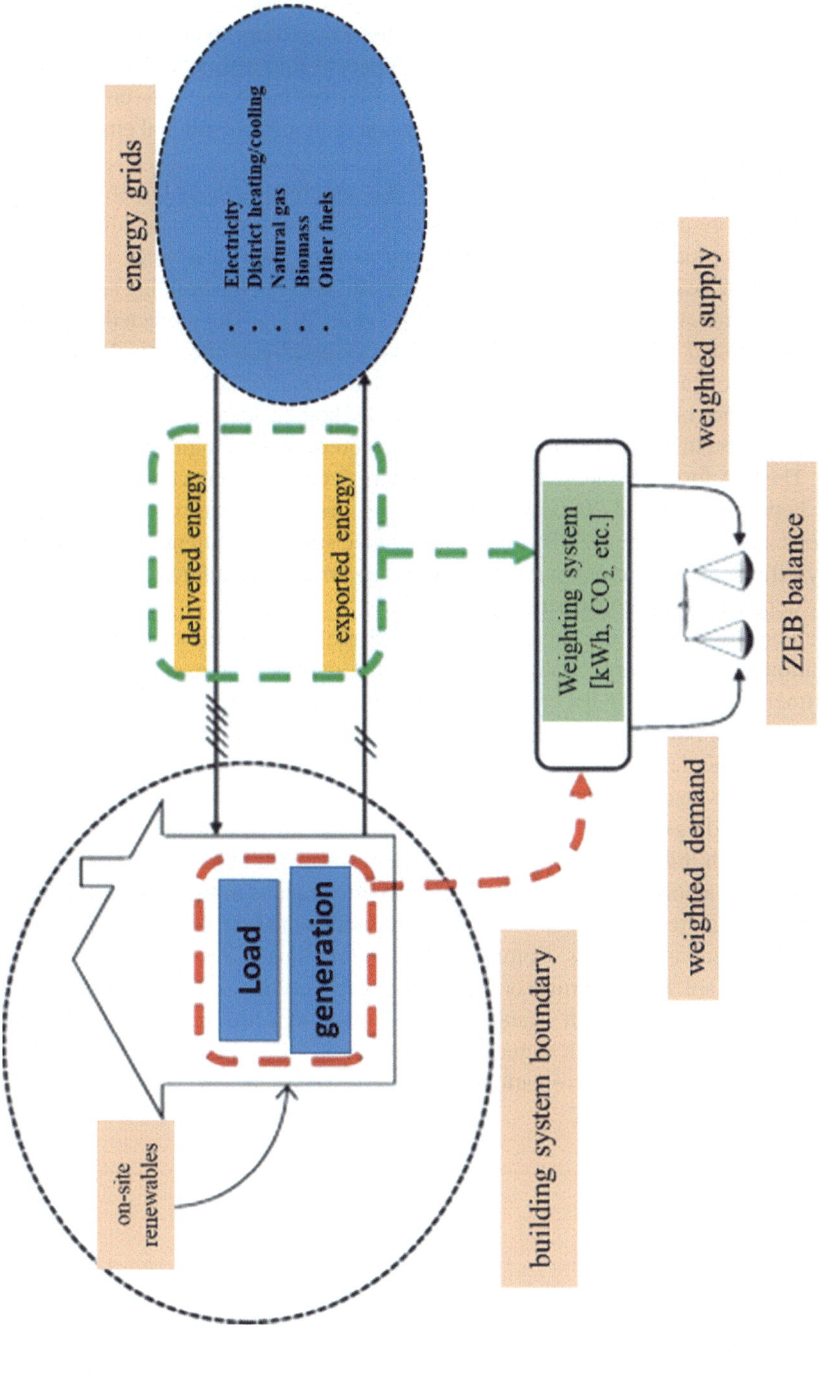

Fig. 5.1 Zero Energy Building (ZEB) Energy Flow Diagram (Sartori et al. 2012)

(Omrany et al. 2022). Consequently, the literature presents a range of definitions and frameworks for NZEB, each tailored to the specific climatic conditions, economic structures, or political frameworks of different countries (Magrini et al. 2020). Additionally, recent evaluations have assessed the impact of four distinct definitions such as net-zero site-energy buildings, net-zero source-energy buildings, net-zero energy-cost buildings, and net-zero emission buildings on reducing operational emission of the building (Shirinbakhsh and Harvey 2024).

In recent decades, the concept of NZEB has emerged as a viable solution to the rising energy consumption and associated CO_2 emissions from building operations. This approach is particularly important as it provides a strategy to reduce dependency on fossil fuels by improving energy efficiency in the building sector and incorporating renewable energy sources (Borghi et al. 2021). The projected increase in energy demand within the construction industry highlights the critical need for energy reduction in buildings. In 2020, the Energy Information Administration (EIA) reported that the construction sector, which includes both residential and commercial buildings, accounted for 29% of total final energy consumption in the world (Bilgen 2014). The EIA further projects a 32% rise in energy consumption in the construction sector between 2015 and 2040, with most of this growth occurring in non-OECD countries. As a result, the NZEB concept has gained significant attention in building energy efficiency regulations worldwide as a key strategy to address the growing energy demands of the construction sector.

The concept of ZEB has increasingly gained global recognition, leading to its integration into the energy policies of various countries. For example, the U.S. Department of Energy (DOE) envisions "market-ready zero energy homes by 2020 and commercial buildings by 2025"(Skidmore et al. 2024). However, despite these clear targets, the definitions of ZEB remain inconsistent and lack standardization. In the United Kingdom, the term "zero carbon homes" specifically refers to new residential buildings, with a political goal of making all new homes zero carbon by 2016, though this goal has been under review since 2006 (Wu and Skye 2021). The absence of a universally accepted definition often results in the term ZEB being applied in commercial contexts without precise understanding, leading to policies based on ambiguous concepts. Commercial interpretations of ZEB can vary, sometimes focusing solely on thermal or electrical energy needs, or allowing buildings to meet ZEB criteria through oversized photovoltaic systems without substantial energy-saving measures. Such vague definitions are insufficient as a foundation for robust regulations and effective national policies.

5.3 Legislation, Codes, Policies and Initiatives of Zero Energy Buildings

The concept of ZEB has transitioned from a futuristic ideal to a viable solution for reducing CO_2 emissions and energy consumption within the construction industry. The growing number of ZEB demonstration projects worldwide and the increasing research interest underscore the expanding focus on ZEB. Internationally, frameworks such as the U.S. Energy Independence and Security Act of 2007 (EISA 2007) and in Europe with the recast of the Directive on EPBD illustrate the global commitment to implementing ZEB. For instance, the Net-Zero Energy Commercial Building Initiative under EISA 2007 aims to achieve net-zero energy in all new commercial buildings by 2030, with intermediate goals of achieving 50% net-zero energy in U.S. commercial buildings by 2040 and complete net-zero energy by 2050 (Saini et al. 2022). Similarly, the EPBD mandates that all new buildings must meet "nearly zero energy" standards by 2020, with public buildings required to comply by 2018.

Globally, various countries have developed unique approaches to ZEB. The U.S. Energy Independence and Security Act defines ZEB based on substantial reductions in energy consumption, balanced energy demand, low carbon dioxide emissions, and economic viability (Yu et al. 2022). In Korea, the Zero Energy Construction Action Plan on Climate Change outlines a strategy for designing zero energy buildings that optimize insulation efficiency, reduce unnecessary energy use, and incorporate sustainable energy generation (Wilberforce et al. 2023). Korea further categorizes ZEB into low-rise, high-rise, and town-scale constructions. Japan has formally established the "Energy-Saving Technologies Strategy 2011" aims to continue reducing structural energy demands to zero, emphasizing the importance of energy efficiency in construction (Li et al. 2020). Meanwhile, Passive House Institute constructed by Germany has pioneered the development of zero-energy greenhouse constructions, setting a heating demand threshold of 15 kWh/(m^2y) to qualify as a "passive dwelling" (Mitchell and Natarajan 2020). The Sino-German Building Energy-Efficiency Collaboration has further promoted passive homes, and Germany has also introduced alternative concepts such as the "self-sufficient solar home" and the "3-L oil structure." Experts from Denmark and France advocate for the "active home" concept, which emphasizes the use of solar radiation and energy storage to enhance internal thermal comfort during winter (Schwarz et al. 2020). Rather than focusing solely on increasing the building envelope to minimize heating and cooling energy consumption, this approach integrates solar energy more actively to maintain a comfortable indoor environment year-round. This method leverages natural energy sources to improve sustainability and energy efficiency in residential buildings.

Today, many nations have committed to establishing transition objectives for their building sectors toward achieving net-zero energy. These goals are documented in Table 5.1, which outlines the various national objectives and definitions for NZEB. Since the concept was first defined by Esbensen and Korsgaard as buildings capable of enduring winter conditions without relying on "artificial" energy, the philosophy of NZEB has evolved (Esbensen and Korsgaard 1977). There is now a growing emphasis on a holistic approach to energy neutrality in the built environment, taking into account the embodied energy of buildings. Despite variations in terminology and requirements, the common objective across these definitions is to offset the annual energy consumption associated with building operations by improving the energy efficiency of building envelopes, relying on renewable energy sources (either on-site or off-site), and promoting the use of energy-efficient appliances and equipment. This shift has become a central focus of study in the built environment due to the increasing legislative emphasis on adopting NZEB and the urgent need to reduce energy consumption in the construction sector.

Since then, the concept of NZEB has undergone significant refinement and has gained widespread acceptance globally. Table 5.2 provides an overview of the current legislation and regulations related to NZEB across various countries. The majority of these policies have been established in Europe and the United States, contributing to the rapid and extensive development of NZEB projects in these regions. However, there is a noticeable lack of relevant policies in hot and humid climates, highlighting the urgent need for a comprehensive evaluation of NZEB in these areas.

The primary ZEB-related policies in the three regions are summarized in Table 5.3. In the United States, the federal government has stepped back from leading climate action, allowing local governments to emerge as the primary drivers of climate change initiatives. This shift has led to the establishment of new, robust carbon reduction goals, with a strong focus on architectural practices, resulting in the implementation of ZEB regulations within various states. In contrast, China has adopted a top-down development model to promote ZEBs, with the national government setting the direction and local governments following suit, particularly in colder regions where ZEB initiatives have been actively advanced.

The EPBD serves as the cornerstone of energy legislation within the EU, guiding each member state to develop its own pathway towards achieving ZEB in compliance with the EPBD requirements. However, as previously mentioned, the terminology is not standardized across the board leading to the adoption of various definitions, such as Net Zero Site Energy, Net Zero Source Energy, Net Zero Emissions, Net Zero Energy Expenses, Net Zero Energy Commercial, and Almost Net Zero Energy Building (Cielo and Subiantoro 2021). The specific interpretation a country adopts is largely influenced by its political goals and unique circumstances. The definition of NZEB adopted here follows the one established by the World Green Building Council, describing a "highly energy-efficient building with all remaining operational energy use supplied by renewable energy, preferably generated on-site but also from off-site sources, to achieve annual net zero carbon emissions." While numerous corporations, businesses, and non-governmental organizations actively support this initiative, the construction of NZEB is seeing significant growth, particularly in Europe and the United States.

Table 5.1 Targets and definitions for realizing net-zero energy buildings (Omrany et al. 2022)

Source	Country	Legislation year	Target year	Target	Definition
Energy Performance in Buildings Directive (EPBD)	EU countries	May 2010	2021	All new EU buildings must be nearly zero energy 2019: all new public buildings must be nearly zero energy	A nearly zero energy building is defined as one with very high energy efficiency, where the very low energy requirement is primarily met by renewable energy sources, either generated on-site or nearby
Energy policies of IEA countries: The UK 2012 Review	United Kingdom	April 2007	2016	All new-build dwellings must be zero carbon	A Zero-Carbon Home is described as a building that has no net carbon dioxide emissions from energy use, including lighting, heating, and other home services throughout the year
Belgium Federal Government, definition of zero-energy house	Belgium	September 2012	2020	All new buildings must be nearly zero energy	A Zero-Energy House refers to a building designed to use minimal energy, with the remaining energy demand for heating and cooling fully supplied by renewable energy produced on-site
US FEDERAL GOVernment	United States	December 2007	2040	50% of US commercial buildings must be zero energy	An NZEB is defined as a building that greatly reduces its energy consumption and meets the remaining energy needs with sources that do not emit greenhouse gases, ensuring no net GHG emissions while remaining cost-effective

(continued)

Table 5.1 (continued)

Source	Country	Legislation year	Target year	Target	Definition
NASA, Net Zero Energy Buildings Roadmap	United States	January 2010	2020	All new NASA buildings must be zero energy	An NZEB is characterized by an ambitious energy efficiency goal, which is met through reduced energy demand and complemented by renewable energy sources to cover any remaining energy use
The California Public Utilities Commission (CPUC)	The State of California, US	September 2008	2030	All new residential construction must be zero energy All new commercial construction must be zero energy 50% of commercial buildings must be retrofitted to become NZEBs 50% of new major renovations of state buildings must be NZEBs	An NZEB is defined as an energy-efficient building where the total annual energy consumption is less than or equal to the energy generated on-site from renewable sources

(continued)

Table 5.1 (continued)

Source	Country	Legislation year	Target year	Target	Definition
Energy Council	Australia	February 2019	2030	Residential and commercial buildings should be zero energy (carbon)	A Zero Energy (and Carbon) Ready Building is a building with a highly efficient thermal shell and appliances, designed to consume very little energy and capable of achieving zero energy (and carbon) usage when paired with renewable or decarbonized energy systems, either on-site or off-site
Ministry of Economy, Trade and Industry	Japan	April 2014	2050	All newly constructed houses must be zero emissions	An NZEB is described as a building where the annual net energy consumption from primary sources is reduced to zero

5.4 Technological Options and Strategies Towards Zero Energy Buildings

The path from a standard building to a zero-energy building can be approached through two key strategies which are reducing energy demand through energy-saving techniques and increasing energy supply by integrating renewable energy systems (Wu and Skye 2021). Energy-saving techniques encompass a broad spectrum of passive and active strategies, including improving building insulation, utilizing natural ventilation, and integrating energy-efficient appliances and lighting systems. These measures work in tandem to significantly reduce the baseline energy consumption of the building. Meanwhile, increasing energy supply involves harnessing renewable energy sources such as solar, wind, and geothermal energy to offset any remaining energy demand. These strategies are illustrated in Fig. 5.2.

While various definitions of ZEB exist, certain core principles are commonly recognized. ZEB are expected to align with bioclimatic architecture, utilize energy-efficient lighting, incorporate significant renewable energy sources, and employ advanced water management and demand response technologies, often controlled by Internet of Things (IoT) systems. Bioclimatic architecture ensures that buildings are designed with consideration of the local climate and environmental conditions, using natural resources like sunlight and wind to reduce energy consumption. For instance,

Table 5.2 Summary of current policies for NZEBs in different countries and regions (Feng et al. 2019)

Country/region	Organization	Program	Content	Year
Europe	Directive on Energy Performance of Buildings (EPBD)	ZEBRA 2020	By 2020, all new buildings are required to meet nearly zero energy standards	2010
Belgium	Brussels Capital Region Ministry of Environment	Brussels Passive House Law 2011	From 2015, new constructions or significant renovations of homes, offices, and schools must adhere to the passive house standard	2011
Germany	EPDB	Act on the Promotion of Renewable Thermal Energy	Aim on achieving an almost climate-neutral building stock by 2050	2010
France	Ministry of Environment, Energy, and the sea	Energy Transition for Green Growth	By 2020, new building regulations require energy-positive standards	2015
Denmark	The Ministry for Climate, Energy, and Building	Building Class 2020	By 2018, energy use must be reduced to 50% of current regulations, with a goal for all new buildings to be zero energy by 2020	2015
USA	Office of the Law Review Counsel	The Energy Independence and Security Act of 2007	The target is for 50% of all new federal buildings to be zero energy by 2040 and all new commercial buildings by 2050	2007
	U.S. Department of Energy (DOE); The California Public Utilities Commission (CPUC)	The Building Technologies Program; Zero Net Energy Action Plan	The goal is for all new residential buildings to be NZEBs by 2020 and all new commercial buildings by 2030, with 50% of major retrofits reaching NZEB status by 2025 and all by 2050	2015
	The New York State Energy Research and Development Authority	Ultra-low energy buildings in high-density urban environments	Incremental targets have been set to achieve very low or nearly zero energy use in buildings by 2025	2014
Canada	The City Planning Division of Toronto	Zero Emissions Buildings Framework	The city plans to adopt near zero emissions standards for new city-owned developments by 2026	2018

(continued)

Table 5.2 (continued)

Country/region	Organization	Program	Content	Year
UK	Ministry of Housing, Communities and Local Government	National Planning Policy Framework	All new homes built after 2016, and all other new buildings constructed from 2019 onward, must be zero carbon	2012
Japan	Ministry of Economy, Trade and Industry (METI)	Strategic Energy Plan 2014	By 2020, all newly constructed buildings and standardized houses must meet zero-energy standards, with the goal for half of all existing buildings to be zero-energy by 2030	2015
Korea	National energy program & Zero energy building certification	Building Energy Efficiency Program	By 2025, new buildings should meet energy efficiency standards, covering most new residential and non-residential buildings, with the aim for 50% of energy saving rate	2012
China	Ministry of Housing and Urban–Rural Development	The 13th Five-Year Plan for building energy conservation and green building development	Pilot zero-energy projects are being implemented in cities across China, with the goal for new buildings to be zero energy by 2020	2017
Association of Southeast Asian Nations (ASEAN)	The ASEAN Member States	ASEAN Energy Awards	The ASEAN Energy Awards recognized initiatives in 2019 aimed at reducing energy consumption, particularly in the building sector	2019

buildings in colder climates can leverage passive solar heating, while those in warmer regions can incorporate natural shading and cross-ventilation to stay cool. The integration of IoT systems further enhances energy efficiency by enabling real-time monitoring and control of energy use, allowing building systems to adjust dynamically to changing conditions. The introduction of new technologies into buildings should be carefully evaluated for their compatibility with the ZEB concept, with a proposed assessment method shown in Fig. 5.3.

There are various strategies to reduce energy consumption in buildings, typically categorized into passive and active approaches as demonstrated in Fig. 5.4. Passive strategies focus on the design and orientation of the building, utilizing natural resources to minimize energy needs. This includes features such as natural ventilation, strategic window placement for daylighting, and the use of materials

Table 5.3 The ZEB regulations and codes in three regions (Wang et al. 2021b)

Region	Name	Main Content	Year
The US National	Energy Conservation and Production Act	Provides incentives for conservation and renewable energy, offering loan programs for energy conservation in public and commercial buildings	1976
	Energy Independence and Security Act	Mandates that new federal buildings be fossil fuel-free by 2030, with designs for new federal buildings required to be zero net energy by 2030	2007
	International Energy Conservation Code (IECC)	Managed by the International Code Council (ICC), the IECC provides a Zero Energy Appendix for residential and commercial buildings, updated every three years	Updated every three years (latest 2021)
	ASHRAE 90.1	Serves as the basis for building codes, offering minimum energy efficiency requirements for new and renovated or retrofitted commercial buildings	Updated every three years (latest 2019)
	California long-term energy efficiency strategic plan	Requires all new residential buildings to be zero net energy by 2020 and all new commercial buildings to achieve zero net energy by 2030	2008
EU	State ZEB policy/EPBD	Mandates that by December 31, 2020, all new buildings occupied and owned by public authorities must be nearly zero-energy buildings, with a clear path for future building standards	2010
	Amending EPBD	Sets a clear trajectory toward low and zero-emission building stock in the EU by 2050, supported by national strategies to decarbonize buildings	2018
China	The 13th five-year plan for the development of building energy efficiency and green building	Targets the development of 10 million square meters (m^2) of ultra-low energy and nearly zero-energy buildings by 2020	2016

with high thermal mass to regulate indoor temperatures. On the other hand, active approaches involve the use of mechanical and technological systems such as energy-efficient HVAC systems, smart lighting, and advanced control systems to manage energy use. Both strategies have been well-researched. However, the literature lacks detailed exploration of specific technologies used in residential buildings at both national and international levels. Additionally, methods like halocarbon reduction

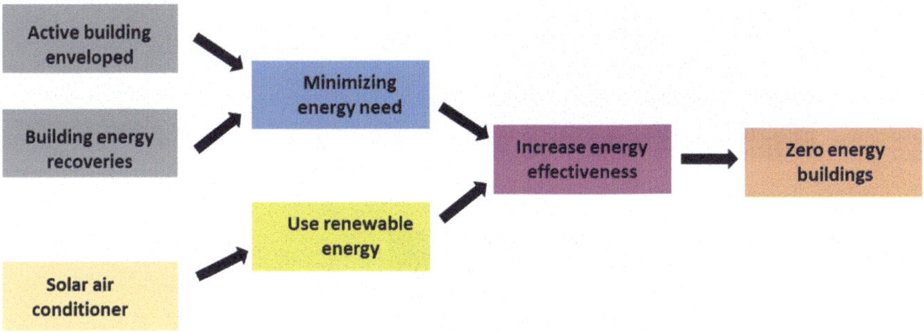

Fig. 5.2 Route of use of technologies towards zero-energy buildings

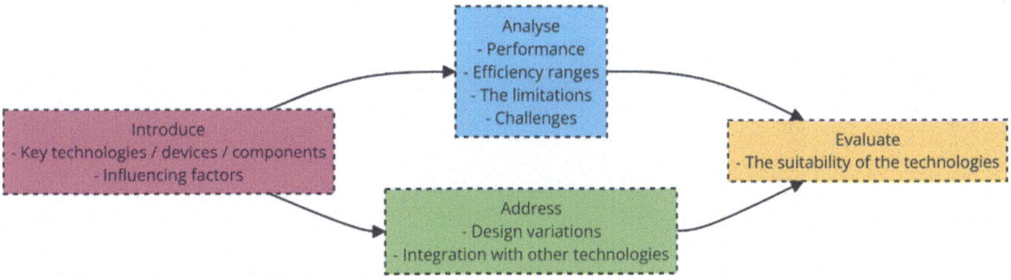

Fig. 5.3 Flowchart for the evaluation of the implementation of new technologies in zero energy buildings

which also contribute to carbon–neutral buildings, are not frequently discussed in existing studies (Hu et al. 2020). Halocarbon reduction focuses on minimizing the use of materials that contribute to greenhouse gas emissions, such as refrigerants used in air conditioning systems. The adoption of low-global-warming-potential (GWP) refrigerants and efficient cooling technologies is critical in ensuring that ZEBs do not inadvertently contribute to climate change through their cooling systems.

5.4.1 Common Technologies for Achieving Zero-Carbon Buildings

Achieving zero-carbon buildings requires the deployment of a wide range of technologies designed to reduce energy consumption, improve energy efficiency, and integrate renewable energy systems. These technologies span various aspects of building design and operations, from the building geometry to the efficiency of its power systems. Below is an in-depth look at the common technologies applied in each major category to achieve zero-carbon performance.

(a) **Building Geometry**

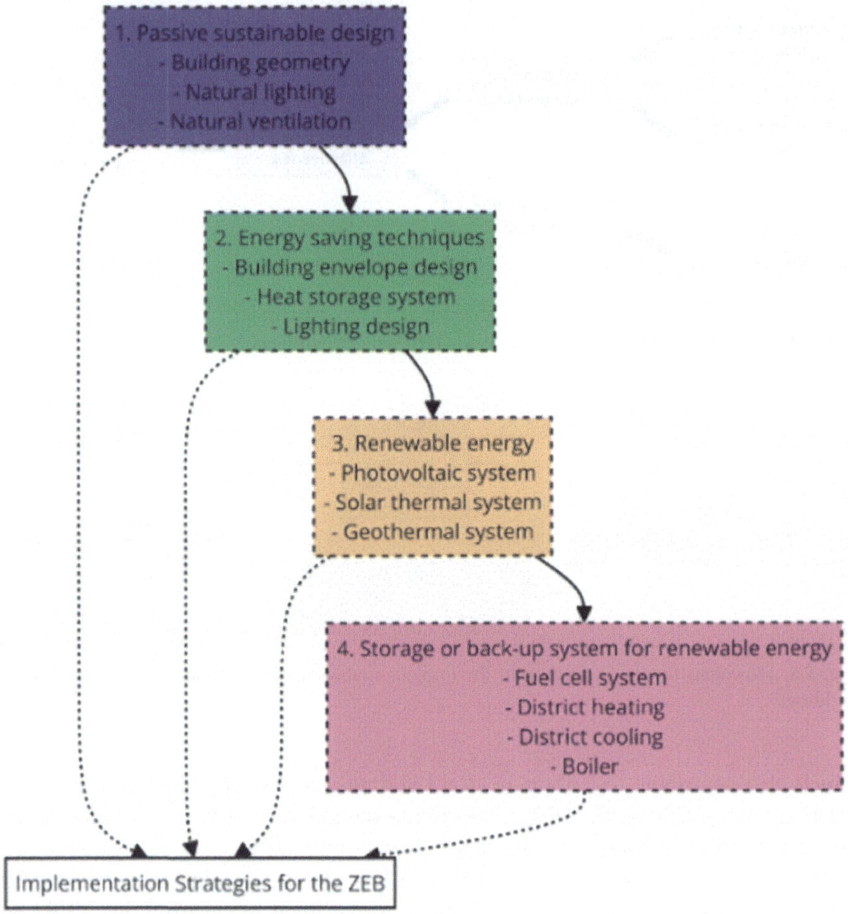

Fig. 5.4 Passive and active strategies to achieve zero energy buildings

The geometry of a building significantly influences its overall energy efficiency and carbon footprint. Optimizing building geometry can reduce the need for mechanical heating, cooling, and lighting, thereby lowering energy consumption and supporting zero-carbon goals (Torabi and Evins 2024). The shape, size, orientation, and spatial configuration of a building all determine how well it interacts with its surrounding environment and natural resources like sunlight, wind, and shade. Building orientation is one of the most critical aspects of geometry. By strategically positioning a building to maximize or minimize solar gains, depending on the climate, energy consumption for heating and cooling can be drastically reduced (Zhao and Du 2020). In colder climates, buildings are often oriented to capture the warmth of the sun during winter, with large, south-facing windows. In contrast, in hot climates, minimizing solar exposure through orientation and shading devices reduces the need for air conditioning.

Another key geometric factor is the form or compactness of the building. Buildings with compact shapes such as cubes, have less surface area exposed to the external environment, which reduces heat loss in the winter and heat gain in the summer. The surface area-to-volume ratio is a crucial design consideration in reducing the energy needed to maintain indoor thermal comfort. Larger surface areas, particularly in irregularly shaped buildings, can result in greater energy loss through walls, roofs, and windows. Daylighting optimization is another benefit of thoughtful building geometry. Skylights, clerestories, and atriums allow natural light to penetrate deep into interior spaces, reducing the need for artificial lighting during the day (Susa-Páez and Piderit-Moreno 2020). The integration of daylighting strategies not only improves energy performance but also enhances the indoor environment, providing occupants with natural light, which has been linked to improved health and productivity.

Finally, the geometry of the building can facilitate natural ventilation. By designing cross-ventilation pathways or strategically placing windows and vents, buildings can take advantage of prevailing winds to cool spaces naturally, reducing the need for mechanical ventilation systems. Buildings designed with atriums or courtyards also promote better air circulation, ensuring fresh air can move throughout the space without the energy-intensive use of fans or HVAC systems. Optimizing building geometry is a passive design strategy that directly influences the energy performance of the building. When combined with other technologies, such as high-performance glazing and insulation, well-planned building geometry plays a critical role in reducing the energy demands of zero-carbon buildings.

(b) **Envelope Improvement**

The building envelope which includes walls, roofs, windows, and floors, acts as the barrier between the internal environment and the external climate. A well-designed envelope can significantly reduce energy consumption, making it a critical component in achieving zero-carbon buildings. Various technologies can be applied to improve the thermal performance of the building envelope, reducing the need for mechanical heating and cooling systems (Kalbasi and Afrand 2022). One of the most effective strategies for improving the building envelope is the installation of thermal insulation. Insulation materials, such as fiberglass, foam, or mineral wool, are placed within walls, roofs, and floors to limit the transfer of heat between the interior and exterior of the building. By maintaining stable indoor temperatures, thermal insulation reduces the demand on heating and cooling systems, ultimately lowering energy consumption. High-performance insulation materials such as aerogels and phase change materials are becoming increasingly popular in zero-carbon designs due to their superior insulating properties.

Air tightness is another key factor in envelope performance. Gaps and cracks in the building envelope can lead to air leaks, which result in significant energy loss as conditioned air escapes and outdoor air enters. Improving the air tightness of a building involves sealing these gaps with materials such as caulks, foams, or weather stripping (Ji et al. 2024). In combination with effective insulation, a well-sealed envelope helps maintain indoor temperatures and reduces the reliance on mechanical systems to compensate for energy losses. In addition to insulation and air tightness, the design of the roof plays a significant role in energy efficiency. Reflective roofs are designed to reflect more sunlight and absorb less heat than standard roofs, making them especially useful in warm climates. Materials such as white membranes or reflective coatings can reduce the heat absorbed by the building, lowering cooling costs. Green roofs provide similar benefits but with added environmental advantages. By covering the roof with vegetation, green roofs improve insulation which reduce the urban heat island effect, and provide stormwater management. Green roofs are particularly effective in urban areas, where large expanses of impermeable surfaces contribute to overheating and environmental stress.

The materials used in the building envelope also contribute to its energy performance. Advanced sustainable materials, such as cross-laminated timber (CLT) or composite panels, offer excellent thermal properties while being environmentally friendly (Thongcharoen et al. 2021). These materials are often sourced from renewable or recycled resources, reducing the embodied carbon of the building. By combining insulation, air tightness, reflective or green roofs, and sustainable materials, the building envelope can be transformed into an energy-saving system that supports zero-carbon objectives. Envelope improvements play a vital role in minimizing energy consumption of the building. Through the strategic use of insulation, air sealing, reflective or green roofs, and advanced materials, the building envelope can significantly reduce heating and cooling demands, making zero-carbon buildings both more feasible and efficient.

(c) **Windows/Glazing Improvement**

Windows are often considered a weak point in a building thermal performance due to their high potential for heat transfer compared to walls or roofs. However, technological advancements in window and glazing design have significantly improved their energy efficiency, making them an integral part of zero-carbon building strategies. The right choice of windows can drastically reduce heating, cooling, and lighting demands, improving the overall performance of the building envelope. Selective glazing is one of the key technologies used in energy-efficient windows (Hafnaoui et al. 2024; Yi et al. 2021). This type of glazing is coated with materials that allow visible light to pass through while reflecting infrared and ultraviolet rays. By blocking solar heat gain while maintaining daylighting, selective glazing reduces the need for air conditioning in hot climates, without compromising natural light. In colder climates, the glazing can be designed to allow more solar gain, helping to reduce heating costs during winter.

Advanced glazing systems such as double or triple-pane windows filled with inert gases like argon or krypton offer excellent insulation properties. These gases have lower thermal conductivity than air, reducing the amount of heat that passes through the window. Multi-pane windows also reduce condensation, which can degrade both the window frame and the surrounding walls (Heydari et al. 2021). The use of low-emissivity coatings on the glass further enhances thermal performance by reflecting heat back into the building during the winter months, keeping the indoor environment warmer with less energy consumption. One of the most innovative technologies in this category is electrochromic window glazing, often referred to as "smart glass."

Electrochromic windows can change their tint in response to an electrical signal, allowing occupants to control the amount of light and heat entering the building (Huang et al. 2024). During periods of high sunlight, the glass can darken to reduce heat gain, while in cooler conditions, the glass can remain clear to maximize solar heat gain. This dynamic adjustment helps to maintain indoor comfort and reduce energy consumption for cooling and heating.

In addition to their energy-saving benefits, these advanced window systems contribute to occupant comfort and well-being by improving indoor daylighting and reducing glare. By incorporating these technologies into the design of zero-carbon buildings, windows and glazing can shift from being an energy liability to becoming an asset that supports the overall energy efficiency of the building. The improvement of windows and glazing through selective coatings, advanced multi-pane designs, and smart glass technologies significantly enhances the thermal and lighting performance of buildings. These technologies help to reduce heating, cooling, and lighting loads, making windows a crucial component in the path to achieving zero-carbon buildings.

(d) **Internal conditions**

The internal conditions of a building such as occupancy levels, internal heat loads, and design parameters, play a crucial role in determining the building overall energy consumption. Managing these conditions effectively is essential for achieving a zero-carbon building, as it reduces unnecessary energy use while maintaining occupant comfort. Various technologies and strategies are employed to control and optimize these internal conditions, ultimately improving the energy efficiency of the buildings. Occupancy control systems are one of the most effective tools for reducing energy consumption. These systems use sensors to detect when rooms or spaces are occupied, automatically adjusting lighting, heating, cooling, and ventilation systems based on real-time occupancy levels (Trofimova et al. 2021). For instance, lights can be dimmed or turned off when a room is unoccupied, and HVAC systems can reduce their output to save energy. This ensures that energy is only used when and where it is needed, significantly lowering overall energy consumption. Occupancy control is particularly effective in office buildings, conference rooms, and other commercial spaces where rooms may be intermittently used throughout the day.

Another important factor is managing internal heat loads, which are generated by appliances, lighting, and even building occupants. By reducing internal heat loads particularly in warm climates, the cooling demands of a building can be minimized. For example, using energy-efficient appliances and low-energy lighting such as LED bulbs can significantly reduce the amount of heat generated inside the building (Zaferanchi and Sozer 2024). This not only lowers electricity consumption for lighting but also reduces the energy needed to cool the space. Internal design conditions also refer to how spaces within the building are arranged and used. Spaces that require higher levels of heating or cooling such as kitchens or data centres can be strategically located in parts of the building with better natural ventilation or shading, reducing the need for artificial temperature control. Additionally, zoning systems can be used to divide the building into different areas that can be heated or cooled independently, allowing more precise control of energy use in different parts of the building. This helps ensure that energy is not wasted by heating or cooling unoccupied or less-used areas. By carefully managing internal conditions, buildings can significantly reduce energy consumption while maintaining a comfortable environment for occupants. These technologies, combined with efficient design strategies, are key components in the pursuit of zero-carbon buildings.

(e) **Heating, Ventilation, and Air Conditioning (HVAC)**

HVAC systems are often among the largest consumers of energy in a building. For zero-carbon buildings, it is essential to employ technologies that not only reduce energy consumption but also integrate renewable energy sources where possible. A range of HVAC technologies are designed to minimize the carbon footprint of heating, cooling, and ventilating spaces, improving both efficiency and sustainability. One of the most effective strategies is natural ventilation, which uses wind and temperature differences between indoor and outdoor spaces to circulate air through the building without the need for mechanical systems (Rey-Hernández et al. 2020). By designing the building to allow for cross-ventilation, warm air can escape naturally and cooler air can enter which reducing the need for air conditioning, particularly in mild climates. Passive cooling techniques, such as shading and natural ventilation can drastically lower energy consumption in buildings that otherwise rely on mechanical systems for ventilation.

In climates where heating is necessary, technologies like biomass boilers and heat pumps provide low-carbon alternatives to conventional fossil fuel-based systems. Air source heat pumps (ASHP) and ground source heat pumps (GSHP) are particularly effective in reducing energy use for heating and cooling (Nikitin et al. 2021). These systems extract heat from the air or ground and transfer it into the building. They are highly efficient and can be powered by renewable electricity, further reducing carbon emissions. Ground source heat pumps leverage the stable temperatures found below the earth surface, making them extremely energy-efficient year-round. For ventilation systems, mechanical ventilation with heat recovery (MVHR) is a highly efficient solution. These systems capture heat from the exhaust air and use it to warm incoming fresh air, reducing the energy required to maintain indoor temperatures (Kowalski et al. 2021). This technology is particularly effective in colder climates, where fresh air is needed but outdoor temperatures are low. By recovering heat that would otherwise be lost, MVHR systems can significantly reduce heating costs.

Another renewable heating technology used in zero-carbon buildings is solar thermal. Solar thermal systems use solar collectors to capture the energy from the sun, which is then used to heat water or provide space heating. This reduces the reliance on conventional heating systems and can be particularly effective in regions with high solar radiation. Solar thermal systems are often integrated into buildings alongside district heating and cooling networks, which provide centralized heating or cooling to multiple buildings. District systems are highly efficient because they can leverage large-scale renewable energy sources such as geothermal or biomass to provide heating and cooling at a reduced carbon cost. Zero-carbon buildings also integrate smart HVAC controls, which adjust the system based on real-time conditions (Chen et al. 2020a). These advanced control systems monitor internal conditions and occupancy levels, automatically optimizing the HVAC system performance. This reduces energy wastage by ensuring the system operates only when needed and at the appropriate levels. Advanced HVAC technologies such as natural ventilation, heat pumps, MVHR systems, and solar thermal solutions are key to minimizing energy consumption in zero-carbon buildings. By combining these technologies with renewable energy sources and intelligent control systems, buildings can achieve significant reductions in their carbon footprint while maintaining a comfortable indoor environment.

(f) **Power Systems**

In zero-carbon buildings, power systems play a critical role in minimizing reliance on fossil fuels by generating and utilizing renewable energy. These systems are designed to produce electricity on-site, reduce peak energy demand, and increase energy independence. By integrating renewable energy technologies, zero-carbon buildings can significantly lower their overall carbon emissions and achieve energy neutrality. One of the most common and effective power technologies used in zero-carbon buildings is solar photovoltaic (PV) systems. Solar PV panels convert sunlight into electricity, providing a clean, renewable source of power for the building (Nematchoua et al. 2021). Depending on the location, solar PV systems can generate a significant portion or even all of the building electricity needs. PV systems are highly scalable, meaning they can be designed for both residential and large commercial buildings. Additionally, modern PV systems are often paired with battery storage allowing buildings to store excess energy generated during the day for use at night or during periods of low solar availability.

Another innovative technology is hybrid photovoltaic-thermal (PV/T) systems, which combine the generation of electricity and heat in one system. PV/T systems capture the heat produced by the PV panels, which would otherwise be wasted, and use it for heating water or space heating (Braun et al. 2020). This increases the overall efficiency of the system and provides multiple forms of renewable energy from a single installation. In addition to solar technologies, micro-wind turbines offer another renewable power option for zero-carbon buildings, particularly in areas with consistent wind patterns. As shown in Fig. 5.5, micro-wind turbines are smaller-scale versions of the large turbines used in wind farms and are designed to be installed on rooftops or open spaces. While they typically generate less electricity than PV panels, they can complement solar systems especially in regions where wind speeds are higher during times when solar energy is less available.

For buildings located in urban environments or areas without consistent access to wind or solar resources, fuel cells and micro-combined heat and power (CHP) systems offer efficient alternatives. Fuel cells generate electricity through an electrochemical reaction, often using hydrogen as fuel which produces only water and heat as by-products. Fuel cells can be an excellent option for zero-carbon buildings because they produce zero direct emissions and can run continuously, providing a steady source of power. Micro-CHP systems combine electricity generation with

Fig. 5.5 Micro-wind turbines units

the recovery of heat for space or water heating making them highly efficient. These systems can be powered by natural gas or renewable biogas, reducing their carbon footprint. To further enhance the efficiency of power systems in zero-carbon buildings, voltage optimizers are used to reduce the voltage supplied to electrical systems, which helps minimize energy waste. By ensuring that appliances and systems operate at their optimal voltage levels, buildings can reduce unnecessary energy consumption and improve overall electrical efficiency. Power systems such as solar PV, hybrid PV/T systems, micro-wind turbines, fuel cells, and micro-CHP are vital components of zero-carbon buildings. These technologies allow buildings to generate their own clean, renewable energy and, when combined with energy storage solutions, significantly reduce reliance on the grid. By integrating multiple renewable energy sources, zero-carbon buildings can achieve a balance between energy consumption and generation, moving closer to total energy independence.

(g) **Lighting**

Lighting systems are a major component of energy consumption in buildings, and optimizing lighting efficiency is crucial for achieving zero-carbon buildings. Advancements in lighting technology have made it possible to reduce the energy required for lighting while maintaining high-quality illumination. By incorporating energy-efficient lighting solutions and smart controls, buildings can significantly lower their overall energy use and contribute to zero-carbon goals. The most widely used and energy-efficient lighting technology today is LED (Light Emitting Diode) lighting. LED are vastly superior to traditional incandescent and fluorescent bulbs in terms of both energy consumption and longevity. LED lights use up to 75% less energy and last up to 25 times longer than incandescent lighting, making them the ideal choice for energy-conscious buildings. In addition to their efficiency, LED lights emit less heat, reducing the cooling load in buildings and contributing further to energy savings.

In zero-carbon buildings, efficient lighting strategies extend beyond just the choice of bulbs. The overall design and layout of lighting systems play a key role in reducing energy consumption. Task lighting focuses light only where it is needed, rather than illuminating entire rooms or spaces unnecessarily. By providing focused lighting for specific activities, task lighting reduces the overall amount of light required, cutting down on energy use. Another important aspect of lighting in zero-carbon buildings is the integration of smart lighting controls (Metallidou et al. 2020). These systems use sensors and timers to adjust lighting based on occupancy or daylight levels. For example, occupancy sensors can automatically turn lights on or off when people enter or leave a room, ensuring that energy is not wasted in unoccupied spaces. Daylight sensors adjust the brightness of artificial lighting based on the amount of natural light entering a space, reducing the need for artificial lighting during the day. Dimming systems also allow lights to operate at reduced levels when full brightness is not necessary, further optimizing energy use.

Daylighting design is another critical component of lighting strategies in zero-carbon buildings. By maximizing the use of natural light through features such as large windows, skylights, and light shelves, buildings can reduce their reliance on artificial lighting during the daytime. Proper daylighting not only lowers energy consumption but also creates a healthier and more comfortable indoor environment for occupants. In conclusion, energy-efficient lighting technologies such as LED lights combined with smart lighting controls and thoughtful daylighting strategies, play a key role in reducing the energy demand of buildings. These lighting solutions, when implemented effectively, help lower overall energy consumption, supporting the achievement of zero-carbon goals in buildings.

(h) **Appliances**

Appliances particularly in residential and commercial buildings, account for a substantial portion of energy consumption. To achieve zero-carbon buildings, it is essential to integrate energy-efficient appliances and systems that minimize power usage without compromising functionality. Advances in appliance design, as well as the use of smart technologies, have significantly improved the energy efficiency of common appliances. These improvements make them vital contributors to reducing the overall carbon footprint of a building. Low-energy appliances are designed to consume significantly less electricity than traditional appliances while maintaining or improving performance. Energy Star-rated refrigerators, washing machines, dishwashers, and HVAC equipment use up to 50% less energy than their conventional counterparts (Reddy and Raghavaiah 2022). These appliances are designed with advanced technologies such as improved insulation, variable-speed compressors, and heat pump systems, which allow them to operate efficiently with minimal energy input. By integrating low-energy appliances into zero-carbon buildings, energy consumption from essential building functions such as cooling, refrigeration, and laundry can be drastically reduced.

Smart appliances are another important innovation that contributes to energy savings in zero-carbon buildings. These appliances are equipped with sensors and communication technology that allow them to interact with building management systems or smart home networks. For example, a smart refrigerator can adjust its cooling cycles based on usage patterns, reducing its energy consumption during periods of lower demand. Smart washing machines can be programmed to run during off-peak hours when electricity rates are lower, or when renewable energy sources like solar power are available, further reducing both costs and carbon emissions. Internet of Things (IoT) devices enable appliances to be controlled remotely, giving users the flexibility to optimize energy use based on real-time conditions or energy availability (Ala'raj et al. 2022). In addition to reducing electricity consumption, zero-carbon buildings also aim to minimize the environmental impact of cooking and food preparation. Clean cooking technologies such as induction cooktops, use electromagnetic energy to heat cookware directly, reducing energy losses compared to traditional gas or electric stoves. Induction cooktops are more efficient and faster while also being safer and easier to control, making them an ideal choice for sustainable, low-energy kitchens.

Another key innovation in the push towards zero-carbon appliances is the development of smart grids and demand response systems, which allow appliances to adjust their energy usage based on grid conditions. During periods of high renewable energy generation, appliances can increase their activity to take advantage of clean power, while during times of high grid demand, they can reduce their power consumption, helping to balance energy supply and demand. The use of low-energy appliances, smart appliances, and clean cooking technologies is crucial for reducing energy consumption in zero-carbon buildings. By incorporating these advanced systems into everyday building operations, zero-carbon buildings can significantly reduce their reliance on the grid and minimize overall energy demand, while still providing comfort and convenience for occupants.

5.5 Case Study of Net-Zero Buildings

The evolution of net-zero energy buildings (NZEB) in hot and humid climates has shown significant progress, with various projects demonstrating the potential to achieve energy neutrality under challenging environmental conditions. These buildings utilize a combination of design strategies and advanced technologies to balance energy consumption and production effectively. One significant achievement is the adaptation of these buildings to extreme climates, proving that even in regions with high heat and humidity, it is possible to reach near-zero or net-positive energy statuses (Zahir et al. 2023). In these climates, managing indoor air quality and temperature control becomes crucial as improper ventilation can lead to high energy consumption for cooling and humidity regulation. NZEB in such climates are particularly innovative in their use of natural ventilation and moisture control technologies to maintain comfort levels without excessive reliance on mechanical systems.

One of the primary approaches observed in these NZEB is the extensive use of passive design features. Common strategies include the implementation of advanced building envelopes, which significantly improve thermal insulation, and the strategic use of solar shading to reduce heat gain. Daylighting and natural ventilation are also frequently employed to minimize the reliance on artificial lighting and mechanical cooling systems. In many cases, passive solar design principles are applied where buildings are oriented and designed to maximize or minimize solar gain depending on seasonal requirements. Features such as overhangs, louvers, and light shelves help regulate sunlight penetration into the building, reducing cooling loads. These passive strategies also incorporate reflective materials in roofs and walls to reduce the absorption of solar radiation, further minimizing heat gain. These passive strategies are complemented by active systems such as photovoltaic panels which are typically installed on rooftops or integrated into building facades to harness renewable energy.

The performance of these buildings varies, but many have achieved impressive results with annual energy consumption intensities below 100 kWh per square meter of floor space. Some buildings have even reached net-positive energy status, meaning they generate more energy than they consume lead to further emphasizing the potential of well-designed NZEB in hot and humid climates (Feng et al. 2019). In particular, net-positive buildings demonstrate the capability of feeding excess energy back into the grid, contributing to a more resilient energy network. These buildings serve as models for future construction, showing how smart integration of renewable energy and energy-saving techniques can push beyond net-zero targets (Himeur et al. 2021). Despite these successes, the case studies also reveal that not all NZEB inherently operate as energy efficient buildings. In some cases, high energy use intensity is still present due to increased demand for cooling and dehumidification in tropical climates (Dezfouli et al. 2022). However, these buildings manage to maintain their net-zero status by maximizing the production of renewable energy such as through larger PV systems or the integration of wind energy in suitable locations. In some instances, buildings with high energy use intensity still manage to achieve NZEB status by relying heavily on renewable energy sources. This points to the importance of a balanced approach that emphasizes both reducing energy demand and increasing renewable energy supply.

Moreover, previous case studies illustrate that the integration of advanced technologies and design strategies must be carefully tailored to the local climate (Wamsler et al. 2020). For example, buildings in regions with high humidity levels benefit significantly from technologies that manage moisture and enhance indoor air quality such as advanced HVAC systems and humidity control devices. In such cases, the use of desiccant-based systems or dehumidification technologies becomes essential for maintaining indoor comfort without excessive energy consumption. Technologies like heat exchangers which cool and dehumidify incoming air have also shown potential in hot and humid climates. These case studies provide valuable insights into the practical implementation of NZEB in hot and humid climates, offering lessons that can guide future projects towards more sustainable and energy-efficient outcomes.

5.5.1 Case Studies in Europe

The state of NZEB across Eastern Europe varies significantly, with each country adopting different strategies and facing unique challenges. The following summaries provide an overview of the NZEB implementation in several Eastern European countries, highlighting their climatic conditions, regulatory frameworks, and key technological approaches.

(a) **Bulgaria**: This country implemented the national NZEB plan in 2015, adapting it to the varying climates continental in the north and Mediterranean in the south (Attia et al. 2022). This plan establishes a minimum energy efficiency standard of 95 kWh/m^2 per year for residential buildings, without imposing specific requirements for renewable energy sources. Mechanical ventilation is not a mandatory feature, and the development of NZEB benchmarks did not follow a cost-optimality approach.
(b) **Croatia**: Experiencing both rainy continental and moderately warm Mediterranean climates, this nation introduced its national NZEB plan in 2015 (Vujnović and Dović 2021). Primary energy thresholds for residential buildings range between 50 and 80 $kWh/m^2.a$, with at least 20% of the annual energy onsite required to come from renewable sources. Air permeability standards are enforced, and the use of heat pumps, particularly when paired with photovoltaic systems, is highly effective.
(c) **Czechia**: With warm summers and cold, snowy winters, this country has established a detailed NZEB plan that compares heating and cooling demands against a reference building. Primary energy demand ranges from 30 to 90 $kWh/m^2.a$. Mechanical ventilation is not mandatory, and there are significant seasonal variations in energy demand. The proliferation of NZEBs in rural areas is slow due to a lack of sophisticated certification mechanisms and a shortage of qualified labour.
(d) **Estonia**: This country with its oceanic temperate climate featuring warm summers and severe winters, has implemented a comprehensive national plan setting primary energy limits between 100 and 145 $kWh/m^2.a$ for residential buildings. Advanced building envelope air permeability requirements are in place, and heat pumps are widely adopted. The cost-optimality approach has been mandatory since 2010 and has been revised multiple times.
(e) **Hungary**: In this continental climate, known for its hot summers and cold winters, the national NZEB plan is still under development. Existing guidelines set primary energy use between 85 and 100 $kWh/m^2.a$, although mechanical ventilation is not required. The cost-optimality approach has been employed since 2016 to define NZEB requirements.
(f) **Latvia**: A country with an oceanic temperate climate, experiencing warm summers and fairly severe winters, has a national plan focusing on limiting non-renewable primary energy use. Thresholds range from 95 to 170 $kWh/m^2.a$, and strict air permeability standards are enforced. While the cost-optimality approach is not widely adopted, technologies like photovoltaic and solar thermal panels, combined with ground source heat pumps, are common.
(g) **Lithuania**: In this climate, similar to that of neighbouring regions, detailed national plans mandate primary energy use between 50 and 100 $kWh/m^2.a$, with a requirement that 50% of energy must come from renewable sources. Although the cost-optimality approach is not commonly used, and mechanical ventilation is not mandatory, significant progress has been made in implementing NZEB.
(h) **Poland**: With its cold climate that combines maritime and continental influences, has established a national plan with primary energy requirements set between 65

and 75 kWh/m^2.a. Renewable energy must cover residual energy requirements, and while the cost-optimality approach guides national requirements, there are inconsistencies in the implementation of energy performance certificates (EPC).

(i) **Romania**: Characterized by a temperate continental climate, transitioning to an oceanic climate on the eastern coast, the national plan of this country sets primary energy thresholds between 93 and 217 kWh/m^2.a, without minimum renewable energy share requirements. Mechanical ventilation and standardized air permeability testing are not required, although the cost-optimality approach is established.

(j) **Slovakia**: In this country, where a continental climate prevails with warm, dry summers and cold winters, national regulations set low primary energy thresholds of 32–54 kWh/m^2.a. Mechanical ventilation and heat recovery systems are mandatory, and renewable energy sources are required to cover at least one energy use item, such as heating or domestic hot water (DHW). The construction sector has widely adopted new technologies, and NZEB practices are well integrated, especially since 2018.

5.5.2 Case Studies of Net Zero Buildings in Cold Regions

In cold climates, achieving net-zero energy status presents unique challenges due to the significant heating demands and extended winter seasons. The harsh climate conditions such as cold winters make it necessary for buildings to use more energy to maintain thermal comfort which in turn complicates the path to achieving net-zero energy goals. Innovative approaches are therefore required to manage energy efficiency while ensuring that buildings are resilient to these extreme weather conditions. Case studies from regions like China, United States, and the European Union demonstrate a range of strategies and technologies that have been employed to overcome these challenges (Wang et al. 2021b). In cold regions, the key is not only to reduce energy consumption but to ensure that energy demand for heating is met through sustainable and renewable sources. This involves the integration of both passive and active energy-saving measures.

A comprehensive evaluation of over 400 zero energy building (ZEB) projects in cold regions reveals that these buildings often rely on a combination of passive and active technologies to meet their energy needs. Passive strategies such as improving the thermal insulation of the building envelope, maximizing solar heat gain, and reducing thermal bridging are common across successful ZEB in cold regions. These strategies are complemented by active technologies like heat pumps, mechanical ventilation with heat recovery (MVHR), and renewable energy systems to minimize energy demand while maintaining indoor comfort. In these regions, high-performance building envelopes are critical with substantial insulation and airtightness playing a central role in reducing heating requirements. The U-values of external walls and roofs in many of these buildings are significantly lower than the maximum limits set by regional ZEB standards, indicating a strong emphasis on minimizing heat loss.

Renewable energy technologies also play a key role in these ZEB. Solar photovoltaic (PV) systems are the most commonly utilized renewable energy source in the United States and the European Union, while ground source heat pumps (GSHP) are more prevalent in China, where building density is higher and the efficiency of GSHP systems can be maximized (Luo et al. 2023). GSHP systems which leverage the relatively stable temperatures underground to provide efficient heating are particularly effective in cold climates where air source heat pumps might struggle to operate efficiently. In China, GSHP systems are integrated into dense urban environments, taking advantage of the proximity of buildings to maximize the efficiency of the technology (Luo et al. 2023). Additionally, biomass energy is more widely used in the European Union, driven by stringent carbon emission reduction policies (Jåstad et al. 2020). In rural areas of Europe, biomass boilers are often used as a renewable alternative to conventional heating systems, providing a carbon–neutral source of heat that is compatible with the goal of reducing reliance on fossil fuels.

Despite the variations in technology adoption, one commonality across these cold-region ZEB is the significant reliance on passive design strategies. These include optimizing building orientation, utilizing thermal mass for heat storage, and implementing advanced glazing systems to maximize solar heat gain while minimizing heat loss (Chen et al. 2020b). In particular, thermal mass materials such as concrete or stone are used to absorb and store heat from the sun during the day and release it at night, reducing the need for mechanical heating systems. High-performance glazing such as triple-glazed windows with low emissivity coatings, further enhances the energy efficiency of buildings in cold climates by reducing heat loss through windows. Natural ventilation is less emphasized in these regions due to the need to maintain indoor warmth. But when it is used, it is typically coupled with mechanical ventilation systems that incorporate heat recovery to maintain energy efficiency. Heat recovery systems are particularly valuable in cold climates, where they capture the warmth from outgoing air and use it to preheat incoming fresh air leads in reducing the amount of energy needed to heat the building.

The actual performance of these ZEB varies, with some buildings achieving net-zero energy more effectively than others. In the United States, a larger proportion of ZEB meet or exceed their energy targets compared to those in the European Union and China, where higher building densities and climatic challenges make it more difficult to achieve net-zero status. However, even in these cases, the buildings often demonstrate significant reductions in energy use compared to conventional buildings, underscoring the effectiveness of the ZEB approach in cold climates.

5.5.3 *Case Studies of Net Zero Energy Buildings in Hot and Humid Climate Zones*

The advancement of NZEB has been notably rapid in affluent nations, while developing regions face significant challenges in achieving similar progress. In hot and humid climates, the high demand for cooling and dehumidification creates significant energy challenges. Conventional air conditioning systems can account for a large proportion of the overall energy use of the building. Furthermore, managing indoor air quality and preventing mould growth are critical issues in these climates, adding another layer of complexity to NZEB design and operation. According to data from the International Energy Agency (IEA) Solar Heating and Cooling Programme (SHC) Task 40, the overwhelming majority of NZEB projects over 90% of nearly 300 listed are located in developed countries such as those in the European Union and the United States (Sakiliba et al. 2020; Kaitouni et al. 2024). In contrast, only a small fraction of these projects about 11, are situated in hot and humid climates and just three of these offer detailed data on their energy performance. The disparity in the distribution of NZEB projects is primarily driven by economic factors. The high initial costs of NZEB construction including the investment in renewable energy technologies and high-performance building materials, pose a barrier in regions where economic development priorities differ from those in affluent nations.

This stark imbalance is largely attributable to economic constraints, as the high initial costs and prolonged payback periods present formidable barriers to the adoption of NZEB in developing countries. In these regions, economic policies and subsidies for renewable energy technologies are often limited or non-existent, making it difficult for developers and homeowners to justify the higher upfront costs associated with NZEB projects. However, there is growing recognition that investing in energy-efficient buildings can yield long-term savings through reduced energy bills and lower maintenance costs. To accelerate the deployment of NZEB in hot and humid climates, it is imperative that these regions prioritize passive design strategies that require lower upfront investments and opt for cost-effective solutions with quicker returns on investment. For example, shading systems, reflective roofing materials, and natural ventilation can be implemented with minimal cost while significantly reducing the need for active cooling. Passive cooling techniques such as cross-ventilation, stack effect ventilation, and evaporative cooling also offer low-cost alternatives to traditional air conditioning in these climates. Evaluating the financial feasibility of NZEB technologies is especially crucial in these areas to ensure their practical adoption and sustainability. Additionally, international funding mechanisms such as climate finance initiatives, could play a key role in supporting the transition to NZEB in developing regions by providing the necessary financial resources to offset initial capital costs.

Although research on NZEB in hot and humid climates remains limited, there have been some notable efforts. For instance, the New Building Institute (NBI) in the United States has documented 482 zero energy projects, including both certified and upcoming buildings, in North America as part of their "Going to Zero" initiative.

This initiative has spurred significant interest in net-zero buildings, particularly in warmer states like California and Florida, where the need for energy-efficient cooling systems is paramount. Similarly, the IEA SHC Task 40 initiative has provided valuable resources, including an interactive map of over 300 NZEB worldwide and a detailed report on 30 selected NZEB projects. Within the IEA SHC Task 40 framework, three NZEB located in hot and humid climates have been identified and analysed in a case study report. These case studies have shown that integrating renewable energy systems such as solar photovoltaic and energy storage, alongside passive design strategies like solar shading and reflective materials is key to achieving energy efficiency in hot and humid climates. Additionally, the NBI list includes 20 NZEB from similar climatic zones. Beyond these larger projects, smaller-scale NZEB case studies and research on energy-efficient buildings in hot and humid regions have been documented in various locations, including Darwin, Australia; Kuala Lumpur, Malaysia; Hong Kong; and several sites in the United States. These projects illustrate the potential of NZEB to significantly reduce energy consumption in hot and humid climates, while also providing valuable insights into the challenges of balancing energy efficiency with thermal comfort in these regions.

References

Ade-Ojo OC (2022) Awareness of the LEED requirements for green housing development among built-environment professionals in Nigeria. Built Environ Project Asset Manage 12(4):521–536

Ahmed A et al (2022) Assessment of the renewable energy generation towards net-zero energy buildings: a review. Energy Build 256:111755

Alaraj M et al (2022) Data-driven based HVAC optimisation approaches: a systematic literature review. J Build Eng 46:103678

Al-Shamkhee D et al (2022) Passive cooling techniques for ventilation: an updated review. Renewable Energy and Environ Sustainab 7:23

Asghari A et al (2021) Evaluation of urban sustainable neighbourhood with zero energy buildings in Valiasr neighbourhood of Tabriz. Sustainable City 4(2):91–106

Attia S et al (2022) Overview and future challenges of nearly zero-energy building (nZEB) design in Eastern Europe. Energy Build 267:112165

Bandeiras F et al (2020) Towards net zero energy in industrial and commercial buildings in Portugal. Renew Sustain Energy Rev 119:109580

Barrutieta X et al (2023) The zero building: an exemplary nearly zero energy office building (NZEB) and its potential to become a positive energy building (PEB). Archit Sci Rev 66(3):214–225

Bertoldi P et al (2021) How to finance energy renovation of residential buildings: review of current and emerging financing instruments in the EU. Wiley Interdisciplinary Rev: Energy and Environ 10(1):e384

Bilgen S (2014) Structure and environmental impact of global energy consumption. Renew Sustain Energy Rev 38:890–902

Braun R et al (2020) System design and feasibility of trigeneration systems with hybrid photovoltaic-thermal (PVT) collectors for zero energy office buildings in different climates. Sol Energy 196:39–48

Brown D et al (2024) A systematic review of barriers to implementing net zero energy buildings in Australia. J Cleaner Prod pp 142910

Chen Y et al (2020a) Transfer learning with deep neural networks for model predictive control of HVAC and natural ventilation in smart buildings. J Clean Prod 254:119866
Chen Y et al (2020b) Experimental investigation of demand response potential of buildings: Combined passive thermal mass and active storage. Appl Energy 280:115956
Christopher S et al (2023) Renewable energy potential towards attainment of net-zero energy buildings status–a critical review. J Clean Prod 405:136942
Cielo D, Subiantoro A (2021) Net zero energy buildings in New Zealand: challenges and potentials reviewed against legislative, climatic, technological, and economic factors. J Build Eng 44:102970
D'Agostino D et al (2021a) Assessing nearly zero energy buildings (NZEBs) development in Europe. Energ Strat Rev 36:100680
D'Agostino D et al (2021b) Data on nearly zero energy buildings (NZEBs) projects and best practices in Europe. Data Brief 39:107641
Del Borghi A et al (2021) Carbon-neutral-campus building: design versus retrofitting of two university zero energy buildings in Europe and in the United States. Sustainability 13(16):9023
Dezfouli M, Sopian K, Kadir K (2022) Energy and performance analysis of solar solid desiccant cooling systems for energy efficient buildings in tropical regions. Energy Convers Manage: X 14:100186
Digitemie WN, Ekemezie IO (2024) A comprehensive review of building energy management systems (BEMS) for improved efficiency. World J Adv Res Rev 21(3):829–841
Echlouchi K, Ouardouz M, Bernoussi A-S (2022) Eco-district, an ideal framework to initiate large-scale urban energy renovation in Morocco. J Ecolog Eng 23(9)
Elaouzy Y, El Fadar A (2022) Energy, economic and environmental benefits of integrating passive design strategies into buildings: a review. Renew Sustain Energy Rev 167:112828
Esbensen TV, Korsgaard V (1977) Dimensioning of the solar heating system in the zero energy house in Denmark. Sol Energy 19(2):195–199
Faraj K et al (2020) Phase change material thermal energy storage systems for cooling applications in buildings: A review. Renew Sustain Energy Rev 119:109579
Feng W et al (2019) A review of net zero energy buildings in hot and humid climates: experience learned from 34 case study buildings. Renew Sustain Energy Rev 114:109303
Fragkos P et al (2021) Energy system transitions and low-carbon pathways in Australia, Brazil, Canada, China, EU-28, India, Indonesia, Japan, Republic of Korea. Russia and the United States. Energy 216:119385
Galvin R (2022) Net-zero-energy buildings or zero-carbon energy systems? How best to decarbonize Germany's thermally inefficient 1950s–1970s-era apartments. J Build Eng 54:104671
González-Torres M et al (2022) A review on buildings energy information: trends, end-uses, fuels and drivers. Energy Rep 8:626–637
Hafnaoui R et al (2024) Smart switchable glazing systems in Saudi Arabia: A review. Energy Build pp 114555
Heizer M, Shelide B (2020) Challenges to statewide net-zero energy buildings: the Oregon experience. ASHRAE Transactions 126(2)
Heydari A, Sadati SE, Gharib MR (2021) Effects of different window configurations on energy consumption in building: optimization and economic analysis. J Build Eng 35:102099
Himeur Y et al (2021) Artificial intelligence based anomaly detection of energy consumption in buildings: a review, current trends and new perspectives. Appl Energy 287:116601
Hu S, Cabeza LF, Yan D (2020) Review and estimation of global halocarbon emissions in the buildings sector. Energy Build 225:110311
Huang R et al (2024) Self-powered electrochromic smart window helps net-zero energy buildings. Nano Energy 129:109989
Jain P, Goswami B (2021) Energy efficiency in South Asia: trends and determinants. Energy 221:119762
Jåstad EO et al (2020) The role of woody biomass for reduction of fossil GHG emissions in the future North European energy sector. Appl Energy 274:115360

Ji Y, Duanmu L, Hu S (2024) Measurement and analysis of airtightness safeguard measures for typical ultra-low energy buildings. Energy Built Environ 5(3):348–363

Kaitouni SI et al (2024) Design parameters influencing the energy performance and indoor comfort of net zero energy building "NZEB" designed for semi-arid urban areas: Digital workflow methodology, sensitivity analysis and comparative assessment. Sol Energy 268:112264

Kalbasi R, Afrand M (2022) Which one is more effective to add to building envelope: phase change material, thermal insulation, or their combination to meet zero-carbon-ready buildings? J Clean Prod 367:133032

Karvonen A, Cook M, Haarstad H (2020) Urban planning and the smart city: projects, practices and politics. Urban Planning 5(1):65–68

Khan MH et al (2022) Modeling and optimization of smart building energy management system considering both electrical and thermal load. Energies 15(2):574

Kowalski P, Szałański P, Cepiński W (2021) Waste heat recovery by air-to-water heat pump from exhausted ventilating air for heating of multi-family residential buildings. Energies 14(23):7985

Kuwahara R, Kim H, Sato H (2022) Evaluation of zero-energy building and use of renewable energy in renovated buildings: a case study in japan. Buildings 12(5):561

Li Y et al (2020) Techno-economic performance analysis of zero energy house applications with home energy management system in Japan. Energy and Buildings 214:109862

Luo J et al (2023) An overview of the recent development of the Ground Source Heat Pump (GSHP) system in China. Renew Energy 210:269–279

Magrini A et al (2020) From nearly zero energy buildings (NZEB) to positive energy buildings (PEB): the next challenge-The most recent European trends with some notes on the energy analysis of a forerunner PEB example. Dev Built Environ 3:100019

Mariano-Hernández D et al (2021) A review of strategies for building energy management system: Model predictive control, demand side management, optimization, and fault detect and diagnosis. J Build Eng 33:101692

Marszal AJ et al (2011) Zero energy building—a review of definitions and calculation methodologies. Energy Build 43(4):971–979

Mazzucato M, Macfarlane L (2023) Mission-oriented development banks: The case of KfW and BNDES. UCL Institute for Innovation and Public Purpose, Working Paper Series (IIPP WP 2023–13). Available at: https://www.ucl.ac.uk/bartlett/public-purpose/wp2023-13

Merabet GH et al (2021) Intelligent building control systems for thermal comfort and energy-efficiency: a systematic review of artificial intelligence-assisted techniques. Renew Sustain Energy Rev 144:110969

Metallidou CK, Psannis KE, Egyptiadou EA (2020) Energy efficiency in smart buildings: IoT approaches. IEEE Access 8:63679–63699

Mitchell R, Natarajan S (2020) UK Passivhaus and the energy performance gap. Energy Build 224:110240

Monstvilas E et al (2023) Impact of the EPBD on changes in the energy performance of multi-apartment buildings in Lithuania. Sustainability 15(3):2032

Moran P, O'Connell J, Goggins J (2020) Sustainable energy efficiency retrofits as residenial buildings move towards nearly zero energy building (NZEB) standards. Energy Build 211:109816

Mushtaha E et al (2021) The impact of passive design strategies on cooling loads of buildings in temperate climate. Case Stud Thermal Eng 28:101588

Nematchoua MK, Nishimwe AM-R, Reiter S (2021) Towards nearly zero-energy residential neighbourhoods in the European Union: a case study. Renew Sustain Energy Rev 135:110198

Nguyen TMP, Davidson K, Coenen L (2020) Understanding how city networks are leveraging climate action: experimentation through C40. Urban Trans 2:1–23

Nikitin A et al (2021) Comparative study of air source and ground source heat pumps in 10 coldest Russian cities based on energy-exergy-economic-environmental analysis. J Clean Prod 321:128979

Omrany H et al (2022) A bibliometric review of net zero energy building research 1995–2022. Energy and Build 262:111996

Park BR, Chung MH (2023) Analysis of the additional energy-saving potential of residential buildings after mandatory zero-energy buildings to achieve carbon neutrality in South Korea. Build Environ 228:109908

Qian F, Shi Z, Yang L (2024) A review of green, low-carbon, and energy-efficient research in sports buildings. Energies 17(16):4020

Reddy BK, Raghavaiah N (2022) Smart energy-efficient techniques for large-scale process industries. Des Dev Efficient Energy Syst 67–100

Rey-Hernández JM et al (2020) Performance analysis of a hybrid ventilation system in a near zero energy building. Build Environ 185:107265

Saini L et al (2022) Net Zero Energy Consumption building in India: an overview and initiative toward sustainable future. Int J Green Energy 19(5):544–561

Sakiliba SK et al (2020) The energy performance and techno-economic analysis of zero energy bill homes. Energy and Build 228:110426

Sartori I, Napolitano A, Voss K (2012) Net zero energy buildings: a consistent definition framework. Energy and Buildings 48:220–232

Schwarz M, Nakhle C, Knoeri C (2020) Innovative designs of building energy codes for building decarbonization and their implementation challenges. J Clean Prod 248:119260

Shirinbakhsh M, Harvey LDD (2024) Feasibility of achieving net-zero energy performance in high-rise buildings using solar energy. Energy and Built Environment 5(6):946–956

Skidmore C, Girling G, McWhirter S (2024) Building the Future: Lessons for a Buildings Breakthrough. M-RCBG Associate Working Paper Series

Susa-Páez A, Piderit-Moreno MB (2020) Geometric optimization of atriums with natural lighting potential for detached high-rise buildings. Sustainability 12(16):6651

Thongcharoen N et al (2021) Development of structural insulated panels made from wood-composite boards and natural rubber foam. Polymers 13(15):2497

Tomrukcu G, Ashrafian T (2024) Climate-resilient building energy efficiency retrofit: evaluating climate change impacts on residential buildings. Energy Build 316:114315

Torabi M, Evins R (2024) Towards net-zero carbon buildings: Investigating the impact of early-stage structure design on building embodied carbon. Int J Life Cycle Ass pp 1–16

Trofimova P et al (2021) Post-occupancy evaluation of indoor air quality and thermal performance in a zero carbon building. Sustainability 13(2):667

Vujnović N, Dović D (2021) Cost-optimal energy performance calculations of a new nZEB hotel building using dynamic simulations and optimization algorithms. J Build Eng 39:102272

Wamsler C et al (2020) Environmental and climate policy integration: targeted strategies for overcoming barriers to nature-based solutions and climate change adaptation. J Clean Prod 247:119154

Wang R et al (2021b) A comprehensive evaluation of zero energy buildings in cold regions: actual performance and key technologies of cases from China, the US, and the European Union. Energy 215:118992

Wang WH, Moreno-Casas V, Huerta de Soto J (2021) A free-market environmentalist transition toward renewable energy: the cases of Germany, Denmark, and the United Kingdom. Energies 14(15):4659

Wilberforce T et al (2023) A review on zero energy buildings—Pros and cons. Energy Built Environ 4(1):25–38

Willingham SA (2020) A case study on the financial implications of the zero net energy (ZNE) Goal in California

Wu W, Skye HM (2021) Residential net-zero energy buildings: review and perspective. Renew Sustain Energy Rev 142:110859

Yan B, Hao F, Meng X (2021) When artificial intelligence meets building energy efficiency, a review focusing on zero energy building. Artif Intell Rev 54(3):2193–2220

Yi Z et al (2021) Energy saving analysis of a transparent radiative cooling film for buildings with roof glazing. Energy and Built Environment 2(2):214–222

Yu F et al (2022) Review of the US policies, codes, and standards of zero-carbon buildings. Buildings 12(12):2060

Zaferanchi M, Sozer H (2024) Effectiveness of interventions to convert the energy consumption of an educational building to zero energy. Int J Build Pathol Adaptat 42(4):485–509

Zahir MH et al (2023) Challenges of the application of PCMs to achieve zero energy buildings under hot weather conditions: a review. J Energy Storage 64:107156

Zhao J, Du Y (2020) Multi-objective optimization design for windows and shading configuration considering energy consumption and thermal comfort: a case study for office building in different climatic regions of China. Sol Energy 206:997–1017

Chapter 6
Eco-Cities: Sustainable Urban Living

Contents

Abstract As global energy demand rises alongside rapid urbanization, the need for sustainable urban development becomes increasingly critical. This chapter explores the concept of eco-cities, focusing on sustainable practices that integrate renewable energy, zero-waste management, and water conservation. It also examines the role of innovative urban planning in mitigating the environmental impacts of mass urbanization. Eco-cities prioritize resilience, circular economies, and carbon–neutral development, addressing the urgent need to decouple economic growth from resource depletion. Real-world examples are highlighted to illustrate the application of these sustainable principles, emphasizing their importance in achieving long-term environmental and economic resilience. Additionally, the chapter addresses the challenges of implementing these practices in rapidly growing urban areas and the necessity of adapting urban planning to accommodate future growth. It underscores the importance of collaboration between governments, industries, and communities to foster sustainable urban living. The discussion concludes with strategies for advancing sustainable cities, particularly in the context of developing countries where urbanization pressures are most intense.

S. Riffat et al., *Sustainable Energy Technologies and Low Carbon Buildings*, Lecture Notes in Energy 45, https://doi.org/10.1007/978-3-031-78853-6_6

6.1 Urbanization and Sustainable City Development

The rising global energy demand along with concerns about climate change and the dwindling supply of fossil fuels has spurred significant interest in developing sustainable cities. Generally, these sustainable cities are powered by renewable energy sources such as solar, wind, and biomass (Aşchilean et al. 2021). Currently, urban areas account for approximately 75% of global energy consumption and are responsible for nearly 70% of greenhouse gas emissions. This disproportionate consumption of energy and resources by cities is primarily due to the concentration of industrial activities, transportation networks, and dense populations within urban centres. The need for sustainable city models is not just a response to climate change but also a proactive approach to mitigating the long-term environmental and social impacts of urbanization.

Mass urbanization is one of the most pressing challenges of our time. The pace at which cities are expanding is particularly notable in developing countries like China, India, and parts of sub-Saharan Africa. China has been experiencing rapid urbanization over the past few decades, with rural-to-urban migration seeing approximately 15 million people relocating annually (Du 2023). This migration trend is expected to continue, with the country planning to build 400 new cities and towns over the next 20 years. This large-scale urban development in China is driven by economic growth and industrialization, but it also brings environmental concerns, particularly in terms of energy consumption, air pollution, and resource depletion.

While cities cover only 2% of land surface, they consume 75% of natural resources and generate 75% of waste of the world (Rezaie et al. 2022). Urban areas are major contributors to carbon dioxide emissions due to the concentration of transportation networks, industrial activities, and energy-intensive buildings. As cities grow, they require more infrastructure such as roads, bridges, water systems, and power grids which, in turn, increases resource consumption and the demand for energy. Today, half of the global population resides in urban areas, a stark contrast to the 15% that lived in cities a century ago. Urbanization has not only reshaped the demographic landscape but also placed significant pressure on ecosystems, natural resources, and existing infrastructure, leading to challenges such as water scarcity, pollution, and the degradation of green spaces.

Figure 6.1 shows the number of cities with populations exceeding one million has skyrocketed from 83 in 1950 to over 400 today. Megacities defined as urban areas with populations exceeding 10 million, have also seen dramatic growth, particularly in Asia, Africa, and Latin America. These megacities face unique challenges, including traffic congestion, inadequate housing, and environmental degradation. As a result, cities have become the primary contributors to carbon dioxide emissions and other greenhouse gases, amplifying global climate change.

Fig. 6.1 Urbanization: In 1950, there were 83 cities with a population of over 1 million. Now there are more than 400 cities over 1 million

6.1.1 Challenges of Urbanization: Environmental and Social Impacts

The rapid pace of urbanization can lead to social instability, which in turn hampers the ability of cities to achieve environmental sustainability and economic prosperity. In many rapidly growing cities particularly in the Global South, the lack of affordable housing and essential services like clean water and sanitation has led to the proliferation of informal settlements and slums. These areas are often overcrowded, lack proper infrastructure, and are highly vulnerable to environmental risks such as flooding, heatwaves, and pollution (Ngcamu 2023). Additionally, the inequality in access services and resources can fuel social unrest and create deep economic divides between urban residents. Urban sprawl, the unplanned and uncontrolled expansion of urban areas into surrounding rural land is another significant challenge of rapid urbanization. Sprawl leads to the fragmentation of ecosystems, loss of agricultural land, increased car dependency, and inefficient resource use (Chettry 2023). The dispersed nature of sprawling cities also makes it difficult and costly to provide infrastructure and public services such as transportation, water supply, and waste management. As cities grow outward, the need for roads and private car use increases, leading to higher levels of air pollution and greenhouse gas emissions. To address these challenges, urban planners and policymakers are advocating for "smart growth" strategies that promote compact, mixed-use urban development. These strategies focus on creating walkable neighbourhoods with access to public transportation, reducing the need for long commutes and minimizing the environmental footprint of urban expansion. Compact cities are not only require more energy-efficient but also foster social interaction and community engagement by promoting diverse and vibrant neighbourhoods.

6.1.2 The Need for Sustainable City Models

Given the environmental and social challenges posed by rapid urbanization, there is an urgent need for a new model of sustainability that offers stronger incentives for energy conservation, reduced consumption, and environmental protection. Sustainable cities also known as eco-cities or green cities that are designed to maximize energy efficiency, minimize resource consumption, reduce pollution, and avoid urban sprawl. The concept of sustainable cities is based on principles of circular economy, resilience, and environmental stewardship. These cities aim to operate within the ecological limits of the planet, ensuring that resources are used efficiently and waste is minimized through recycling, composting, and reusing materials. A key characteristic of sustainable cities is high density and social diversity, where economic and social activities intersect, and communities are organized around neighbourhoods (Bibri et al. 2020). High-density urban areas reduce per capita energy and resource consumption by concentrating housing, employment, and services in a smaller geographic footprint.

In addition, high-density cities are more conducive to public transportation systems, which can reduce reliance on private cars and lower greenhouse gas emissions (He et al. 2023). These cities should be designed to meet the changing needs of their inhabitants while promoting well-being, equity, and inclusivity. Resilience is another critical aspect of sustainable cities. As climate change increases the frequency and intensity of extreme weather events such as floods, heatwaves, and storms, cities must be able to adapt to and recover from these shocks. This requires the integration of green infrastructure such as parks, wetlands, and green roofs which can mitigate the effects of climate change by absorbing stormwater, reducing urban heat island effects, and improving air quality. Moreover, resilient cities are designed with redundancy in critical systems, ensuring that disruptions to energy, water, or transportation networks do not cripple urban life (Li 2024).

(a) **Green Roofs and Walls**

Green roofs and walls are transformative elements of green infrastructure, offering numerous environmental and social benefits in urban areas. As shown in Fig. 6.2, green roofs involve the installation of vegetated layers on the rooftops of buildings, effectively turning underutilized spaces into productive, eco-friendly zones (Liberalesso et al. 2020). These vegetated layers absorb rainwater, reduce runoff, and help mitigate the urban heat island effect by lowering rooftop temperatures. Additionally, green roofs serve as insulation for buildings, reducing the need for heating in

Fig. 6.2 Installation of a green roof with vegetated layers on the rooftops of buildings

winter and cooling in summer, thus improving energy efficiency. The cooling effect provided by plants can lower building temperatures, decreasing the energy demand for air conditioning and contributing to reduced greenhouse gas emissions.

In cities like Toronto and Chicago, policies have been enacted to encourage or mandate the installation of green roofs on new developments (Loder 2014). Green walls or vertical gardens are another popular green infrastructure solution, integrating vegetation directly onto building facades. These walls not only enhance an aesthetic appeal of the building but also provide similar environmental benefits such as air purification and temperature regulation. Green walls are particularly effective in filtering particulate matter from the air, thus improving air quality in urban environments. This integration of green roofs and walls into city landscapes demonstrates how architecture and nature can coexist to create healthier, more sustainable cities.

(b) **Urban Parks and Green Corridors**

Urban parks and green corridors are vital components of green infrastructure that enhance urban liveability and environmental sustainability. Parks provide essential green spaces for recreation, relaxation, and social interaction, while also supporting urban biodiversity by creating habitats for wildlife (Peng et al. 2021). These spaces improve public health by offering residents opportunities for physical activity and mental relaxation. Parks also contribute to environmental sustainability by acting as natural cooling systems that reduce the urban heat island effect. Trees and vegetation in parks absorb carbon dioxide and release oxygen, which improves air quality and helps mitigate the impacts of climate change.

Green corridors such as tree-lined pathways or linear parks are designed to connect different parts of the city, allowing for both wildlife movement and easier access to green spaces for residents. These corridors often serve as pedestrian and cycling routes, encouraging sustainable modes of transportation (Wang and Pei 2020). A prominent example is High Line in New York, a former elevated railway track transformed into a public park. The High Line not only serves as a green space but also links neighbourhoods and enhances biodiversity in the heart of the city. By integrating urban parks and green corridors into the fabric of cities, planners can create more sustainable, resilient, and socially vibrant urban environments.

(c) **Permeable Pavements**

Permeable pavements are an innovative green infrastructure solution designed to manage stormwater more effectively in urban areas. Traditional concrete and asphalt surfaces are impermeable, causing rainwater to run off into drains and potentially overwhelming sewage systems during heavy rainfall. This can lead to flooding, water pollution, and increased strain on water treatment facilities. Permeable pavements, on the other hand, are designed to allow water to seep through the surface and into the ground below (Zhuge 2020). By doing so, they help reduce stormwater runoff, mitigate flooding risks, and promote groundwater recharge. In Portland, Oregon, permeable pavements have been installed in streets and parking lots to manage stormwater sustainably while reducing the environmental footprint of urban infrastructure.

The integration of permeable pavements also contributes to improved water quality, as the water that passes through these surfaces is filtered by underlying soil layers before it reaches natural water bodies. This filtration process helps remove pollutants and sediment, reducing the contamination of rivers and lakes. Moreover, permeable pavements are increasingly being used in conjunction with green roofs, rain gardens, and other stormwater management techniques to create a more comprehensive approach to sustainable water management in cities. By incorporating permeable pavements, urban planners can address both flooding concerns and water quality issues while creating more resilient and adaptable cities.

(d) **Constructed Wetlands**

Constructed wetlands are engineered systems designed to mimic the water filtration functions of natural wetlands. These green infrastructure installations are highly effective in treating stormwater and wastewater in urban areas. Constructed wetlands act as natural filters where vegetation, soil, and microorganisms work together to remove pollutants and sediment from water. This process helps improve the quality of water before it is released back into natural water bodies or reused within the city (Haron and Feisal 2020). Constructed wetlands are especially beneficial in cities with combined sewer systems, where stormwater and wastewater are transported in the same pipes, often leading to overflow during heavy rains. By using constructed wetlands, cities can treat excess water in a natural, sustainable way, reducing the risk of water pollution.

Beyond water management, constructed wetlands provide additional benefits such as enhancing biodiversity and offering aesthetic and recreational value. In cities like

Singapore and Melbourne, constructed wetlands have been integrated into urban landscapes to improve water management while also creating green spaces that residents can enjoy (Neo et al. 2022; Pettigrove et al. 2023). These wetlands attract birds, insects, and other wildlife, helping to reintroduce natural ecosystems into densely populated urban environments. By blending ecological functionality with urban design, constructed wetlands demonstrate how cities can use natural systems to address environmental challenges while creating healthier, more liveable urban spaces.

6.1.3 Sustainable City Development in China: A Case Study

China is currently responsible for about half of construction activity of the world. As a result of rapid industrialization and urbanization, China has experienced significant environmental challenges, including air pollution, water scarcity, and loss of biodiversity (Bian et al. 2024). However, the country is also emerging as a leader in sustainable city development. Over the next two decades, China plans to build 400 new cities and towns, many of which are being developed under the framework of the low-carbon economy initiatives (Cheng and Hu 2010). The "Go West" initiative highlights the shift in focus towards developing interior regions instead of coastal areas, addressing the growing demand for urbanization in less developed parts of the country. The low-carbon demonstration zones in China, which include five provinces and eight cities, serve as experimental hubs for testing and scaling up sustainable urban practices.

These zones are implementing cutting-edge technologies such as smart grids, energy-efficient buildings, and electric transportation systems. The goal is to create cities that are not only environmentally sustainable but also economically competitive and socially inclusive. The nation is also advancing rapidly towards a low-carbon economy, having designated several cities as low-carbon demonstration projects. In these projects, renewable energy sources like solar and wind power are being integrated into urban grids, and energy-efficient technologies such as LED street lighting, building automation systems, and electric vehicle infrastructure are being deployed (Wang et al. 2020). The efforts in sustainable urban development provide valuable insights into how rapidly growing nations can balance the demands of urbanization with the need for environmental sustainability. The success of these initiatives depends on strong government policies, public–private partnerships, and the ability to integrate innovative technologies at scale.

6.1.4 *Principles of Sustainable Cities*

Creating sustainable urban environments requires designers to thoroughly understand and effectively implement the core principles of green urbanism. These principles serve as the foundation for developing cities that are not only environmentally friendly but also resilient, liveable, and economically viable. By aligning urban development with ecological principles, cities can minimize their environmental impact while promoting social and economic well-being. The key principles of sustainable cities include:

(a) **Climate and Context**

Each location has unique conditions related to orientation, solar radiation, rain, humidity, topography, shading, lighting, noise, and air pollution. These factors greatly influence how buildings and infrastructure perform over time, directly impacting energy consumption, comfort, and sustainability (Mishra and Sadhu 2023). In order to develop sustainable urban environments, it is crucial to analyse these environmental variables carefully. By tailoring urban design and architecture to the specific climate and environmental conditions of a location, cities can significantly reduce their dependence on energy-intensive systems such as heating, ventilation, and air conditioning (HVAC) while enhancing indoor comfort and air quality. For instance, tropical regions which experience high temperatures and humidity year-round can incorporate natural cooling strategies such as cross-ventilation, wind corridors, and open floor plans to promote airflow (Soebiyan et al. 2023). In these climates, buildings are often designed with large overhangs, deep eaves, and louvers to provide shade and reduce direct solar heat gain, making interiors cooler and more comfortable without the need for artificial cooling. Shading devices, such as awnings, pergolas, and vertical screens, can also minimize the impact of solar radiation while allowing natural light to enter.

In contrast, cities located in colder climates need to optimize their designs to maximize solar heat gain in the winter months while minimizing heat loss. Here, passive solar design principles are essential. Buildings can be oriented to face south or north to capture the energy from the sun for natural heating (Xue et al. 2022). The use of high-performance windows and well-insulated building envelopes can further enhance energy efficiency by reducing heat loss through walls, roofs, and windows. Additionally, thermal mass materials such as concrete, brick, or stone that absorb and store heat, can be strategically used to retain heat during the day and release it at night, further reducing the need for active heating systems. Topography also plays a critical role in urban design. Hills and valleys can influence wind patterns, temperature variations, and the movement of water through the urban landscape (Toland and Cate Christ 2020). Planners must account for these factors to maximize natural ventilation, avoid cold air pooling in lower-lying areas, and minimize the risk of flooding. For cities in mountainous regions, tiered or stepped building designs can follow the natural contours of the land, reducing the need for major earthworks and preserving the ecological balance. Similarly, cities located in windy regions can leverage natural wind patterns to cool buildings, but must also design windbreaks or barriers to prevent strong winds from causing discomfort or damage to structures.

Rainfall patterns particularly in monsoon or tropical regions dictate the need for effective stormwater management and flood protection. Understanding the local context allows planners to design flood resilient infrastructure in areas prone to heavy rainfall or rising sea levels (Song et al. 2024). In regions with frequent heavy rains, cities can integrate green infrastructure such as bioswales, rain gardens, and permeable pavements to manage stormwater runoff, reduce flooding risks, and recharge groundwater supplies. Urban wetlands and green roofs are also valuable in absorbing excess rainwater and providing a buffer against extreme weather events. In coastal cities or regions prone to sea-level rise, planners must also design with long-term climate resilience in mind. This includes elevating infrastructure above projected flood levels, building seawalls or levees, and adopting adaptive architecture, such as floating buildings or amphibious structures, that can withstand periodic flooding. These measures not only protect urban areas from immediate climate impacts but also help cities adapt to future environmental changes.

Noise and air pollution are additional environmental factors that influence the liveability and sustainability of urban areas. High-density cities often suffer from elevated levels of noise and air pollution due to traffic congestion, industrial activities, and dense populations. Proper urban planning that considers noise barriers, tree planting, and the creation of green spaces can help mitigate these issues (Oquendo-Di Cosola et al. 2022). Green corridors and vertical gardens can also absorb pollutants, improve air quality, and provide thermal insulation for buildings, contributing to overall urban health. By designing urban spaces that respond to these climate and context factors, cities can not only minimize their environmental footprint but also enhance the health, comfort, and resilience of their populations. When cities are designed in harmony with the local environment, they can harness natural resources more effectively, reduce the costs and environmental impacts of energy consumption, and create built environments that are adaptive to changing climatic conditions. This approach supports the broader goals of sustainable development, fostering cities that are both economically viable and ecologically balanced.

(b) **Renewable Energy for Zero CO_2 Emissions**

This principle addresses the integration of energy supply systems and the optimization of energy consumption and operations. The overarching goal is to transform cities into hubs of clean energy production, moving away from reliance on fossil fuels and drastically reducing their carbon footprints. Achieving this requires both technological innovation and systemic changes in how cities generate, distribute, and consume energy. It advocates for a significant shift towards renewable energy sources, accelerating the transition away from fossil fuels, and reimagining cities as net energy producers rather than merely consumers. By harnessing renewable resources like sunlight, wind, and geothermal energy, cities can provide clean, sustainable power to their residents and industries while minimizing their impact on the environment. One of the cornerstones of this strategy is the integration of renewable energy technologies into the urban fabric itself. Renewable energy technologies, such as solar panels, wind turbines, and geothermal energy systems, should be incorporated into the urban fabric, powering homes, businesses, and public infrastructure (Ang et al. 2020). For instance, solar photovoltaic (PV) panels can be installed on rooftops, building facades, or even on parking structures, allowing cities to generate electricity directly at the point of consumption. Large-scale solar farms can also be established on city outskirts or in underutilized spaces such as brownfields. Similarly, small-scale wind turbines can be incorporated into urban areas where wind conditions are favourable, particularly along coastal cities or in elevated regions where wind speeds are higher.

In addition to harnessing solar and wind energy, geothermal systems offer a reliable, low-carbon source of energy, especially for heating and cooling. Geothermal heat pumps which utilize the stable temperatures found just below the surface of the earth, can be used in both residential and commercial buildings to reduce the energy needed for climate control (Senova et al. 2022). This makes them particularly effective in regions with extreme seasonal temperature fluctuations. Hydropower, in some cases, may also be feasible, particularly in cities located near rivers or along coastlines with tidal energy potential. Essential components of this strategy include on-site electricity generation and the implementation of smart grids that incorporate local energy production and storage solutions. On-site energy generation reduces transmission losses associated with centralized power plants and enables cities to be more self-sufficient. Smart grids in combination with energy storage technologies such as batteries, play a crucial role in ensuring that the energy generated from renewable sources can be effectively distributed and used when demand is high. These grids allow cities to balance supply and demand dynamically, integrating a variety of energy sources and improving the efficiency of the entire urban energy system.

Smart grids represent a key innovation in urban energy management. These grids are equipped with advanced sensors, automated controls, and two-way communication systems that enable real-time monitoring and optimization of energy flow. Smart grids also support the integration of decentralized energy resources, such as solar panels on individual homes or electric vehicle (EV) charging stations, enabling cities to better manage the variability of renewable energy sources like solar and wind. In addition, smart grids allow for demand-response programs, where electricity usage can be adjusted based on price signals or grid conditions, further enhancing energy efficiency and grid stability. By decentralizing energy generation and promoting community-scale renewable energy systems, cities can enhance their energy resilience while reducing greenhouse gas emissions. Decentralized energy systems such as microgrids, enable local communities to generate, store, and use their own energy independently of the central grid (Bramareswara Rao and Padma 2019). These systems can operate in conjunction with the main grid or in isolation during power outages, providing an added layer of resilience against disruptions caused by extreme weather events or other emergencies. Microgrids also facilitate the integration of diverse energy sources, such as solar, wind, and battery storage, creating a more flexible and adaptable energy system that is less reliant on large, centralized power plants.

Community-scale renewable energy projects, such as shared solar installations or district heating systems powered by renewable sources, allow neighbourhoods and local businesses to pool resources and collectively benefit from clean energy (Syed et al. 2020). These initiatives are particularly effective in promoting energy equity, ensuring that even low-income households have access to affordable renewable energy. Additionally, cities can incentivize the adoption of renewable energy systems by offering tax breaks, subsidies, or low-interest loans for solar panel installations, energy-efficient retrofits, and electric vehicle infrastructure. Energy storage technologies are critical in enabling cities to maximize the benefits of renewable energy systems. Since renewable energy sources like solar and wind are intermittent, battery storage allows cities to store excess energy produced during periods of low demand and deploy it during peak consumption times or when renewable energy generation is low. Advancements in battery technologies such as lithium-ion and solid-state batteries are improving the efficiency and capacity of energy storage systems, making them more accessible for urban applications.

Moreover, transitioning to renewable energy is not only beneficial for reducing carbon emissions but also for enhancing air quality in cities. Fossil fuel-based energy production is a major source of urban air pollution, contributing to respiratory illnesses, heart disease, and premature deaths. By replacing coal, oil, and natural gas with clean energy sources, cities can significantly reduce their levels of harmful pollutants such as particulate matter, nitrogen oxides, and sulphur dioxide, leading to healthier living conditions for their residents. Electrification of urban transportation systems powered by renewable energy, further accelerates the reduction of urban carbon emissions. Cities can promote the widespread adoption of electric vehicles (EV) by developing charging infrastructure powered by renewables. This shift not only reduces the carbon footprint in transportation sector but also lowers air pollution and noise levels in densely populated areas. Public transportation systems such as electric buses, trams, and metro systems, can also be powered by clean energy, making mass transit an even more sustainable option for urban mobility.

The transitioning to renewable energy for zero CO_2 emissions is a multifaceted strategy that involves the integration of various technologies, the optimization of energy systems, and the decentralization of power generation. By adopting renewable energy technologies, cities can dramatically reduce their greenhouse gas emissions which enhance their energy resilience, and improve the overall quality of life for their residents. As cities move toward net-zero emissions, they will not only mitigate the impacts of climate change but also create more sustainable, equitable, and liveable environments for future generations.

(b) **Zero-Waste City**

Sustainable waste management redefines waste as a valuable resource. In a zero-waste city, the goal is not simply to manage waste efficiently but to eliminate it altogether by treating it as a resource that can be continuously reused, recycled, or repurposed. This transformation requires a systemic change in how products are designed, consumed, and discarded, moving from a linear economy to a circular economy. Cities should implement zero-waste systems modelled on natural processes, focusing on reducing consumption, enhancing recycling efforts, reusing materials, and composting organic waste to generate energy (Amaral et al. 2020). In a circular economy, products are designed with their end-of-life in mind, ensuring that materials can be easily disassembled, repaired, and repurposed, rather than disposed of in landfills. A circular economy approach is vital, where products are designed to have longer life cycles and materials are continuously repurposed. By extending the lifecycle of products and minimizing the need for virgin resources, cities can drastically reduce their environmental footprint while simultaneously fostering economic resilience. Key to the success of zero-waste cities is the development of strong waste reduction policies. Cities can promote responsible consumption by incentivizing the reduction of single-use plastics and encouraging the adoption of reusable alternatives (Shakir et al. 2024a). They can also implement extended producer responsibility (EPR) programs which hold manufacturers accountable for the disposal or recycling of their products. In many cases, manufacturers can be required to take back products at the end of their lifecycle, creating a closed-loop system that reduces the burden on municipal waste management services.

A critical component of a zero-waste city is the emphasis on waste segregation at the source. This means households, businesses, and industries must separate recyclables organic waste and non-recyclable waste to ensure that each waste stream can be processed efficiently. For instance, organic waste which constitutes a significant portion of urban waste can be composted or processed through anaerobic digestion to produce biogas. It is a renewable energy source that can be used for electricity generation, heating, or transportation fuel (Binhweel 2024). Composting organic materials also produces nutrient-rich fertilizer which can be used to support urban agriculture, thus closing the nutrient loop. Advanced waste management technologies are essential for processing and repurposing the various waste streams produced by cities. Technologies such as anaerobic digestion and waste-to-energy systems allow cities to convert organic waste into electricity or heat, reducing the reliance on landfills and minimizing environmental pollution. Anaerobic digestion is particularly effective in urban settings where food waste, agricultural waste, and sewage sludge are abundant. This biological process breaks down organic matter in the absence of oxygen, producing biogas, which can then be used to generate electricity, heat buildings, or power vehicles.

Waste-to-energy (WTE) facilities which convert non-recyclable waste into energy through incineration or gasification, also play a role in reducing landfill use (Liang et al. 2022). While WTE technologies have faced criticism for potential air pollution, advances in emission controls have made modern WTE plants much cleaner and more efficient. These facilities can recover valuable energy from waste that cannot be recycled, providing a more sustainable alternative to landfill disposal. In some cities, WTE plants provide district heating to neighbourhoods, creating an integrated system where waste is used to heat homes and buildings. The success of a zero-waste city also depends on robust recycling infrastructure. Recycling programs must be designed to accommodate a wide range of materials, including paper, metals, glass, plastics, and electronic waste. In addition to traditional recycling, cities can promote industrial symbiosis where the waste products of one industry become the raw materials for another.

This approach aims to minimize landfill use and create a circular economy where waste products are continuously repurposed, contributing to environmental sustainability and resource efficiency. Landfills are not only a source of pollution that leaching harmful chemicals into soil and water, but also a major emitter of methane, a potent greenhouse gas (Cusworth et al. 2024). Minimizing landfill use not only reduces these emissions but also preserves valuable land resources. Education and community engagement play a vital role in the success of zero-waste cities. Public awareness campaigns can help citizens understand the importance of waste reduction, proper waste sorting, and participation in recycling programs. Schools, community centres, and local governments can organize workshops and events to educate residents about sustainable consumption habits such as buying in bulk, repairing items instead of replacing them, and composting at home.

Furthermore, cities can support the development of reuse and repair markets, where discarded items can be refurbished and resold. Reuse centres or community repair workshops, where people can learn to fix broken household items or electronics, not only reduce waste but also foster community ties and job creation. These initiatives help shift the mindset from one of disposability to one of resource stewardship. Zero-waste cities also involve collaboration with businesses and industries to implement more sustainable practices. Retailers can reduce packaging waste by offering bulk bins or refill stations, while restaurants and food vendors can switch to compostable or reusable packaging. Retailers can also implement take-back schemes for electronics, clothing, or furniture, ensuring that products are returned for recycling or refurbishment rather than being discarded.

Zero-waste cities are built on the principles of sustainability, circularity, and resource efficiency. By treating waste as a valuable resource, cities can reduce their reliance on landfills, lower their greenhouse gas emissions, and conserve natural resources. Through innovative waste management technologies, strong waste reduction policies, and active community participation, cities can achieve a zero-waste future while fostering economic growth and environmental sustainability.

(d) **Water Management**

Effective water management focuses on minimizing water consumption, optimizing the use of water resources, safeguarding water quality, and preserving aquatic ecosystems. Water is a critical resource for cities and with increasing urbanization, climate change, and population growth, managing water sustainably has become a key challenge for many urban areas around the world. Given the increasing challenges of water scarcity in many urban areas, cities must adopt strategies to maximize the efficient use of water resources. This involves not only reducing water consumption but also rethinking how water is sourced, distributed, used, and treated within urban settings. One of the most pressing issues facing cities today is water scarcity. Many regions around the world are already experiencing severe water stress due to over-extraction of groundwater, pollution, and the effects of climate change. Rising temperatures and irregular rainfall patterns are reducing the availability of freshwater in many regions, forcing cities to seek alternative solutions to meet their water needs. In response, cities must implement comprehensive water management strategies that address both demand and supply.

Cities should implement strategies such as rainwater harvesting, wastewater recycling, and stormwater management to enhance water sustainability. Rainwater harvesting systems collect and store rainwater from rooftops, streets, and other surfaces, which can then be used for non-potable purposes such as irrigation, toilet flushing, and cooling systems (Lúcio et al. 2020). This reduces reliance on municipal water supplies, especially during periods of drought. Additionally, rainwater harvesting helps to reduce the burden on stormwater infrastructure by capturing runoff that would otherwise flow into sewers, contributing to urban flooding. Wastewater recycling and reuse are essential components of sustainable water management. Treated wastewater can be used for various non-potable applications, such as irrigation, industrial processes, or landscape maintenance, reducing the need for fresh water. In some cases, cities are adopting more advanced approaches, such as the use of treated wastewater also known as "greywater" for potable purposes, a practice known as direct potable reuse (DPR). DPR requires high levels of filtration and purification to ensure the water meets drinking standards, but it offers a sustainable way to address water scarcity in regions where freshwater resources are severely limited.

Green infrastructure solutions such as permeable pavements, rain gardens, and constructed wetlands, can help cities manage stormwater and reducing the risk of flooding while improving water quality. Permeable pavements allow rainwater to infiltrate the ground, reducing surface runoff and recharging groundwater supplies. This reduces the strain on urban drainage systems and prevents overflows during heavy rainfall events, which are becoming more frequent due to climate change. Constructed wetlands are engineered ecosystems designed to mimic the natural filtration processes of wetlands. These systems use plants, soil, and microorganisms to remove contaminants from stormwater or wastewater, improving water quality before it is released back into rivers and lakes. Constructed wetlands also provide additional benefits such as carbon sequestration, biodiversity conservation, and aesthetic and recreational opportunities for urban residents. By incorporating green infrastructure into urban water management, cities can not only improve water quality but also enhance their resilience to climate change impacts, such as flooding and drought.

Stormwater management is another critical component of urban water sustainability. As cities expand and impervious surfaces like roads and buildings replace natural landscapes, rainwater has fewer opportunities to be absorbed into the ground. This increases surface runoff, which can overwhelm drainage systems and lead to flooding, particularly in low-lying areas. To mitigate this, cities can implement sustainable urban drainage systems (SUDS) that slow down the flow of water, promote infiltration, and reduce flood risks. SUDS include features such as bioswales, green roofs, detention basins, and retention ponds that capture and store stormwater during heavy rain events. Green roofs are particularly effective in managing stormwater in dense urban environments. By covering rooftops with vegetation, green roofs absorb rainwater, reduce runoff, and lower urban temperatures by mitigating the heat island effect. In addition, they provide insulation for buildings, reducing energy demand for heating and cooling, and contribute to biodiversity by creating habitats for pollinators and birds.

These measures not only reduce the strain on freshwater resources but also contribute to the resilience of urban environments against water scarcity and climate change impacts. As climate change leads to more frequent and severe droughts in some regions, while causing more intense storms and flooding in others, cities must adopt water management strategies that are both adaptive and flexible. This requires an integrated approach that considers the entire water cycle from capturing and storing rainwater to treating and reusing wastewater and addresses both supply-side and demand-side issues. Another important aspect of sustainable water management is the protection and restoration of natural water bodies and aquatic ecosystems. Rivers, lakes, wetlands, and aquifers provide critical ecosystem services including water purification, flood control, and habitat for wildlife. However, these ecosystems are often degraded by urban development, pollution, and over-extraction of water. Cities must prioritize the restoration and protection of natural waterways by reducing pollution, preventing the discharge of untreated sewage, and ensuring that urban development does not encroach on these vital ecosystems. Riparian buffer zones which are vegetated areas along the banks of rivers and streams can help filter pollutants, reduce erosion, and provide wildlife corridors.

Water conservation programs are essential for reducing per capita water consumption in cities. Public awareness campaigns can encourage residents and businesses to adopt water-saving practices such as installing low-flow fixtures, fixing leaks, and reducing outdoor water use. Some cities have implemented water pricing schemes that charge higher rates for excessive water use, incentivizing conservation while ensuring that basic water needs remain affordable for all citizens. Cities can also incentivize water-efficient technologies, such as smart irrigation systems and water-efficient appliances, to further reduce water demand. The sustainable water management requires a holistic, integrated approach that addresses water scarcity, quality, and resilience. By implementing innovative technologies, green infrastructure, and water reuse strategies, cities can enhance their water security, reduce their environmental impact, and improve the quality of life for their residents. As water becomes an increasingly scarce and valuable resource, cities that invest in sustainable water management practices will be better equipped to meet the challenges of the future.

(d) **Landscape, Gardens, and Urban Biodiversity**

Sustainable cities emphasize the importance of well-designed parks and public gardens, which serve as vital spaces for leisure and recreation. These green spaces provide urban residents with opportunities for physical activity, social interaction, and relaxation, improving both mental and physical well-being (Negret et al. 2022). Access to these areas is essential for enhancing quality of life, particularly in dense urban environments where natural spaces can be limited. Studies have shown that exposure to green spaces reduces stress, enhances cognitive function, and can even improve mental health (Zhang et al. 2020). Well-designed parks and gardens also promote social cohesion by providing communal areas where residents can gather and engage in activities such as sports, picnics, or cultural events. Beyond their social and recreational benefits, urban green spaces play a critical role in mitigating climate change by acting as carbon sinks and improving air quality. Trees, shrubs, and other vegetation absorb carbon dioxide from the atmosphere, sequestering carbon and helping to offset the greenhouse gas emissions produced by cities. In addition to sequestering carbon, trees and plants filter pollutants from the air, including particulate matter and nitrogen dioxide. This improves urban air quality and helps reduce the incidence of respiratory illnesses. Green spaces also contribute to stormwater management, as vegetation and permeable surfaces absorb rainwater, reducing runoff and preventing flooding in urban areas.

Urban biodiversity is a critical aspect of sustainability, as it ensures the health and resilience of ecosystems within cities. Trees, urban forests, and green roofs help reduce the urban heat island effect, which can make cities significantly hotter than surrounding rural areas. The urban heat island effect occurs when concrete, asphalt, and other impervious surfaces absorb and retain heat, raising temperatures in cities by several degrees compared to rural areas (Karimi et al. 2023). By increasing vegetation cover, cities can mitigate this effect, cooling the air and reducing the energy demand for air conditioning in buildings. Urban forests, which consist of trees and wooded areas within cities, are vital for both human health and environmental sustainability. These forests provide shade, reduce the need for cooling in the summer, and support biodiversity by offering habitat for a wide range of species. Planting diverse species of trees in urban forests enhances their resilience to pests, diseases, and climate change, ensuring that these green areas continue to provide ecosystem services in the future. Cities like New York, Singapore, and Tokyo have made significant investments in expanding their urban forests, recognizing the multiple benefits they provide in terms of air quality, climate resilience, and aesthetic appeal (Agnihotri 2022).

Prioritizing the enhancement of natural environments, planning based on ecological principles, and increasing urban vegetation are critical strategies for preserving biodiversity. Ecological urban planning involves designing cities to coexist with and support natural ecosystems, rather than displacing them. This approach recognizes the interconnectedness of human and natural systems, ensuring that urban development does not come at the expense of local flora and fauna. Cities can adopt planning strategies that preserve natural habitats, such as wetlands, rivers, and woodlands, which act as refuges for wildlife and contribute to the overall health of the urban ecosystem. Incorporating green corridors continuous stretches of vegetation that connect parks, gardens, and natural areas can facilitate the movement of species, preventing habitat fragmentation and supporting biodiversity in urban environments.

In cities that are vulnerable to the impacts of climate change, enhancing urban biodiversity can also serve as a strategy for building resilience. Green infrastructure, such as wetlands and mangroves in coastal cities, not only provides habitats for wildlife but also acts as a buffer against storm surges, flooding, and sea-level rise. For example, cities in tropical regions have successfully implemented mangrove restoration projects to protect coastal areas from erosion while supporting biodiversity and carbon sequestration (Ellison et al. 2020). Similarly, restoring riparian habitats along rivers and streams in urban areas can prevent flooding and improve water quality, benefiting both wildlife and city residents. These efforts help counteract the loss of species and ecosystems, contributing to the overall ecological health and resilience of urban areas. Urban expansion often results in habitat loss, fragmentation, and pollution, all of which threaten the survival of native species. However, by intentionally creating and maintaining biodiverse spaces within cities, urban areas can become part of the solution to global biodiversity loss rather than contributing to it.

Cities can also engage their residents in the protection and enhancement of urban biodiversity through initiatives such as tree-planting campaigns, citizen science projects, and educational programs. Encouraging community participation in urban greening efforts fosters a sense of environmental stewardship and builds connections between people and nature. In many cities, volunteers participate in monitoring local wildlife, planting trees, or maintaining community gardens, helping to create a culture of sustainability and environmental awareness. In conclusion, integrating landscape, gardens, and urban biodiversity into the design and planning of sustainable cities not only improves quality of life for residents but also enhances the ecological and climate resilience of urban areas. By prioritizing green spaces, protecting ecosystems, and promoting biodiversity, cities can create vibrant, healthy, and sustainable environments that benefit both people and nature. These green spaces act as lungs for the city, providing clean air, sequestering carbon, and supporting a diverse range of species, while also offering residents a connection to the natural world.

(e) **Sustainable Transport and Public Spaces**

Effective access to fundamental transport services is crucial for reducing dependence on automobiles and minimizing unnecessary travel. In many cities, private car use remains a dominant form of transportation, contributing to traffic congestion, air pollution, and high carbon emissions. By prioritizing public transportation and non-motorized options, cities can significantly reduce their environmental footprint and improve the overall quality of life for their residents (Mansoor et al. 2022). Sustainable cities prioritize public transportation systems, such as buses, trams, and subways, alongside non-motorized transport options like cycling and walking. Public transportation systems are not only more energy-efficient than private vehicles but also help to reduce road congestion, lower pollution levels, and provide equitable access to mobility for all citizens. One of the most effective strategies for creating sustainable urban transport is the development of integrated, multi-modal public transportation networks. These networks combine various forms of transport into a cohesive system that enables seamless transfers between modes. For instance, many cities have implemented integrated ticketing systems that allow passengers to switch between buses, trains, and trams using a single fare card, making public transport more convenient and accessible (Zhang et al. 2021). By designing transportation systems that are easy to use, affordable, and widely available, cities can encourage residents to choose public transport over private cars, significantly reducing carbon emissions and road congestion.

Investing in high-quality, reliable public transportation is also key to promoting inclusivity and reducing social inequities in cities. Public transit systems connect residents to employment opportunities, education, healthcare, and social services, making it easier for people from all economic backgrounds to access essential services. For lower-income communities, affordable and efficient public transport can be a lifeline, reducing the need for car ownership and the associated costs. Additionally, cities can prioritize the electrification of public transit fleets, transitioning from diesel-powered buses to electric or hybrid vehicles, which produce fewer emissions and reduce noise pollution. The integration of non-motorized transport options such as the development of bicycle-friendly infrastructure, free bike rental programs, and well-designed public spaces is essential. Creating infrastructure that supports active transportation, such as protected bike lanes, pedestrian-friendly streets, and ample sidewalks, encourages residents to walk or cycle for short trips. This reduces the reliance on cars for daily commutes or errands. In cities like Copenhagen, Amsterdam, and Portland, cycling has become a major mode of transportation due to extensive bike lane networks and dedicated bicycle traffic signals, as shown in Fig. 6.3. These cities also provide ample parking facilities, making cycling both safe and convenient.

For pedestrians, well-designed public spaces are crucial to creating vibrant and liveable cities. Cities should invest in pedestrian-friendly streetscapes, with wide sidewalks, safe crosswalks, and traffic-calming measures such as speed bumps and raised intersections (Filion et al. 2023). Urban planners can also prioritize the creation of car-free zones or pedestrianized streets, where motorized vehicles are restricted, allowing people to walk freely and safely. These pedestrian zones often found in city centres, not only reduce traffic-related pollution but also enhance the aesthetic appeal of the area, boosting local economies by attracting shoppers, diners, and tourists. Investing in electric vehicle (EV) infrastructure and promoting ride-sharing platforms

Fig. 6.3 Bicycle parking facility in Amsterdam

can also contribute to reducing the environmental impact of urban transportation (Luo and Qiu 2020). The electrification of the transportation sector is essential for reducing the greenhouse gas emissions produced by internal combustion engine vehicles. Cities can support this transition by installing electric vehicle charging stations in public areas, such as parking lots, shopping centres, and office buildings, making it more convenient for EV owners to charge their cars. Additionally, policies that incentivize the purchase of electric vehicles such as tax credits, rebates, and reduced tolls, can accelerate the adoption of EV, further reducing the carbon footprint of urban transportation.

Beyond private electric vehicles, cities can electrify their public transportation fleets. Electric buses, trams, and light rail systems can provide efficient, low-emission public transit options while reducing the noise and air pollution typically associated with diesel or gasoline-powered vehicles. Some cities have also invested in electric car-sharing services, where residents can rent EV for short trips without the need for car ownership. This approach not only reduces the total number of vehicles on the road but also ensures that those vehicles have a smaller environmental impact. Ride-sharing platforms and mobility-as-a-service (MaaS) technologies further contribute to sustainable urban mobility by optimizing vehicle use and reducing the total number of cars on the road. Platforms like Uber, Lyft, and local ride-sharing apps have introduced carpooling services, where passengers traveling in the same direction can share a ride, reducing individual vehicle trips and lowering emissions. MaaS platforms allow users to plan, book, and pay for multiple types of transportation through a single app, streamlining mobility options and promoting the use of shared, sustainable modes of transport. These strategies promote a more sustainable urban environment by encouraging active transportation, reducing carbon emissions, and enhancing the quality of public spaces for community interaction and well-being. Public spaces, such as plazas, parks, and squares, are essential for creating social hubs where residents can gather, interact, and engage in community activities. When well-designed, these spaces encourage foot traffic, reduce car dependency, and create a sense of place and belonging within a neighbourhood. Cities that prioritize pedestrian plazas, green spaces, and outdoor seating areas not only improve their liveability but also foster social cohesion by providing spaces for social interaction, cultural events, and recreational activities.

In addition to environmental benefits, sustainable transport systems contribute to the overall health and well-being of urban residents. Active transportation options, such as cycling and walking, offer physical health benefits by promoting exercise, while reduced air pollution from lower car use improves respiratory health. Public transportation systems reduce the stress associated with traffic congestion and long commutes, providing more reliable and efficient mobility for all residents. Therefore, creating a sustainable transport system and developing public spaces that prioritize active transportation and social interaction is key to fostering healthier, more resilient urban environments. By reducing dependence on private vehicles, investing in public transportation, and encouraging non-motorized transport options, cities can significantly reduce their carbon emissions while improving the quality of life for their residents. Well-designed public spaces that promote walking, cycling, and social engagement are critical components of a sustainable urban future.

(f) Local and Sustainable Materials with Less Embodied Energy

This principle emphasizes the importance of utilizing advanced material technologies and optimizing supply chains to reduce the embodied energy in construction. Embodied energy refers to the total energy required to extract, manufacture, transport, and install building materials. In the context of sustainable urban development, minimizing embodied energy is crucial because construction processes, from material extraction to end-of-life disposal, contribute significantly to carbon emissions. By prioritizing materials with lower embodied energy and leveraging advanced building technologies, cities can dramatically reduce their overall environmental footprint. By prioritizing locally sourced materials and leveraging regional expertise, such as the use of native timber, cities can minimize environmental impact and support local economies. Locally sourced materials not only reduce transportation-related emissions but also often reflect materials that are well-adapted to the local climate and environmental conditions. For example, timber harvested from nearby forests can be a more sustainable and climate-appropriate building material than steel or concrete in certain regions. Moreover, sourcing materials locally fosters regional economic growth, supports local industries, and reduces the need for long-distance transportation, further lowering the environmental impact.

In many regions, traditional building materials have evolved based on local availability and environmental considerations. For instance, in desert regions, adobe or rammed earth construction has been used for centuries due to its natural insulating properties and suitability to the climate. Similarly, cities in temperate or forested regions may use timber extensively, as it is abundant, renewable, and has a lower embodied energy than materials like steel or concrete (Hamadyk et al. 2020). Leveraging these regional materials and construction techniques can help cities build sustainably while maintaining cultural and historical connections to their natural environments. Using materials with lower embodied energy, such as recycled steel or low-carbon concrete can significantly reduce the overall carbon footprint of buildings and infrastructure. Concrete is one of the most widely used construction materials but has a high embodied energy due to the production of cement, which accounts for a significant portion of global carbon dioxide emissions. To mitigate this, innovations in low-carbon concrete such as using industrial by-products like fly ash or slag as partial cement replacements can help reduce the carbon intensity of concrete without compromising its structural integrity (Subramaniam and Sathiparan 2023). Recycled steel, which requires far less energy to produce than virgin steel is another example of how using alternative materials can lower the carbon impact of construction projects.

The construction industry is increasingly turning to sustainable building materials that not only reduce embodied energy but also enhance the thermal efficiency and lifespan of buildings. Examples of these materials include cross-laminated timber (CLT), which has gained popularity for its strength and renewable properties, and bio-composite materials which are derived from natural fibres and biodegradable polymers. CLT is making headway in sustainable architecture, offering a renewable alternative to concrete and steel while also sequestering carbon, as trees absorb CO_2 during their growth. Bio-composites made from agricultural by-products such as straw, hemp, or flax, can be used for insulation, flooring, or interior finishes, reducing reliance on synthetic, petroleum-based materials. Incorporating recycled materials into the construction process is another strategy for reducing embodied energy. Recycled aggregates from construction and demolition waste can be used in road building, concrete production, or landscaping, diverting materials from landfills and lowering the demand for new raw materials. Glass, plastic, and rubber are also being repurposed into construction materials such as insulation or paving, contributing to a more circular economy where materials are continuously reused and repurposed rather than discarded.

The focus on sustainable materials not only reduces carbon footprints but also enhances the resilience and sustainability of urban development. Buildings constructed with sustainable, low-embodied-energy materials tend to be more durable, energy-efficient, and adaptable to changing climate conditions. For instance, buildings made from natural materials like wood, straw, or clay tend to perform better in terms of thermal insulation, reducing the need for heating or cooling systems, which in turn lowers operational energy demands over the lifetime of the building. Additionally, materials that are locally sourced or produced can be more easily replaced or repaired, reducing the need for long supply chains or high-embodied-energy replacements. The integration of advanced building technologies such as 3D printing with eco-friendly materials, further supports the transition to greener construction practices. 3D printing technologies are revolutionizing the construction industry by enabling the precise use of materials, reducing waste, and allowing for greater flexibility in design. This method allows buildings to be constructed more efficiently using exactly the amount of material needed for a specific design, thereby minimizing material waste. Moreover, 3D printing with sustainable materials such as recycled plastics or bio-based composites, provides an innovative way to further lower the embodied energy of construction projects.

In addition, modular construction where prefabricated building components are manufactured off-site and then assembled on location, can significantly reduce both material waste and construction time. Modular construction techniques can also leverage low-embodied-energy materials while allowing for greater energy efficiency through precise design and manufacturing processes. The use of prefabrication also reduces the need for on-site construction activities, which lowers the environmental impact associated with heavy machinery, labour, and waste production. Moreover, designing buildings with deconstruction and reuse in mind known as design for disassembly (DfD), helps extend the life cycle of building materials (Ostapska et al. 2024). Buildings constructed with DfD principles allow for components to be easily taken apart and reused or recycled at the end of their life, significantly reducing the embodied energy of future construction projects. This approach contributes to a circular economy, where building materials are not treated as waste but as valuable resources that can be reused in new construction.

Sustainable urban development also involves assessing the life cycle of materials from extraction to disposal, to ensure that each phase of material use is optimized for environmental performance. Life Cycle Assessment (LCA) tools can help cities evaluate the embodied energy of materials and make informed decisions about which materials to prioritize in construction projects. LCA analysis takes into account the full environmental impact of materials, including extraction, transportation, manufacturing, and eventual disposal or recycling. By choosing materials with lower life-cycle impacts, cities can reduce the total carbon emissions associated with their infrastructure and buildings. The use of local and sustainable materials with less embodied energy is a critical strategy for reducing the environmental impact of urban development. By sourcing materials locally, incorporating recycled and low-embodied-energy alternatives, and utilizing advanced construction technologies, cities can build resilient, energy-efficient structures that support both environmental sustainability and local economies. As cities continue to grow and evolve, adopting sustainable materials and construction practices will be essential for creating urban environments that are not only functional and efficient but also environmentally responsible.

(g) **Density and Retrofitting of Existing Districts**

Promoting densification in city centres through mixed-use urban infill, regeneration of central areas, and the development of green transit-oriented projects is essential. Densification refers to the strategic increase in the number of people, businesses, and services in a given urban area, particularly in central or established districts, without expanding the geographic footprint of the city. By promoting compact urban development, cities can reduce urban sprawl, minimize the conversion of natural landscapes into built environments, and optimize the use of existing infrastructure (Abdrabo et al. 2021). High-density cities are inherently more resource-efficient, as they reduce the need for extensive transportation networks and allow for the efficient delivery of services such as water, electricity, and waste management. Densification allows cities to accommodate growing populations while minimizing the environmental impact. In many rapidly urbanizing regions, uncontrolled urban sprawl has led to increased car dependency, higher energy consumption, and loss of valuable agricultural or green spaces. By concentrating development in central areas, cities can provide more efficient public transportation options, reduce commute times, and lower overall energy use. Mixed-use development where residential, commercial, and recreational spaces are integrated within the same area ensures that people can live closer to work, schools, shops, and services, reducing the need for long-distance travel and promoting walkability.

Transit-oriented development (TOD) is an important strategy for densification, involving the creation of high-density, mixed-use communities around major public transit hubs. TOD projects encourage the use of public transportation by making it more accessible and convenient, while also reducing the reliance on private vehicles. In cities like Tokyo, New York, and Singapore, TOD has been instrumental in creating walkable, transit-rich neighbourhoods that reduce traffic congestion, air pollution, and greenhouse gas emissions. These areas are typically characterized by pedestrian-friendly streetscapes, easy access to buses, subways, or trains, and a range of housing options catering to different income levels. Integrating green spaces, parks, and recreational facilities into TOD projects further enhances the liveability and sustainability of these high-density areas. This approach includes retrofitting inefficient buildings to modern standards and systematically reducing the overall carbon footprint of the urban environment. Retrofitting is a key component of sustainable urban development, as it focuses on upgrading existing buildings, many of which are often energy-inefficient or rather than demolishing them and constructing new ones. The construction and demolition of buildings generate significant amounts of waste and emissions, while retrofitting allows cities to preserve the embodied energy in existing structures. This approach making the process more sustainable. Retrofitting is also more cost-effective in the long run, as it reduces energy costs for building owners and tenants, while improving indoor comfort and reducing carbon emissions.

One of the most effective retrofit strategies is improving the insulation of the building. Many older buildings lose a significant amount of energy through poorly insulated walls, roofs, and windows, resulting in higher heating and cooling demands. By upgrading insulation, cities can drastically reduce the energy required to maintain comfortable indoor temperatures. This can include installing high-performance insulation materials in walls, roofs, and floors, as well as sealing gaps to prevent drafts and heat loss. Similarly, replacing old windows with energy-efficient double or triple-glazed units can improve thermal performance, reduce energy consumption, and improve indoor comfort by minimizing heat transfer. Another critical retrofit measure is the integration of renewable energy systems into existing buildings. Solar panels can be installed on rooftops to generate clean electricity on-site to reduce reliance on fossil fuels and lowering energy costs for building occupants. Other renewable energy technologies, such as geothermal heat pumps or solar thermal systems for hot water can also be retrofitted into buildings to further reduce their carbon footprint. Incorporating smart energy management systems that monitor and optimize energy use can help ensure that retrofitted buildings operate at peak efficiency, reducing energy waste and maximizing the benefits of renewable energy systems.

Retrofitting also provides an opportunity to upgrade buildings to meet modern sustainability standards, including improving water efficiency, indoor air quality, and lighting systems. Installing water-efficient fixtures, such as low-flow faucets and toilets, can reduce water consumption, while rainwater harvesting systems can be added to collect and reuse water for irrigation or other non-potable uses. Upgrading lighting systems to LED technology not only reduces energy consumption but also improves lighting quality and longevity, contributing to overall building performance. In addition to individual building retrofits, entire districts can be regenerated through comprehensive retrofitting strategies. Urban regeneration projects involve revitalizing underutilized or degraded areas, transforming them into vibrant, mixed-use neighbourhoods that support social and economic development while reducing environmental impacts. These projects often prioritize affordable housing, public spaces, and sustainable infrastructure to create inclusive, liveable communities that can adapt to future challenges, such as climate change or population growth.

Density and retrofitting are not only about environmental benefits but also about creating more equitable and inclusive cities. High-density development can help address the housing shortages that many cities face by increasing the supply of affordable housing in central areas, close to employment and services. Mixed-use developments provide opportunities for small businesses to thrive in vibrant urban areas, while retrofitting older buildings can help preserve historical and cultural assets, maintaining the unique character of neighbourhoods while modernizing them for future needs. Retrofitting existing buildings with energy-efficient technologies, such as improved insulation, energy-efficient windows, and renewable energy systems, can significantly reduce their energy consumption and emissions. Older buildings, particularly those built before modern energy codes were established, tend to be highly inefficient in terms of energy use. Retrofitting these buildings allows cities to bring them up to current standards, making them more energy-efficient and comfortable while reducing their environmental impact.

By focusing on these strategies, cities can enhance their sustainability to improve resource efficiency, and create vibrant, liveable communities that meet the demands of growing urban populations. Densification and retrofitting are crucial for making cities more compact, efficient, and resilient to future challenges, including climate change, population growth, and resource constraints. High-density, mixed-use neighbourhoods reduce the environmental impact of cities while enhancing the quality of life for residents by providing better access to transportation, services, and amenities. Additionally, retrofitting existing buildings allows cities to transition to a more sustainable built environment without the need for extensive new construction, which can be both costly and resource-intensive.

(h) **Green Buildings and Districts Using Passive Design Principles**

Implementing low-energy building designs that adhere to best practices in passive design principles is crucial for reducing overall energy consumption. Passive design involves designing buildings in such a way that they naturally regulate indoor temperatures and lighting, minimizing the need for mechanical heating, cooling, and artificial lighting (Li et al. 2021). This approach is particularly important in the context of sustainable urban development, as buildings are responsible for a significant portion of energy use in cities. By reducing energy demand of the building at the design stage, passive strategies contribute to lower operational energy costs, reduced greenhouse gas emissions, and improved indoor comfort for occupants. Passive design strategies focus on harnessing the natural elements such as sunlight, wind, and thermal mass, to create comfortable indoor environments with minimal reliance on energy-intensive systems. Optimizing building orientation is one of the foundational principles of passive design. In certain countries, buildings are typically oriented with large windows facing south to maximize exposure to sunlight during the winter months, reducing the need for artificial heating. In contrast, shading devices such as overhangs, louvers, or vegetation can be used to block direct sunlight in the summer, preventing overheating and reducing the demand for air conditioning. By carefully positioning buildings based on local climate conditions, cities can enhance energy efficiency and occupant comfort.

Using natural ventilation is another key passive design strategy. Buildings designed to capture prevailing breezes and allow for cross-ventilation can remain cool in hot weather without the need for air conditioning. This is particularly effective in tropical and temperate climates, where mechanical cooling can account for a large portion of energy consumption (Elshafei et al. 2021). Features like operable windows, wind towers, and courtyards can be incorporated into building designs to facilitate the natural flow of air, thereby reducing indoor temperatures and enhancing air quality. For buildings in colder climates, natural ventilation can also help reduce indoor humidity levels and improve indoor air quality, creating healthier living environments (Shakir et al. 2024b). Incorporating thermal mass into building designs helps regulate indoor temperatures by absorbing, storing, and gradually releasing heat (Wonorahardjo et al. 2020). Materials like concrete, brick, and stone have high thermal mass and can store heat during the day, releasing it at night when temperatures drop. This passive strategy is particularly useful in climates with significant temperature variations between day and night, as it reduces the need for mechanical heating and cooling. Thermal mass can be strategically placed in floors or walls, allowing buildings to maintain more consistent temperatures and reducing the energy demand for climate control.

This includes the adoption of compact solar architecture and the retrofitting of existing structures to enhance their energy efficiency. Compact solar architecture refers to the design of buildings that efficiently capture and use solar energy for heating, cooling, and lighting. Solar architecture also involves designing compact, dense buildings that maximize solar gain without requiring excessive energy for heating or cooling (Wang et al. 2021). By incorporating photovoltaic panels and solar water heaters, buildings can generate a portion of their energy needs on-site, further reducing their reliance on external energy sources. Compact building designs also reduce the amount of exterior surface area exposed to the elements, minimizing heat loss in colder climates and heat gain in warmer climates. Retrofitting existing structures with passive design principles is a highly effective way to improve the energy efficiency of older buildings. Many older buildings were constructed before modern energy codes were established and may have poor insulation, inefficient windows, and energy-intensive HVAC systems. Retrofitting these buildings with passive design strategies such as installing energy-efficient windows and optimizing natural ventilation, can significantly reduce energy consumption and improve occupant comfort without the need for major structural changes. In addition, retrofitting existing buildings avoids the environmental impacts associated with demolition and new construction, making it a more sustainable option.

High-performance glazing refers to advanced window technologies that improve thermal insulation, reduce heat gain, and minimize energy loss. Double or triple-glazed windows with low-emissivity coatings are designed to reflect heat back into the building during the winter while blocking heat from entering during the summer. These windows also reduce drafts and improve the overall energy efficiency of buildings by minimizing heat transfer through the glass. By incorporating high-performance glazing, buildings can maximize the benefits of natural light without compromising energy efficiency. By focusing on passive design, cities can create buildings and districts that minimize reliance on artificial heating and cooling, thereby lowering carbon emissions and promoting long-term sustainability. Passive design strategies not only reduce the operational energy use of individual buildings but also contribute to the overall sustainability of entire districts. When implemented at the neighbourhood or district scale, passive design principles can enhance urban microclimates, create comfortable public spaces, and reduce the energy demands of entire communities. District energy systems such as district heating and cooling networks, can further support passive design by distributing renewable energy across multiple buildings, increasing overall efficiency.

In addition to reducing carbon emissions, passive design promotes resilience by creating buildings that can maintain comfortable indoor conditions during power outages or extreme weather events. For example, a building designed with sufficient insulation, shading, and natural ventilation will remain habitable even if the power grid fails, providing a level of energy independence and resilience that is crucial in the face of climate change. Implementing passive design principles in green buildings and districts is essential for reducing energy consumption, lowering carbon emissions, and enhancing the overall sustainability and resilience of urban environments. By focusing on natural strategies for heating, cooling, and ventilation, cities can create energy-efficient, comfortable buildings that contribute to long-term environmental and economic sustainability. As urban areas continue to grow, adopting passive design strategies will be critical for meeting the demands of urban populations while reducing the ecological footprint of the built environment.

(i) **Liveability, Healthy Communities, and Mixed-Use Programs**

Developing cities with mixed-use and mixed-income areas fosters social inclusion and revitalizes city centres. Mixed-use developments are designed to combine different functions such as residential, commercial, recreational, and institutional spaces within close proximity. This approach encourages a more integrated urban experience, where people can live, work, shop, and engage in leisure activities without needing to commute long distances. By facilitating these interactions within a neighbourhood, cities create more vibrant, walkable, and efficient communities, which help reduce traffic congestion, pollution, and energy use associated with transportation. Meanwhile, mixed-income development promotes economic diversity and inclusion by ensuring that housing is accessible to people of various income levels, reducing the socioeconomic divide within cities (Sanga 2024). The integration of residential and commercial spaces also enhances the local economy, as businesses benefit from a built-in customer base, while residents have easy access to goods, services, and employment opportunities. Mixed-use developments promote local businesses by creating pedestrian-friendly environments where small shops, restaurants, and cafes can thrive. This dynamic foster community engagement and encourages local spending, which strengthens neighbourhood economies and supports small enterprises. In turn, local economies become more resilient as they are less reliant on distant commercial centres.

Mixed-use developments not only reduce the need for long commutes but also contribute to more sustainable modes of transport by encouraging walking, cycling, and public transit use. When homes, offices, and amenities are clustered together, it becomes easier for residents to engage in daily activities without relying on cars. This compact urban form significantly lowers energy consumption and carbon emissions by decreasing the reliance on private vehicles, making cities more sustainable overall. In addition, mixed-use developments can be complemented by transit-oriented development (TOD) strategies, where high-density neighbourhoods are designed around efficient public transportation networks, further reducing the environmental impact of urban sprawl. Creating connected, compact communities that implement mixed-use concepts and prioritize housing affordability is essential for promoting liveability and well-being. By offering affordable housing options within these developments, cities ensure that diverse populations including low- and middle-income residents can benefit from living in vibrant, well-served urban areas. Affordable housing can be integrated with market-rate housing, fostering social inclusion and reducing the risk of segregating communities based on income levels. Affordable housing initiatives within mixed-use developments help combat the issue of housing shortages, which is particularly acute in many urban areas, and ensure that city centres remain accessible to all citizens, not just the wealthy.

Housing affordability and accessibility are critical components of liveability, as they provide residents with stable, secure, and healthy living conditions. When people live in safe, affordable housing in proximity to their workplaces, schools, and essential services makes their overall quality of life improves. Reducing the financial stress associated with high housing costs allows individuals and families to allocate resources to health, education, and other personal development areas. Additionally, mixed-use developments that prioritize affordable housing promote more diverse, inclusive communities, where people from different socioeconomic backgrounds can live, work, and interact. Incorporating green spaces, pedestrian-friendly streets, and community centres into these developments enhances the quality of life for residents and fosters social interaction. Green spaces, such as parks, gardens, and playgrounds, are integral to creating healthy, liveable neighbourhoods. They provide residents with areas for recreation, exercise, and relaxation, contributing to both physical and mental well-being. Access to nature within urban environments is known to reduce stress, improve mood, and encourage outdoor activities which are essential for maintaining a healthy lifestyle. Urban green spaces also serve as gathering places where neighbours can interact, promoting social cohesion and fostering a sense of community.

Pedestrian-friendly streets are a hallmark of liveable cities, as they prioritize walking and cycling over car use, creating safer, more enjoyable public spaces. Wide sidewalks, safe crosswalks, street trees, benches, and bicycle lanes all contribute to making streets accessible and welcoming to people of all ages and abilities. By reducing car traffic and increasing the walkability of neighbourhoods, pedestrian-friendly streets enhance air quality, lower noise pollution, and make cities more attractive and healthier places to live. These streets also support local businesses by encouraging foot traffic, which helps small shops and restaurants thrive. Community centres, libraries, and recreational facilities within mixed-use developments provide essential services and opportunities for social engagement, education, and recreation.

These amenities serve as focal points for community life, offering residents spaces to gather, learn, and participate in various activities. By fostering social interaction, community centres strengthen the social fabric of neighbourhoods, helping to reduce isolation and build strong support networks among residents. The inclusion of these shared spaces is essential for creating a sense of belonging and fostering a collective identity within urban neighbourhoods.

These strategies help build vibrant neighbourhoods that support diverse populations, encourage social interaction, and ensure access to essential services, contributing to the overall health and sustainability of the urban environment. When residents live in close proximity to schools, healthcare facilities, shops, offices, parks, and public transit, their quality of life improves. They can spend less time commuting and more time engaging in community activities, pursuing personal interests, and participating in civic life. These connected, compact communities provide the foundation for healthier, more inclusive, and more sustainable cities, where social, economic, and environmental needs are addressed in an integrated manner. In summary, mixed-use and mixed-income developments are key to creating liveable, healthy, and sustainable urban environments. By integrating residential, commercial, and recreational spaces, these neighbourhoods reduce travel distances, promote active lifestyles, and foster social inclusion. Incorporating green spaces, pedestrian-friendly infrastructure, and community facilities into these developments ensures that residents can enjoy a high quality of life, with access to essential services and opportunities for social interaction. These strategies not only make cities more vibrant and resilient but also create diverse, inclusive, and equitable communities that support long-term urban sustainability.

(j) **Local Food and Short Supply Chains:**

Prioritizing local food production, regional supply chains, and urban agriculture, including initiatives like "eat local" and "slow food," is vital for enhancing sustainability. Food production and consumption are integral to urban sustainability. Cities that prioritize local food systems can reduce their environmental footprint, enhance food security, and strengthen community resilience (Langemeyer et al. 2021). Local food production reduces the need for long-distance transportation, lowering the carbon emissions associated with importing food from distant regions or countries. By shortening the supply chain, cities can also reduce food waste, as fresher products with shorter transportation times are less likely to spoil before reaching consumers. Urban agriculture is a key component of this approach, offering cities the opportunity to produce food within or near urban centres, reducing reliance on imported goods (Orsini et al. 2020). Urban agriculture such as rooftop gardens, vertical farms, and community gardens, can help cities reduce their reliance on imported food, lower transportation emissions, and provide fresh, healthy produce to local communities. Rooftop gardens utilize underutilized spaces on top of buildings to grow fruits, vegetables, and herbs, often using lightweight soil or hydroponic systems. These gardens not only provide fresh produce but also help insulate buildings, reduce urban heat island effects, and manage stormwater runoff. Vertical farming, which involves growing crops in stacked layers within controlled environments, is another innovative method for urban food production, allowing cities to grow large amounts of produce in compact spaces year-round, regardless of weather conditions.

Community gardens, where residents collectively grow food in shared plots of land are also gaining popularity in cities. These gardens foster a sense of community, provide educational opportunities about sustainable agriculture, and give urban residents access to fresh, locally grown food. Community gardens also encourage healthy eating habits, as participants can directly engage in growing their own fruits and vegetables, promoting better nutrition and reducing dependence on processed foods. Local food production reduces the environmental impact of food transportation, a major contributor to greenhouse gas emissions. Transporting food over long distances requires significant amounts of energy for refrigeration, packaging, and fuel, all of which contribute to climate change. By growing food closer to where it is consumed, cities can cut down on these emissions and reduce the use of plastic packaging that is often necessary for long-haul transport. Additionally, local food systems are less vulnerable to disruptions in global supply chains, making cities more resilient to shocks like pandemics, natural disasters, or geopolitical tensions that could impact food imports.

These efforts reduce the environmental impact of food transportation, support local economies, and promote healthier, more resilient communities. Local food systems are not only more sustainable but also provide economic benefits by keeping money within the community. Farmers markets, local food cooperatives, and farm-to-table restaurants create economic opportunities for small-scale farmers and producers, while also offering consumers access to fresher, higher-quality products. By buying locally, consumers support regional farmers and businesses, helping to create jobs and boost local economies. Encouraging the development of local food markets and farm-to-table initiatives also strengthens the connection between urban residents and local farmers. Farmers markets are a direct way for urban consumers to purchase food from nearby farms, fostering relationships between producers and consumers. These markets offer transparency about where food comes from and how it is grown, giving consumers more control over the quality and sustainability of their food choices. Farm-to-table initiatives emphasize sourcing ingredients from local farms and producers for use in restaurants and food service establishments. These initiatives promote the consumption of local, seasonal food and reduce the environmental footprint of the foodservice industry. These initiatives also offer economic benefits to both farmers and restaurants, as locally sourced food tends to be fresher and of higher quality, attracting customers who value sustainable, farm-fresh ingredients.

Local food systems are also integral to fostering food security in urban areas, especially as cities face the growing challenges of climate change and population growth. By investing in urban agriculture and local supply chains, cities can reduce their reliance on imported foods that may be affected by global market fluctuations, climate-related crop failures, or supply chain disruptions. Local food production ensures that fresh, nutritious food remains available to urban residents, particularly in times of crisis. In regions vulnerable to extreme weather events, local farms and urban agriculture initiatives can help stabilize food supplies and ensure that urban populations have access to sufficient, healthy food. Incorporating urban agriculture into city planning is essential for building resilient, self-sufficient food systems. Zoning policies that designate spaces for community gardens, urban farms, and greenhouses can encourage the growth of local food production in cities. Municipal governments can also offer incentives, such as tax breaks or grants, to developers and property owners who incorporate rooftop gardens or vertical farming systems into their buildings. In cities where land is scarce, innovative solutions like vertical farming, hydroponics, and aquaponics can maximize food production in limited space, further enhancing urban food resilience.

The "slow food" movement which encourages consumers to prioritize quality, sustainability, and local sourcing over fast, industrialized food production is another important aspect of creating sustainable food systems. Slow food emphasizes the cultural and environmental value of food, promoting traditional farming practices, biodiversity, and the preservation of local culinary traditions. By encouraging consumers to "eat local" and support slow food initiatives, cities can promote a shift away from industrialized agriculture. This shift addresses the issues of prioritizing quantity over quality, which often leads to environmental degradation and loss of biodiversity. Slow food also encourages people to reconnect with their food, understanding where it comes from, how it is produced, and its impact on the environment. By fostering a strong connection between urban areas and local food sources, cities can ensure food security and contribute to a more sustainable and self-sufficient urban ecosystem. Urban food systems that prioritize local production and short supply chains are less reliant on external factors, making them more resilient to economic or environmental disruptions. These systems also encourage healthier eating habits by providing urban residents with access to fresh, locally grown produce, reducing the prevalence of processed and unhealthy foods in urban diets. By shortening the distance between producers and consumers, cities can create a more transparent, efficient, and equitable food system that benefits both people and the planet.

Furthermore, local food systems can play a significant role in reducing urban food deserts in areas where access to affordable, nutritious food is limited. Urban agriculture initiatives, such as community gardens and farmers' markets, can help bring fresh food to underserved neighbourhoods, improving the health outcomes of residents in those areas. Policies that promote equitable access to local food, such as subsidizing farmers markets or offering incentives for grocery stores to source locally, can help ensure that all urban residents benefit from sustainable, healthy food options. In conclusion, prioritizing local food production and short supply chains is essential for creating sustainable, resilient, and self-sufficient urban ecosystems. By investing in urban agriculture, supporting local farmers, and encouraging the development of farm-to-table and slow food initiatives, cities can reduce their environmental impact, strengthen local economies, and improve the health and well-being of their residents. These efforts not only contribute to more sustainable urban living but also build stronger, more connected communities that are better equipped to meet the challenges of the future.

(k) **Cultural Heritage, Identity, and Sense of Place**

Every city possesses a unique environment shaped by its materials, history, and the desires of its population, which must be carefully considered in urban design. Cultural heritage and identity are at the core character of the city which influencing how residents connect with their surroundings and how the city is perceived both locally and globally. Preserving cultural heritage of the city and architectural character is essential for maintaining a sense of place and community identity. The architectural styles, public spaces, historical landmarks, and cultural traditions found in cities contribute to their unique identity, and safeguarding these elements is crucial for creating cohesive and vibrant urban environments. Preserving built heritage and sustaining a strong cultural identity are essential for maintaining the sense of place. When cities evolve and grow, they often face pressures to modernize, expand, or redevelop, which can threaten historical buildings, traditional neighbourhoods, and cultural sites. However, urban development does not have to come at the expense of cultural heritage. By incorporating historical preservation into planning and development processes, cities can maintain their unique identities while still accommodating growth and change (Bogdan et al. 2022).

Historical buildings, landmarks, and cultural sites serve as physical links to the past of the city, providing continuity for residents and fostering a sense of belonging. These sites act as touchstones for collective memory, reminding people of the evolution and the generations that shaped its history. By preserving these structures, cities can create spaces that promote reflection, education, and pride in local culture, strengthening the social fabric of the community. Whether it is a century-old marketplace, a public square, or an iconic monument, these spaces offer a sense of permanence and contribute to the identity of the city, connecting its past to its future (Coombes and Viles 2021). Integrating historic landmarks and cultural sites into modern urban developments allows cities to balance progress with preservation, fostering community pride and continuity. Rather than viewing historical preservation and urban development as conflicting goals, cities can adopt strategies that blend the old with the new, creating a harmonious urban landscape that honours both heritage and innovation. For example, adaptive reuse projects where old buildings are repurposed for new functions will offer a way to preserve architectural heritage while meeting modern needs. Old factories, warehouses, and civic buildings can be converted into residential spaces, cultural centres, or commercial hubs, breathing new life into historical structures while retaining their original character.

By integrating cultural heritage into new developments, cities create environments that reflect their unique histories and offer a deeper sense of place for residents and visitors alike. Preserving historical streetscapes, public squares, and other heritage sites can enhance the aesthetic and cultural value of a city, making it more attractive for tourism, local businesses, and community activities. In cities such as Rome, Paris, or Kyoto, where historic and modern elements coexist, the preservation of cultural landmarks not only supports tourism but also instils a sense of identity and pride among residents (Bogdan et al. 2022; Giliberto and Appendino 2022). Cultural heritage also plays a crucial role in creating community identity, as it is often tied to local traditions, festivals, and practices that bring people together. The preservation of cultural traditions and community activities, such as local festivals, markets, or religious ceremonies, enhances the cultural life of a city and provides residents with opportunities to celebrate their shared history. These events help reinforce community ties and give residents a sense of continuity, ensuring that cultural traditions are passed down to future generations.

Maintaining a strong cultural identity in cities also requires protecting intangible cultural heritage such as language, music, art, and culinary traditions that reflects the spirit and diversity of urban communities. Cities can promote these cultural expressions by supporting local artists, musicians, and cultural institutions, ensuring that these creative traditions remain integral to the urban experience. Public art, murals, street performances, and cultural festivals all contribute to a vibrant urban environment, creating spaces where residents and visitors can engage with cultural narrative of the city in dynamic and interactive ways. Incorporating cultural heritage into urban planning and design also offers significant social and economic benefits. Historic districts and cultural landmarks attract tourists, provide local employment, and stimulate economic activity by encouraging spending in nearby shops, restaurants, and businesses. Heritage tourism is a growing industry, and cities that preserve and promote their cultural assets can capitalize on this trend, using it as a tool for economic development. In cities like Prague, Florence, and Istanbul, tourism driven by cultural heritage not only boosts the economy but also helps fund the preservation of historical sites, creating a self-sustaining cycle of preservation and economic growth (Kopuz and Bal 2023; Lazzeretti and Oliva 2021).

Beyond economic benefits, preserving cultural heritage fosters a deeper sense of community and social cohesion. When residents feel a connection to their city past and cultural identity, they are more likely to take pride in their community, participate in civic life, and engage in efforts to maintain and improve their surroundings. This sense of belonging is particularly important in an age of rapid urbanization, where cities are often subject to constant change and disruption. By preserving cultural landmarks and traditions, cities can provide stability and continuity, helping residents maintain their connection to the local community. Urban design that incorporates cultural heritage and emphasizes a sense of place can also create more liveable and attractive environments. Public spaces designed with respect for local culture can preserve historic architecture, include public art, or reflect traditional styles through the design of new buildings. These spaces offer residents places to gather, interact, and experience the cultural legacy of their city. These spaces provide more than just aesthetic value, but they serve as social hubs that foster interaction, cultural exchange, and civic engagement.

These efforts not only honour the historical and cultural legacy of the city but also foster community pride and contribute to a cohesive and vibrant urban environment. When residents feel connected to history and culture of their city, they are more likely to take ownership of its future, actively participating in efforts to improve the urban landscape and preserve its unique character. This engagement leads to stronger communities, where people work together to protect and celebrate the distinct qualities that make their city special. In conclusion, preserving a cultural heritage of the city and sense of place is essential for creating a cohesive, vibrant urban environment. By balancing historical preservation with modern development, cities can honour their past while embracing progress, creating spaces that reflect their unique identity and foster community pride. Through the integration of historic landmarks, cultural traditions, and community-driven design, cities can ensure that their heritage remains a living part of the urban experience, enriching the lives of residents and visitors alike.

(l) **Urban Governance, Leadership, and Best Practices**

Transforming cities into sustainable communities requires robust urban governance and visionary leadership. Urban governance refers to the frameworks, policies, and decision-making processes that guide the development and management of cities. Strong governance is essential for ensuring that urban growth follows sustainability principles and that resources are allocated effectively to meet the needs of residents. Visionary leadership is critical for driving change and inspiring innovation in urban sustainability, mobilizing both political will and public support for ambitious initiatives. Leaders who champion sustainability can set the agenda for green development, secure investment in eco-friendly infrastructure, and foster a culture of environmental responsibility within both government and society (Deng et al. 2021). Strong governance structures ensure that urban development follows sustainability principles by establishing clear regulations, performance standards, and accountability measures. These frameworks help cities plan for sustainable growth, manage resources efficiently, and ensure that urban infrastructure and services are designed to minimize environmental impact. Effective governance also involves setting long-term goals for sustainability, such as reducing carbon emissions, increasing energy efficiency, and promoting the use of renewable energy sources. By adopting policies that encourage sustainable construction, transportation, and energy use, urban governments can create a regulatory environment that supports green development.

Urban governance also plays a key role in coordinating the various actors involved in city management, including government agencies, private sector businesses, non-governmental organizations, and local communities. Collaborative governance models are essential for aligning the interests of these stakeholders and ensuring that urban sustainability initiatives are comprehensive and inclusive (Yahia et al. 2021). By facilitating partnerships between the public and private sectors, urban leaders can leverage the expertise, resources, and innovation of businesses and civil society to implement large-scale sustainability projects. For example, public–private partnerships (PPP) can fund renewable energy projects, affordable housing developments, and green transportation networks (Fell and Mattsson 2021). Visionary leadership can mobilize resources and public support for ambitious sustainability initiatives, driving cities toward greener futures. Leaders who prioritize sustainability must inspire action by outlining clear, achievable goals and communicating the benefits of sustainable urban development to the public. They play a critical role in engaging citizens, building political coalitions, and securing investment in projects that promote long-term environmental, social, and economic well-being. When urban leaders make sustainability a central focus of their policies and governance, they create a ripple effect, influencing other cities, regions, and even national governments to adopt similar approaches.

This encompasses the provision of efficient public transportation systems, the creation of accessible and high-quality public spaces, and the development of affordable housing. Sustainable urban governance requires a focus on creating cities that are not only environmentally friendly but also liveable and inclusive. Public transportation is a cornerstone of sustainable urban development. By investing in efficient, affordable, and accessible public transit networks, cities can reduce car dependency, lower greenhouse gas emissions, and make transportation more equitable for all residents. Well-designed public transit systems help reduce traffic congestion, improve air quality, and increase mobility, particularly for low-income communities who may not have access to private vehicles. Public spaces are another essential element of sustainable urban design. Accessible, well-maintained parks, plazas, and recreational areas contribute to the social and mental well-being of urban residents, foster community interaction, and provide green spaces that help mitigate the urban heat island effect (Navarrete-Hernandez et al. 2021). Urban governance plays a crucial role in ensuring that public spaces are designed to meet the needs of diverse populations, offering places for leisure, exercise, cultural activities, and social gatherings. These spaces not only enhance quality of life but also promote social cohesion and inclusivity by providing environments where people from different backgrounds can come together.

Affordable housing is critical for creating inclusive and sustainable cities. Urban leaders must work to develop policies that ensure housing is accessible to all residents, regardless of income. In many cities, rising real estate prices and the lack of affordable housing have led to increased inequality and displacement of lower-income communities. Governments can address this issue by investing in affordable housing developments, offering incentives for developers to include low-cost housing units in new projects, and protecting existing affordable housing through rent control or preservation programs. Ensuring that housing remains affordable is essential for reducing homelessness, promoting social equity, and creating communities where people from all walks of life can live and thrive. Collaboration between government agencies, private sectors, and civil society is crucial in implementing effective urban policies that promote sustainability and inclusivity. Urban governance is most effective when it is collaborative and inclusive, bringing together diverse stakeholders to develop and implement sustainability strategies. Government agencies play a central role in coordinating policy efforts, enforcing regulations, and managing public services, while the private sector provides the expertise and capital needed to drive innovation and build new infrastructure. Civil society, including non-governmental organizations (NGO) and community groups, represents the interests of residents, ensuring that the voices of local communities are heard in decision-making processes. By working together, these groups can ensure that urban policies are responsive to the needs of all residents and aligned with sustainability goals.

Additionally, maintaining high standards of urban management and securing strong political backing are crucial for ensuring that these initiatives succeed. High standards of urban management such as transparency, accountability, and effective service delivery are essential for building public trust and ensuring that sustainability initiatives are implemented efficiently. Political will is also key to the success of urban sustainability efforts. Without strong political backing, sustainability projects may lack the necessary funding, legislative support, or public buy-in to succeed. Urban leaders must champion sustainability initiatives and build broad-based support among political stakeholders, business leaders, and the public to ensure their long-term viability. Transparent and participatory urban planning processes, which involve stakeholders at all levels, help ensure that the needs and concerns of the local population are addressed. Participatory governance models encourage citizens to take an active role in shaping their urban environments, giving them a voice in decisions about land use, infrastructure, public services, and environmental protections. When residents are involved in urban planning processes, the resulting policies are more likely to reflect local needs and priorities, making them more effective and equitable. Public consultations, town hall meetings, and participatory budgeting are just a few ways in which cities can engage citizens in governance processes, creating more democratic and inclusive urban spaces.

Together, these elements foster a well-managed urban environment that meets the needs of its residents while promoting sustainability and equity. Good governance is not just about creating a functional urban environment, it is about creating cities that are liveable, resilient, and fair. By ensuring that urban policies address the needs of all residents, particularly vulnerable populations, cities can promote social justice while advancing sustainability goals. Effective urban governance is essential for creating cities that are adaptable to future challenges, such as climate change, economic inequality, and rapid population growth. Benchmarking and sharing best practices across cities can also accelerate progress, allowing for replication of successful models globally. Cities around the world face similar challenges, from managing population growth to reducing carbon emissions, and sharing best practices can help accelerate the transition to sustainability. International networks provide platforms for cities to collaborate, share knowledge, and learn from experiences from others. By benchmarking their performance against global standards and adopting proven models from other cities, urban leaders can implement more effective policies and avoid repeating past mistakes.

In conclusion, strong urban governance, visionary leadership, and best practices are essential for transforming cities into sustainable, inclusive, and resilient communities. By prioritizing collaboration, transparency, and long-term planning, cities can address the complex challenges of urbanization while ensuring a high quality of life for all residents. The integration of efficient public services, affordable housing, and inclusive public spaces is key to creating cities that not only thrive economically but also uphold social and environmental values.

(m) **Education, Research, and Knowledge.**

This principle emphasizes the importance of technical training, skill enhancement, and the exchange of experiences among professionals. In the context of urban sustainability, education and research are foundational pillars that support the development of innovative solutions to address the complex challenges cities face (Kallio et al. 2021). As urban environments evolve and grapple with issues such as climate change, resource depletion, and population growth, continuous learning and knowledge dissemination are essential for ensuring that cities can adapt and thrive in a sustainable manner. Ongoing education and research are essential for developing new solutions to the complex challenges cities face as they strive to become more sustainable. Through education, cities can equip the next generation of urban planners, architects, engineers, and policymakers with the skills and knowledge necessary to design and implement sustainable solutions. Education plays a vital role in fostering a culture of sustainability within urban environments. From primary education to university degrees, sustainability-focused curricula help raise awareness and foster a culture of environmental responsibility among citizens. By teaching students about ecological city theory, climate change, resource conservation, and sustainable design, educational programs cultivate environmentally conscious citizens. These citizens understand the importance of protecting natural resources and reducing environmental impact. This early exposure to sustainability principles encourages individuals to adopt eco-friendly practices, such as reducing waste, conserving energy, and supporting green transportation, which collectively contribute to the sustainability of urban environments.

At the higher education level, universities and technical institutes offer specialized programs that train professionals in sustainable urban planning, green architecture, renewable energy, and environmental engineering. These programs prepare future leaders and experts to design cities that are more energy-efficient, resilient, and environmentally friendly (Sharp and Raven 2021). For example, degree programs in urban sustainability, environmental policy, or sustainable design teach students the technical skills and theoretical knowledge needed to develop smart cities. These programs also promote green building practices and help implement sustainable energy systems. By integrating sustainability into higher education, cities can cultivate a workforce equipped with the expertise needed to drive green innovation and implement sustainable practices across industries. It also involves the dissemination of knowledge through publications focused on ecological city theory and sustainable design. Research and knowledge sharing are key to advancing urban sustainability practices. As cities experiment with new approaches to sustainable development, it is essential to document successes and challenges so that best practices can be replicated in other cities. Academic journals, research papers, and industry publications focused on urban ecology, green technologies, and sustainable design provide valuable insights for professionals, policymakers, and city planners looking to implement sustainable solutions. In addition, public access to research findings encourages transparency and allows citizens to become informed participants in discussions about their future.

Research institutions, universities, and think tanks play a critical role in developing and testing new technologies, urban design strategies, and policies that can guide cities toward a more sustainable future. These institutions serve as hubs of innovation, where experts from diverse fields including architecture, engineering, environmental science, economics, and social science to collaborate to solve urban challenges (Bayuo et al. 2020). Universities often partner with local governments to pilot sustainability projects, such as renewable energy installations, green infrastructure, or smart city technologies. These collaborative projects provide cities with data-driven solutions that can be scaled up for broader implementation. Research institutions also engage in long-term studies that assess the environmental, social, and economic impacts of sustainability initiatives, ensuring that policies are grounded in evidence and can deliver tangible benefits to urban populations. Innovation and experimentation are essential for identifying the most effective strategies for reducing energy consumption, managing resources, and mitigating climate impacts in cities. Universities and research centres are often at the forefront of this work, developing cutting-edge technologies such as advanced renewable energy systems, sustainable building materials, and smart urban infrastructure. For example, research on photovoltaic systems, battery storage, and green roofs has led to technological advancements that allow cities to reduce their reliance on fossil fuels and create more resilient energy systems. As cities adopt these innovations, they contribute to the global effort to combat climate change and reduce greenhouse gas emissions.

By fostering a culture of continuous learning and sharing expertise, cities can advance sustainable practices, encourage innovation, and build the intellectual foundation necessary for developing resilient and environmentally responsible urban environments. Professional development programs, workshops, and conferences that bring together experts from various sectors are essential for keeping urban professionals informed about the latest trends, technologies, and best practices in sustainability. Cities that invest in continuous education for their workforce create a culture of innovation, where professionals are encouraged to explore new ideas and implement forward-thinking solutions. This mindset drives progress in areas such as green building design, sustainable mobility, and circular economy models, ultimately leading to more sustainable urban environments. Cross-sector collaborations, including partnerships between academia, industry, and government, can accelerate the development and deployment of cutting-edge sustainability technologies (Anthony 2024). By fostering collaboration among these key stakeholders, cities can benefit from a diverse range of expertise and resources. Academia contributes research and technical knowledge, industry provides practical experience and financial investment, and governments set the policy frameworks that enable large-scale adoption of sustainable technologies.

Collaborative efforts can also enhance the ability of cities to respond to global challenges like climate change, urbanization, and resource scarcity. International partnerships and knowledge exchanges allow cities to learn from each other experiences and adopt best practices from around the world. Global networks facilitate the sharing of research and policy innovations that contribute to sustainable urban development. By participating in these global networks, cities can leverage international expertise to address local challenges, such as reducing carbon emissions, managing water resources, or increasing resilience to extreme weather events. In addition to formal education and research, public awareness campaigns and community-based educational programs can also play a significant role in promoting sustainable urban practices. Cities can engage residents in sustainability efforts by offering educational programs on energy conservation, recycling, water use, and green living. These initiatives can take the form of workshops, community events, or digital platforms that provide practical advice and resources for adopting more sustainable lifestyles. By empowering citizens with knowledge, cities can create a culture of environmental stewardship that extends beyond government initiatives and becomes ingrained in everyday life.

In conclusion, education, research, and knowledge dissemination are critical for driving sustainable urban development. Through ongoing learning, innovation, and collaboration, cities can equip professionals and citizens with the tools they need to build more resilient and environmentally responsible urban environments. By integrating sustainability into educational programs, fostering cross-sector partnerships, and sharing research findings widely, cities can ensure that they remain at the forefront of global efforts to create sustainable, liveable, and thriving communities for future generations.

(n) **Strategies for Cities in Developing Countries**

Urban planning in developing countries necessitates tailored strategies distinct from those applied in developed regions. Cities in the Global South, particularly in Africa, Asia, and Latin America are experiencing rapid urbanization, which presents both opportunities and significant challenges (Henderson and Turner 2020). Unlike cities in developed countries, many rapidly urbanizing cities in the Global South face issues such as limited financial resources, rapid population growth, and inadequate infrastructure. These challenges require innovative, context-specific solutions that balance economic development with environmental sustainability and social equity. One of the most pressing issues in developing countries is the rise of informal settlements commonly known as slums. These areas often lack access to basic services such as clean water, sanitation, electricity, and adequate housing, which undermines the health and well-being of their residents (Ketadzo et al. 2021). Slums are typically the result of rapid, unplanned urbanization where formal housing and infrastructure development have not kept pace with population growth. Addressing informal settlements is a critical priority for cities in developing regions, as these areas represent a significant portion of the urban population and contribute to economic and social inequality.

Key areas of focus include addressing informal settlements, upgrading urban slums, developing low-cost housing solutions, and designing mass housing typologies. Informal settlements pose significant challenges to sustainable development because they are often located in environmentally sensitive areas, such as floodplains or hillsides, making residents vulnerable to natural disasters. Moreover, the lack of formal infrastructure and services contributes to environmental degradation such as poor waste management, water pollution, and deforestation. Urban upgrading initiatives programs that aim to improve infrastructure, housing, and public services in informal settlements are critical for promoting more inclusive and sustainable urbanization. Rather than displacing slum dwellers, urban upgrading focuses on improving living conditions within these communities. This is done by providing access to essential services such as clean water, electricity, sewage systems, and public transportation. Upgrading also involves enhancing housing structures to ensure they are safe and resilient to natural disasters, as well as providing public spaces, health services, and educational facilities. These efforts not only improve the quality of life for residents but also foster social inclusion, reducing the marginalization of slum communities.

Developing low-cost housing solutions is another key strategy for addressing the housing crisis in many developing countries. As urban populations grow, the demand for affordable housing continues to outpace supply, forcing many low-income residents into informal settlements. Governments and urban planners must prioritize the development of affordable housing that meets the needs of the urban poor while ensuring that these housing projects are sustainable and integrated into the broader urban fabric (Seitz 2021). Mass housing typologies which involve designing and constructing large-scale, low-cost housing units, can help address the housing shortage while providing safe, healthy, and sustainable living environments. These projects should incorporate green building principles, such as energy-efficient designs, locally sourced materials, and adequate access to transportation and public services. In addition to housing and infrastructure challenges, cities in developing countries must also address the environmental impacts of rapid urbanization. Urban slums, which often lack waste management systems, can contribute to significant environmental degradation. Poor waste disposal practices, including dumping waste into rivers and open spaces, lead to pollution, public health risks, and the depletion of natural resources. Implementing waste management systems that include recycling, composting, and waste-to-energy technologies is essential for reducing the environmental footprint of rapidly growing cities. Additionally, strategies for managing water resources are vital, as many cities in developing countries experience water scarcity or face challenges related to water contamination. Rainwater harvesting, wastewater treatment, and infrastructure to prevent flooding can contribute to better water management and increase urban resilience to climate change.

Urban upgrading efforts must be closely integrated with poverty reduction programs to ensure that development promotes social equity and improves living conditions for all residents. Sustainable urbanization cannot be achieved without addressing the root causes of poverty, which include lack of access to education, employment, healthcare, and social services. Holistic approaches that combine physical infrastructure upgrades with programs aimed at improving health, education, and economic opportunities for low-income residents are critical for reducing inequality and fostering more inclusive cities. Poverty reduction programs that focus on skills training, job creation, and small business development can empower slum dwellers to improve their economic prospects while contributing to the overall development of their cities. Financing is one of the biggest obstacles to sustainable urban development in developing countries, where resources are often limited and competing demands are high. Innovative financing models, such as micro-financing, public–private partnerships (PPP), and international development aid, are essential for funding sustainable urban projects in these regions. Micro-financing can provide low-income families with the resources needed to improve their homes or start small businesses, while

PPP can bring together public sector funding with private sector expertise to finance large-scale infrastructure projects, such as transportation networks, water supply systems, and affordable housing developments. In addition, international development organizations, such as the World Bank, UN-Habitat, and regional development banks, play a crucial role in providing technical and financial assistance to cities in developing countries.

Capacity-building programs are equally important in equipping local governments and urban planners with the skills and tools needed to plan and manage sustainable urban growth effectively. Many city officials and planners in developing regions lack the training and expertise required to design and implement comprehensive sustainability strategies. Capacity-building programs, which provide education and technical assistance in areas such as urban planning, environmental management, and disaster resilience can empower local governments. These programs enable governments to take a more proactive role in shaping the future of their cities. By enhancing the technical and managerial skills of urban leaders, these programs can improve the efficiency of municipal governments and increase their ability to deliver sustainable urban services. Additionally, the involvement of local communities in the planning and implementation of urban projects is critical for their success. Participatory urban planning, where residents are actively engaged in decision-making processes, ensures that the voices and needs of marginalized groups such as slum dwellers are heard and addressed. Community-driven initiatives, such as participatory budgeting or grassroots urban upgrading projects, allow residents to have a say in how resources are allocated and what improvements are prioritized. By involving local communities, cities can create more inclusive and equitable urban environments that reflect the needs and aspirations of all residents, not just the wealthy or politically powerful.

Developing cities must also be resilient to the impacts of climate change as many are located in regions prone to extreme weather events such as flooding, droughts, and hurricanes. Climate resilience strategies such as building flood defences, improving drainage systems, and retrofitting buildings to withstand extreme weather are essential for protecting vulnerable communities and ensuring the long-term sustainability of urban development. Disaster risk reduction strategies including early warning systems and evacuation plans are also crucial for safeguarding lives and livelihoods in disaster-prone areas. By integrating climate adaptation into urban planning, cities can enhance their ability to respond to future challenges and ensure that development is sustainable and resilient. Therefore, tailored strategies for cities in developing countries must address the unique challenges posed by rapid urbanization, poverty, and resource constraints. Through innovative solutions such as urban upgrading, low-cost housing, and participatory planning, cities can improve living conditions for all residents while promoting social equity and environmental sustainability. Financing models, capacity-building, and cross-sector collaboration will be key to achieving these goals, ensuring that developing cities can grow sustainably and build resilience to future challenges. By adopting context-specific solutions and leveraging the strengths of local communities, governments, and the private sector, cities in the Global South can create more inclusive, equitable, and sustainable urban environments for their rapidly growing populations.

6.2 Sustainable Buildings and Cities

A sustainable building is more than just an energy-efficient structure. It represents a holistic approach to design, construction, operation, renovation, and repurposing, aiming to minimize the overall environmental impact of the building throughout its entire lifecycle. This includes the selection of materials, the construction process, the use of renewable resources, and strategies to maximize energy efficiency. Sustainable buildings are designed with the intent to not only reduce energy consumption but also conserve water, improve indoor air quality, and optimize the overall well-being of occupants (Zhang et al. 2022). By addressing these key aspects, sustainable buildings contribute to the broader goals of environmental stewardship and resource efficiency at every stage of their lifecycle. Sustainable buildings are meticulously planned and constructed to reduce their ecological footprint, aiming to safeguard the environment and promote sustainable development. This involves integrating energy-efficient systems, such as passive solar design, high-performance insulation, and advanced heating, ventilation, and air conditioning (HVAC) technologies (Pham et al. 2020). By reducing reliance on non-renewable resources and lowering greenhouse gas emissions, sustainable buildings help mitigate the effects of climate change while conserving natural resources. They also prioritize the responsible sourcing of building materials, such as using recycled or locally sourced materials, to reduce the environmental impact of construction.

By integrating advanced technologies and sustainable practices, sustainable buildings enhance the health and well-being of occupants through improved indoor air quality, natural lighting, and thermal comfort. The focus on occupant health is a critical component of sustainable building design. High indoor air quality is achieved through the use of low-emission materials, effective ventilation systems, and the incorporation of natural air flow, reducing the concentration of pollutants and improving the overall comfort of building users. Additionally, the use of natural lighting is maximized through strategic window placement, skylights, and solar tubes, reducing the need for artificial lighting during the day and improving occupant productivity and mental well-being. Thermal comfort is achieved through the use of passive design strategies that regulate indoor temperatures, such as thermal mass materials, shading devices, and natural ventilation. This minimizes the need for energy-intensive heating and cooling systems while ensuring that indoor environments remain comfortable year-round. These strategies not only improve occupant satisfaction but also contribute to significant energy savings, further enhancing the sustainability of the building.

Moreover, sustainable buildings optimize the use of critical resources such as energy and water, employing systems like rainwater harvesting, greywater recycling, and renewable energy sources to minimize waste and consumption. Water conservation is a fundamental element of sustainable building design, with technologies like low-flow fixtures, dual-flush toilets, and water-efficient irrigation systems being commonly used. Rainwater harvesting systems allow buildings to collect and store rainwater for non-potable uses, such as landscape irrigation or toilet flushing.

Greywater recycling systems capture wastewater from sinks, showers, and washing machines for reuse in irrigation or other applications. These water-saving strategies not only reduce the strain on municipal water supplies but also lower utility costs for building occupants. In terms of energy use, renewable energy systems such as photovoltaic panels, solar water heaters, and wind turbines are increasingly being integrated into sustainable buildings. These systems generate clean energy on-site, reducing reliance on fossil fuels and lowering carbon emissions. Combined with energy-efficient appliances, LED lighting, and advanced building management systems, these technologies significantly reduce the operational energy demand of buildings, contributing to a more sustainable urban environment. By using less energy and water, sustainable buildings reduce their environmental impact, lower operational costs, and enhance long-term resource efficiency.

In addition to these environmental benefits, sustainable buildings are also economically advantageous. They are designed to lower operational costs through energy savings, reduced maintenance, and longer building lifespans, making them cost-effective in the long run (Lu et al. 2020). While the initial investment in sustainable technologies and design features may be higher, the long-term savings from reduced energy consumption, lower water bills, and decreased maintenance needs often offset these upfront costs. Moreover, sustainable buildings tend to have higher property values and attract tenants or buyers who prioritize environmental responsibility, further enhancing their economic appeal. As part of a broader strategy for sustainable cities, these buildings contribute to urban resilience by reducing dependency on non-renewable resources and enhancing the overall sustainability of urban environments. Sustainable buildings are a key component of resilient cities, as they are better equipped to withstand the impacts of climate change, such as extreme weather events and resource shortages. Their energy-efficient design and reliance on renewable energy sources make them less vulnerable to fluctuations in energy prices, while their water conservation strategies help ensure a stable supply of clean water even in times of drought or water scarcity. By integrating sustainable buildings into urban planning, cities can create more resilient, adaptive, and sustainable environments that support both environmental and human health. The concept of sustainable buildings is illustrated in Fig. 6.4, which highlights their role in achieving a balanced and sustainable urban ecosystem. Sustainable buildings not only contribute to the reduction of carbon emissions but also promote the health, well-being, and resilience of cities. In doing so, they form a critical part of the solution to the challenges posed by rapid urbanization, climate change, and resource depletion.

Achieving sustainability in building construction necessitates the integration of energy-efficient technologies into the architectural design. Architectural design plays a pivotal role in determining the sustainability of a building by influencing its orientation, material selection, window types and sizes, and ventilation systems. The layout and structure of a building can significantly impact its energy efficiency. For example, a well-designed building can take advantage of natural light and heat, reducing the need for artificial lighting and heating. Various advanced technologies can be incorporated to enhance energy efficiency, including solar water and space heating, photovoltaic panels, ground-source heat pumps, distributed power systems, green roofs,

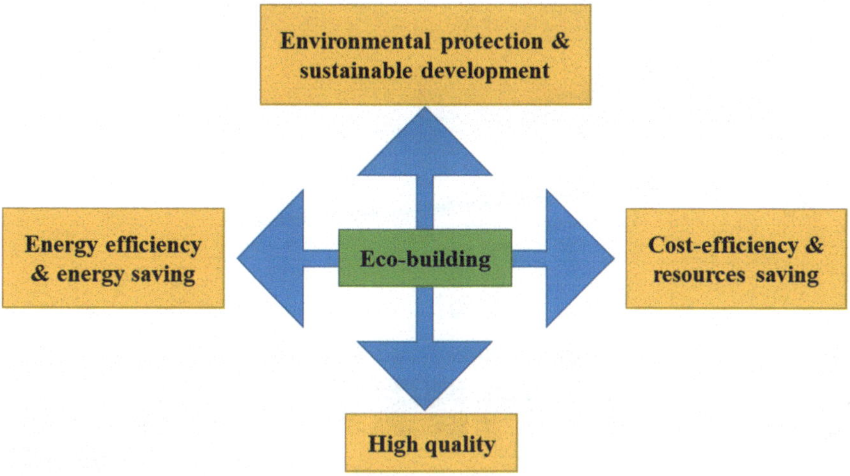

Fig. 6.4 Concept of a sustainable building

net metering, solar light pipes, and wind turbines (Gu 2024). These technologies, when integrated into the design of the building, collectively contribute to reducing the environmental impact and operational costs of buildings while ensuring that they meet the highest standards of sustainability.

An example of a sustainable housing project along with its features is depicted in Fig. 6.5. This passive house, constructed to Level 4 standards of the Code for Sustainable Homes, is part of the Creative Energy Homes Project at the University of Nottingham and is sponsored by BASF. The house features smart metering systems for comprehensive control and scheduling, integrating various systems such as ventilation, heating, internal and external lighting, and automated windows and blinds. Smart metering technology allows occupants to monitor and optimize their energy usage, reducing waste and lowering energy bills. The house has been both occupied and continuously monitored, with projections suggesting that it will exceed the UK Code for Sustainable Homes' Level 4 requirements. Additionally, it has been specifically designed to comply with the rigorous German PassivHaus standards, which are among the strictest energy efficiency standards globally. These standards require that the building achieve a total energy demand for space heating and cooling of less than 15 kWh/m^2 annually and a primary energy consumption of less than 120 kWh/m^2 per year. These stringent energy performance targets have been successfully met, demonstrating the potential for sustainable buildings to achieve significant reductions in energy consumption.

Sustainable buildings are a critical component of the broader effort to create more resilient, efficient, and liveable urban environments. By integrating advanced technologies, resource-efficient practices, and occupant-centred design, these buildings reduce their environmental footprint, lower operational costs, and enhance the quality of life for their users. As cities continue to grow and evolve, sustainable building

Fig. 6.5 Passive solar house at the University of Nottingham

practices will play a key role in shaping the future of urban development, ensuring that cities remain resilient, resource-efficient, and environmentally responsible.

6.2.1 Sustainable Cities

Globally, numerous sustainable cities and towns have been developed, with Europe leading several green city initiatives that prioritize environmental stewardship, energy efficiency, and sustainable urban living. These cities serve as models for how urban development can be aligned with ecological principles to reduce carbon emissions, conserve resources, and enhance the quality of life for residents. One significant initiative includes 11 low-energy demonstration developments across nine EU countries, totaling 645 dwellings. These projects aim to showcase innovative sustainable

Fig. 6.6 Aerial view in showcasing sustainable urban planning in Malmö, Sweden

technologies and urban planning strategies that can be replicated on a larger scale across the continent. Notable examples of European sustainable cities include Malmö (Sweden), Freiburg (Germany), Trondheim (Norway), Tudela (Spain), and Helsingborg/Helsingør (Sweden/Denmark). These cities have made impressive strides in reducing their environmental impact by embracing renewable energy, green building practices, and integrated public transportation systems. In Fig. 6.6, Malmö, Sweden with a population of 277,000 inhabitants, serves as an exemplary model of sustainable urban development. The city adopts an ecological approach to urban planning and building construction, aiming to be an international leader in environmental adaptation of densely built urban areas (Zinkernagel and Neij 2023). Locally produced renewable energy sources, such as solar, wind, water, and biogas from organic waste, are extensively used. Solar heating alone meets 12% of the annual heat demand of the city, demonstrating the significant role that renewable energy can play in urban sustainability. The long-term goal is to be carbon–neutral by 2030, making it a prime example of how cities can align development with ecological principles (Salvia et al. 2021).

In contrast, the discovery of oil in the United Arab Emirates (UAE) has catalyzed rapid development, leading to significant expansions in infrastructure and the built environment. As the UAE experienced an economic boom, cities such as Dubai underwent remarkable transformations, shifting from traditional desert landscapes to modern, high-tech urban centers. As shown in Fig. 6.7a and b, this transformation is evident in skyscrapers, highways, and advanced transportation systems in Dubai which have propelled the city into a global metropolis. However, with this rapid development came environmental challenges. Currently, the UAE holds the distinction of being the largest world CO_2 emitter per capita, primarily due to its reliance on fossil fuels and energy-intensive industries.

In response to these challenges, the Government of Abu Dhabi has launched Masdar City depicted in Fig. 6.7c, as an ambitious and forward-thinking sustainable urban development project. Masdar City located on the outskirts of Abu Dhabi is the

Fig. 6.7 Transformation and sustainability in the United Arab Emirates (UAE): **a** Traditional desert landscape, **b** Modern urban development in Dubai, **c** Masdar City, and **d** Personal rapid transit systems within Masdar City

first city in the world designed to be entirely carbon–neutral, waste-free, and car-free (K'Akumu and Alhamoudi 2023). The city is powered solely by renewable energy sources, including solar energy and bioenergy from organic waste. Masdar City represents a bold vision for the future of urban living, where sustainability is integrated into every aspect of city planning, from transportation to water management. Planned to accommodate 50,000 residents and 1,500 businesses, the city is a model for how new urban developments can be designed with sustainability at their core. In addition to housing residents and businesses, approximately 40,000 workers commute daily to Masdar City using personal rapid transit systems, in addition to road and rail connections that extend beyond the city, as seen in Fig. 6.7 (d). The water supply of the city is managed through a solar-powered desalination plant, while photovoltaic panels generate electricity to power homes and businesses (Ismael 2023). Furthermore, Masdar City has adopted a circular economy approach, converting biological waste into fertilizer and using some of it as a supplementary power source via waste incineration. This integration of renewable energy, waste management, and efficient urban design positions Masdar City as a leading example of how future cities can be both technologically advanced and environmentally sustainable.

6.3 Global Initiatives in Sustainable Energy and Eco-City Development

The rapid increase in global energy demand coupled with the environmental pollution linked to fossil fuel consumption has led to a growing global focus on developing sustainable buildings and eco-cities. Governments around the world are recognizing that transitioning to renewable energy and promoting sustainable urbanization are essential to combating climate change and ensuring a secure energy future. In response, many countries have initiated extensive research and development efforts in sustainable energy technologies, green building practices, and eco-city frameworks. These efforts are crucial for reducing greenhouse gas emissions, conserving natural resources, and improving the resilience of cities to the impacts of climate change. For example, The US Obama Government committed to strategically investing $150 billion over the next decade to ensure that 10% of electricity of the nation would be generated from renewable sources by 2012, with an aim of reaching 25% by 2025 (AMERICA'S NEW 2020). This ambitious investment aimed to accelerate the transition to clean energy, foster innovation in renewable energy technologies, and create jobs in the green energy sector. Similarly, the UK government has committed £100 billion to renewable energy projects, targeting over 30% of electricity, 14% of heat, and up to 10% of transport fuels from renewable sources by 2020 (Raybould et al. 2020). These national initiatives highlight the importance of government investment and policy support in driving the adoption of sustainable energy on a large scale.

As the world largest emitter of greenhouse gases, China has also made substantial progress in sustainability by designating five provinces and eight cities as low-carbon demonstration zones. These zones are part of broader strategy of China to reduce carbon emissions, promote energy efficiency, and accelerate the deployment of renewable energy technologies (Chen et al. 2023). Rapid urbanization in China has posed significant environmental challenges, but its commitment to developing low-carbon cities reflects a growing recognition of the need for sustainable development practices. Meanwhile, European countries like Germany and Sweden continue to lead in eco-city and sustainable building initiatives, with numerous cities setting ambitious targets for reducing carbon emissions and transitioning to renewable energy. Germany, in particular, has been a global leader in renewable energy, with its Energiewende (Energy Transition) policy aiming to phase out nuclear energy and fossil fuels in favour of wind, solar, and other renewable sources (Yang 2022). Sweden has similarly adopted aggressive goals for carbon neutrality and energy efficiency, positioning itself as a global pioneer in sustainable urban development.

The broader adoption of sustainable urban models worldwide will necessitate the development of cost-effective technologies, comprehensive education, and training programs to equip professionals with the skills needed to implement these models. Global collaboration, knowledge sharing, and continued investment in sustainable technologies will be essential for scaling up eco-city initiatives and ensuring that cities around the world can meet the challenges of climate change and rapid urbanization. As cities become the primary drivers of global economic growth, their ability to integrate sustainability into urban planning will determine the future of urban life on the planet.

References

Abdrabo KI et al (2021) A methodological approach towards sustainable urban densification for urban sprawl control at the microscale: case study of Tanta Egypt. Sustainability 13(10):5360

Agnihotri V (2022) Paradigm shifts from social forestry to urban forestry: a synergy to mitigate urban heat island in high-density developments

Amaral RE et al (2020) Waste management and operational energy for sustainable buildings: a review. Sustainability 12(13):5337

AMERICA'S NEW, CE (2020) Jumpstarting US clean energy manufacturing in economic stimulus and infrastructure legislation. Energy

Ang YQ, Berzolla ZM, Reinhart CF (2020) From concept to application: a review of use cases in urban building energy modeling. Appl Energy 279:115738

Anthony B Jr (2024) The role of community engagement in urban innovation towards the co-creation of smart sustainable cities. J Knowl Econ 15(1):1592–1624

Aşchilean I et al (2021) Pairing solar power to sustainable energy storage solutions within a residential building: A case study. Int J Energy Res 45(10):15495–15511

Bayuo BB, Chaminade C, Göransson B (2020) Unpacking the role of universities in the emergence, development and impact of social innovations–A systematic review of the literature. Technol Forecast Soc Chang 155:120030

Bian H et al (2024) China's safe and just space during 40 years of rapid urbanization and changing policies. Landscape Ecol 39(4):74

Bibri SE, Krogstie J, Kärrholm M (2020) Compact city planning and development: emerging practices and strategies for achieving the goals of sustainability. Devel Built Environ 4:100021

Binhweel F et al (2024) Kinetics, thermodynamics, and optimization analyses of lipid extraction from discarded beef tallow for bioenergy production. Separat Sci Technol pp 1–20

Bogdan A et al (2022) Heritage Building Preservation in the Process of Sustainable Urban Development: The Case of Brasov Medieval City, Romania. Sustainability 14(12):6959

Bramareswara Rao S, Padma K (2019) A review on schemes for interconnecting microgrids of urban buildings. Microelectronics, Electromagnetics and Telecommunications: Proceedings of the Fifth ICMEET 2019, pp 431–438

Chen B et al (2023) Can the new energy demonstration city policy promote green and low-carbon development? Evidence from China. Sustainability 15(11):8727

Cheng H, Hu Y (2010) Planning for sustainability in China's urban development: Status and challenges for Dongtan eco-city project. J Environ Monit 12(1):119–126

Chettry V (2023) A critical review of urban sprawl studies. J Geovisualization Spat Anal 7(2):28

Coombes MA, Viles HA (2021) Integrating nature-based solutions and the conservation of urban built heritage: Challenges, opportunities, and prospects. Urban Forestry and Urban Greening 63:127192

Cusworth DH et al (2024) Quantifying methane emissions from United States landfills. Science 383(6690):1499–1504

Deng T, Zhang K, Shen Z-JM (2021) A systematic review of a digital twin city: a new pattern of urban governance toward smart cities. J Manage Sci Eng 6(2):125–134

Du Y (2023) Urbanizing the periphery: infrastructure funding and local growth coalition in China's peasant relocation programs. Changing Asian Urban Geographies. Routledge, pp 77–96

Ellison AM, Felson AJ, Friess DA (2020) Mangrove rehabilitation and restoration as experimental adaptive management. Front Mar Sci 7:327

Elshafei G et al (2021) Towards an adaptation of efficient passive design for thermal comfort buildings. Sustainability 13(17):9570

Fell T, Mattsson J (2021) The role of public-private partnerships in housing as a potential contributor to sustainable cities and communities: a systematic review. Sustainability 13(14):7783

Filion P, Moos M, Sands G (2023) Urban neoliberalism, smart city, and big tech: the aborted sidewalk labs Toronto experiment. J Urban Aff 45(9):1625–1643

Giliberto F, Appendino F (2022) Handling change in historic urban landscapes: an analysis of urban heritage conservation approaches in bordeaux (France), Edinburgh (UK), and Florence (Italy). Landscape as Heritage. Routledge, pp 216–230

Gu C (2024) Research on the application of sustainable building design in civil architecture. Highlights in Science, Engineering and Technology 106:108–116

Hamadyk E, Amado M, De Brito J (2020) Use of timber for the sustainable city growth and its role in the climate change. in IOP Conference Series: Earth and Environmental Science. 2020. IOP Publishing

Haron A, Feisal Z (2020) Constructed wetland parks: a pathway to sustainability for Cairo. J Urban Res 36(1):61–81

He X, Yu Y, Jiang S (2023) City centrality, population density and energy efficiency. Energy Economics 117:106436

Henderson JV, Turner MA (2020) Urbanization in the developing world: too early or too slow? J Econom Perspect 34(3):150–173

Ismael NI (2023) Sustainability criteria used in designing energy-efficient smart cities a study of Masdar City as a model for one of the smart cities that realize the idea of sustainable development. Int J Adv Eng Bus Sci 4(1):42–71

K'Akumu O, Alhamoudi AM (2023) The smart city as policy concept in the Arab world: a critical research into the strategic visions of urban development in Abu Dhabi. J Asian African Stud p 00219096231192316

Kallio KP et al (2021) Skatescape in the making: Developing sustainable urban pedagogies through transdisciplinary education. Sustainability 13(17):9561

Karimi A et al (2023) New developments and future challenges in reducing and controlling heat island effect in urban areas. Environ Dev Sustain 25(10):10485–10531

Ketadzo JA, Nkongolo NV, Akrofi MM (2021) Empirical analysis of the impact of urbanization on groundwater quality within the slums of Accra Ghana. World Water Policy 7(1):112–131

Kopuz AD, Bal A (2023) The conservation of modern architectural heritage buildings in Turkey: İstanbul Hilton and İstanbul Çınar Hotel as a case study. Ain Shams Eng J 14(4):101918

Langemeyer J et al (2021) Urban agriculture—A necessary pathway towards urban resilience and global sustainability? Landsc Urban Plan 210:104055

Lazzeretti L, Oliva S (2021) Rethinking city transformation: florence from art city to creative fashion city. Dislocation: Awkward Spatial Transitions. Routledge, pp 156–173

Li K (2024) The Ideology of sustainable cities. Academ J Manage Soc Sci 6(2):87–91

Li B et al (2021) Sustainable passive design for building performance of healthy built environment in the Lingnan area. Sustainability 13(16):9115

Liang X et al (2022) Conversion of landfilled waste-to-electricity (WTE) for energy efficiency improvement in Shenzhen (China): A strategy to contribute to resource recovery of unused methane for generating renewable energy on-site. J Clean Prod 369:133078

Liberalesso T et al (2020) Green infrastructure and public policies: an international review of green roofs and green walls incentives. Land Use Policy 96:104693

Loder A (2014) 'There's a meadow outside my workplace': a phenomenological exploration of aesthetics and green roofs in Chicago and Toronto. Landsc Urban Plan 126:94–106

Lu Y et al (2020) A critical review of sustainable energy policies for the promotion of renewable energy sources. Sustainability 12(12):5078

Lúcio C, Silva CM, Sousa V (2020) A scale-adaptive method for urban rainwater harvesting simulation. Environ Sci Pollut Res 27:4557–4570

Luo X, Qiu R (2020) Electric vehicle charging station location towards sustainable cities. Int J Environ Res Public Health 17(8):2785

Mansoor U et al (2022) A review of factors and benefits of non-motorized transport: a way forward for developing countries. Environ Dev Sustain 24(2):1560–1582

Mishra V, Sadhu A (2023) Towards the effect of climate change in structural loads of urban infrastructure: a review. Sustain Cities Soc 89:104352

Navarrete-Hernandez P, Vetro A, Concha P (2021) Building safer public spaces: exploring gender difference in the perception of safety in public space through urban design interventions. Landsc Urban Plan 214:104180

Negret HR, Negret R, Montes-Londoño I (2022) Residential garden design for urban biodiversity conservation: experience from Panama City, Panama. Biodiversity Islands: strategies for conservation in human-dominated environments. Springer, pp 387–417

Neo TH et al (2022) Evaluation of active, beautiful, clean waters design features in tropical urban cities: a case study in Singapore. Water 14(3):468

Ngcamu BS (2023) Climate change effects on vulnerable populations in the Global South: a systematic review. Nat Hazards 118(2):977–991

Oquendo-Di Cosola V, Olivieri F, Ruiz-García L (2022) A systematic review of the impact of green walls on urban comfort: temperature reduction and noise attenuation. Renew Sustainable Energy Rev 162:112463

Orsini F et al (2020) Features and functions of multifunctional urban agriculture in the global north: a review. Frontiers in Sustainable Food Systems 4:562513

Ostapska K et al (2024) Design for disassembly: a systematic scoping review and analysis of built structures designed for disassembly. Sustainable Prod Consumpt 48:377–395

Peng J et al (2021) How to quantify the cooling effect of urban parks? Linking maximum and accumulation perspectives. Remote Sens Environ 252:112135

Pettigrove V et al (2023) Catchment sourcing urban pesticide pollution using constructed wetlands in Melbourne Australia. Sci Total Environ 863:160556

Pham A-D et al (2020) Predicting energy consumption in multiple buildings using machine learning for improving energy efficiency and sustainability. J Clean Prod 260:121082

Raybould B et al (2020) An investigation into UK government policy and legislation to renewable energy and greenhouse gas reduction commitments. Clean Technol Environ Policy 22:371–387

Rezaie S et al (2022) Governing the circular economy: how urban policymakers can accelerate the agenda

Salvia M et al (2021) Will climate mitigation ambitions lead to carbon neutrality? An analysis of the local-level plans of 327 cities in the EU. Renew Sustain Energy Rev 135:110253

Sanga N (2024) Low-income housing development in India: Strategies for income mixing and inclusive urban planning. Creating Mixed Communities through Housing Policies. Routledge, pp 35–52

Seitz W (2021) Urbanization in Kazakhstan: desirable cities, unaffordable housing, and the missing rental market. Int J Urban Sci 25(sup1):135–166

Senova A et al (2022) The use of geothermal energy for heating buildings as an option for sustainable urban development in Slovakia. Processes 10(2):289

Shakir MA, Ahmad MI, Abdul Khalil HPS (2024a) Utilization of marine green algae for the development of bioplastics. In: Rafatullah M et al (eds) Multidisciplinary applications of marine resources: a step towards green and sustainable future. Springer Nature Singapore, Singapore, pp 303–319

Shakir MA, Yhaya MF, Ramli SN, Ahmad MI (2024b) Synthesis and performance characterization of green desiccant from cockle shell and marble waste. Clean Technol Environ Policy https://doi.org/10.1007/s10098-024-03031-6

Sharp D, Raven R (2021) Urban planning by experiment at precinct scale: embracing complexity, ambiguity, and multiplicity. Urban Planning 6(1):195–207

Soebiyan V, Koerniawan M, Triyadi S (2023) A review of hybrid ventilation on humid tropics climate. In: IOP conference series: earth and environmental science. 2023. IOP Publishing

Song Y Medda, F, Wang M (2024) The value of resilience bond in financing flood resilient infrastructures: a case study of Towyn. J Sustainable Fin Invest pp 1–24
Subramaniam DN, Sathiparan N (2023) Comparative study of fly ash and rice husk ash as cement replacement in pervious concrete: mechanical characteristics and sustainability analysis. Int J Pavement Eng 24(2):2075867
Syed MM, Hansen P, Morrison GM (2020) Performance of a shared solar and battery storage system in an Australian apartment building. Energy and Buildings 225:110321
Toland A, Christ (2020) Documenting topographic ecologies in Hong Kong: Visual methods for hyper-dense and hyper-topographic urban spaces in landscape architecture. Vis Commun 19(3):391–414
Wang H, Pei Z (2020) Urban green corridors analysis for a rapid urbanization city exemplified in Gaoyou City, Jiangsu. Forests 11(12):1374
Wang Y et al (2020) A review of approaches to low-carbon transition of high-rise residential buildings in China. Renew Sustain Energy Rev 131:109990
Wang P, Liu Z, Zhang L (2021) Sustainability of compact cities: a review of inter-building effect on building energy and solar energy use. Sustain Cities Soc 72:103035
Wonorahardjo S et al (2020) Characterising thermal behaviour of buildings and its effect on urban heat island in tropical areas. Int J Energy Environ Eng 11:129–142
Xue Q, Wang Z, Chen Q (2022) Multi-objective optimization of building design for life cycle cost and CO_2 emissions: a case study of a low-energy residential building in a severe cold climate. In Building simulation. Springer
Yahia NB et al (2021) Towards sustainable collaborative networks for smart cities co-governance. Int J Inf Manage 56:102037
Yang P (2022) Urban expansion of Energiewende in Germany: a systematic bibliometric analysis and literature study. Energy, Sustainab Soc 12(1):52
Zhang Y et al (2020) The association between green space and adolescents' mental well-being: a systematic review. Int J Environ Res Public Health 17(18):6640
Zhang W et al (2021) Assessment of FRP–concrete interfacial debonding with coupled mixed-mode cohesive zone model. J Compos Constr 25(2):04021002
Zhang X et al (2022) Linking urbanization and air quality together: a review and a perspective on the future sustainable urban development. J Clean Prod 346:130988
Zhuge Y (2020) A review of permeable concrete and its application to pavements. Progress in Mechanics of Structures and Materials, pp 601–607
Zinkernagel R, Neij L (2023) Localizing SDGs: the case of city planning in Malmö. Frontiers in Sustainable Cities 5:1154124

Index

S. Riffat et al., *Sustainable Energy Technologies and Low Carbon Buildings*, Lecture Notes in Energy 45, https://doi.org/10.1007/978-3-031-78853-6

The manufacturer's authorised representative in the EU is Springer Nature Customer Service Centre GmbH, Europaplatz 3, 69115 Heidelberg, Germany. If you have any concerns regarding our products, please contact ProductSafety@springernature.com

Printed and bound by CPI Group (UK) Ltd, Croydon, CR0 4YY
01/10/2026
02225913-0014